Human Rights of Older People

IUS GENTIUM

COMPARATIVE PERSPECTIVES ON LAW AND JUSTICE

VOLUME 45

More information about this series at http://www.springer.com/series/7888

Claudia Martin · Diego Rodríguez-Pinzón
Bethany Brown

Human Rights of Older People

Universal and Regional Legal Perspectives

Springer

Claudia Martin
Academy on Human Rights and Humanitarian Law
American University Washington College of Law
Washington, DC
USA

Bethany Brown
HelpAge USA
Washington, DC
USA

Diego Rodríguez-Pinzón
Academy on Human Rights and Humanitarian Law
American University Washington College of Law
Washington, DC
USA

ISSN 1534-6781 ISSN 2214-9902 (electronic)
Ius Gentium: Comparative Perspectives on Law and Justice
ISBN 978-94-017-7184-9 ISBN 978-94-017-7185-6 (eBook)
DOI 10.1007/978-94-017-7185-6

Library of Congress Control Number: 2015942242

Springer Dordrecht Heidelberg New York London

Printed on acid-free paper

Springer Science+Business Media B.V. Dordrecht is part of Springer Science+Business Media (www.springer.com)

Contents

About the Authors

Claudia Martin is Co-Director of the Academy on Human Rights and Humanitarian Law and Professorial Lecturer in Residence at American University Washington College of Law. She teaches international human rights law and Inter-American human rights law courses. Moreover, she oversees the academic coordination of the Academy's prestigious human rights specialized summer program. Ms. Martin has several publications on international law, international and comparative human rights law, and Inter-American human rights law. She has lectured extensively to lawyers, judges, and human rights professors around the world on the jurisprudence of the Inter-American Human Rights System. Her many appointments include serving as a Member of the Editorial Board, Oxford Reports on International Law in Domestic Courts, Oxford University Press and Amsterdam Center for International Law; and as a Contributing editor to Oxford Reports on International Human Rights Law, Oxford University Press.

Diego Rodríguez-Pinzón (J.D. Universidad de Los Andes, LL.M. Washington College of Law, American University, S.J.D. George Washington University Law School) is Professorial Lecturer in Residence and Co-Director of the Academy on Human Rights and Humanitarian Law at American University, Washington College of Law, in Washington DC. He teaches courses in the fields of human rights law, international law, and comparative law. He has written extensively in his areas of expertise, both in English and Spanish. He served as Ad Hoc Judge in the Inter-American Court of Human Rights of the Organization of American States (OAS) from 2007 to 2011. He was also staff attorney at the Inter-American Commission on Human Rights of the OAS and Officer for Latin America at the International Human Rights Law Group (now Global Rights), a Washington DC based nongovernmental organization. He has been invited faculty in the annual summer Study Session of the René Cassin International Institute of Human Rights, in Strasbourg, France, between 2009 and 2012. He has also taught at the Åbo Akademi Institute for Human Rights in Turku, Finland, and at the Interdisciplinary Course of the Inter-American Institute on Human Rights in San Jose, Costa Rica, among other prestigious international programs. As correspondent for the British periodical Butterworths

Human Rights Cases, Professor Rodríguez-Pinzón covers the Americas. He has served as international legal consultant for USAID, the Inter-American Development Bank (IADB), and the Organization of American States (OAS), among other institutions.

Bethany Brown is the Policy Director at HelpAge USA, the US affiliate of HelpAge International. HelpAge works for and with older people in 67 low- and middle-income countries, responding to needs in emergencies, in development, and in international human rights. Before working with HelpAge, she worked for the UNFPA researching older people's rights, in UNDESA's Population Ageing Division, at the US Department of Health and Human Services' Medicare Operations Division, and the AARP Legal Counsel for the Elderly Washington, D.C. Long-Term Care Ombudsman. She is the author of chapters and articles on older people's rights and older people's use of technology. You can read her blog posts on the HelpAge USA blog. She holds a B.A. in Psychology from Barnard College, Columbia University, and a JD from American University Washington College of Law.

Introduction

A. Framing the International Landscape

Great strides have been undertaken since the first legal article on older persons' rights was published in 2002.[1] This work builds on that international foundation. It focuses on descriptions of developments in legal frameworks and policy. It also includes a side-by-side comparison of the work of regional human rights mechanisms, which have picked up some momentum in the past decade. Through its inclusion of law, policy, and current processes, the widest net possible has been cast to collect a descriptive resource for advocates.

In the concluding chapter, some inferences and critiques on these existing laws, policies and processes are drawn. Weaknesses are present in the existing international and regional systems this text addresses; readers will see gaps in the international legal system's protection of older persons' human rights. These shortcomings are discussed in the concluding chapter as well.

Human rights are universal, indivisible and interdependent. This book pays equal regard to the civil and political rights of older persons as well as the economic, social and cultural rights of older persons. These explorations of rights are organized into these two categories because of existing systems of organization, not as a statement of the primacy of one set of rights over another. One of its main objectives is facilitation of comparison of the rights of older persons across different regions and legal frameworks. This sheds light on the state of play of older persons' existing rights around the world. It may leave the reader wanting to know more about where the gaps are: for example, if social security rights are being litigated in Europe for 65 year-olds, does that mean that older persons' participation in political affairs is equally protected? This is an exciting time to research older

[1]Diego Rodríguez-Pinzón & Claudia Martin, The International Human Rights Status of Elderly Persons, 18 Am. U Int'l L. Rev. 915, 969 (2002).

C. Martin et al., *Human Rights of Older People*, Ius Gentium: Comparative Perspectives on Law and Justice 45, DOI 10.1007/978-94-017-7185-6_1

persons' rights because these questions are just starting to be asked and answered. International law informs regional frameworks and vice versa. As different systems grapple with answers, the whole human rights framework stands to benefit.

The movement toward equal rights for persons of all ages is in a state of transformation: old age is making the transition from being simply a question of demographics, social issues, and welfare to being a question of rights. This book has shown that the conversation is changing rapidly as social scientists, legal advocates, and human rights experts learn from each other and explore how to work together. Seven core international human rights treaties already recognize specific groups, and experts from each of these recognized fields are contributing to the shared emerging understanding of older persons' human rights.

Social scientists at the Economic Commission for Latin America and the Caribbean (ECLAC), lawyers at the American Bar Association (ABA), and other powerful professional organizations have crafted thoughtful arguments for involvement in discussions of older persons' rights. ECLAC points to a "Paradigm Shift," which includes, among other things, the movement toward a rights-based approach to aging. The ABA has urged the US Dept. of State to participate in efforts and consultations toward an international and regional human rights instrument on the rights of older persons. These are just a few examples of the influence different disciplines can make when they begin thinking about older persons' rights.

Frameworks for older persons' rights are starting to emerge. The United Nations' Open Ended Working Group toward Strengthening the Human Rights of Older Persons (OEWG) began its work in 2011, as did the Organization of American States' working group drafting an Inter-American Convention on older persons' human rights. The question of what makes older persons a unique group in need of special protection, and the normative gaps that exist to prevent their access to equal rights, and what the articulation of their rights should be are all subjects of inquiry.

One of the main themes running through this work is the cumulative effect of multiple discriminations across the life course. This can be viewed through the lens of vulnerable groups, or culture by culture. Women, for example, are biologically predisposed to longer life expectancy and where available, must often depend on pensions based on lower earnings than men.[2] The region of Latin America and the Caribbean has recognized the importance of promoting and strengthening affirmative actions for elderly women and for increasing their access to resources.[3] Exactly which groups of older persons are at heightened risk for discrimination

[2]U.N. Econ. and Soc. Comm'n for Asia & the Pac., *The Highlights of Progress and Challenges in Implementing the Beijing Platform for Action: Good Practices, Obstacles and New Challenges*, 36, http://www.unescap.org/esid/GAD/Events/HLM-2009/download/BPA09_2E.pdf (last visited Oct. 22, 2014).

[3]Report of the Ninth Session of the Regional Conference on Women in Latin America and the Caribbean, ECLAC, (June 10–12 2004), at 14, http://www.un.org/womenwatch/daw/Review/english/documents/ECLACOutcomeDocument-English.pdf (last visited Oct. 23, 2014).

may vary from region to region and culture to culture. For example, where older men and women experience a bar from formal employment in countries in one region, they may face long-term discrimination in health care in another.

There are two main implementation methods for protecting rights: "soft law," which includes documents such as the United Nations Madrid International Plan of Action on Ageing (MIPAA); and "hard law" which refers to statutes and conventions which link national law with international law and carry expectations of implementation and accountability. In international "soft law" there has been a treatment of issues and strategies, but the discourse is not couched in legal terms or terms of rights, and affected groups are left without redress or enforceability. No sanctions for non-performance or rights infringements are attached.[4] Soft law can, over time create a body that can become customary law.[5] And the normative principles therein …"may lack the supposedly harder edge of a "rule" or an "obligation", but they are certainly not legally irrelevant."[6]

In international "hard law," human rights primary responsibility rests with the state to respect human rights to protect individuals and groups, to make positive actions. State responsibility includes resourcing the implementation and administration of law. New law often requires both paradigm and attitudinal shifts, which can be resource-intensive to achieve.[7] Some such resulting policy changes may be attained through soft law measures, without such resource-intensive outlay.[8] State constitutions may provide protection of older persons' rights,[9] for example, and hard law can change state constitutions. The enforcement of such law depends on domestic legal and political environments.[10] With soft law, even in favorable legal and political environments, enforceability is not a possibility.

At every level, one of the biggest problems older persons face is their invisibility. Worldwide, the United Nations High Commissioner for Refugees (UNHCR) has estimated that older persons make up 8.5 % of the overall refugee population, and in some cases comprise more than 30 %.[11] International human rights training

---

[4]Expert Group Meeting on the Rights of Older Persons, Bonn, Ger., May 5–7, 2009, *Report of the Expert Group Meeting on the Rights of Older Persons*, 14, www.un.org/esa/socdev/ageing/documents/egm/bonn09/report.pdf (last visited Oct. 23, 2014).

[5]*See* A.E. Boyle, *Some Reflections on the Relationship of Treaties and Soft Law*, 48 Int'l & Comp. L.Q. 901, 901–904 (1999).

[6]*Id.* at 907.

[7]*Report of the Expert Group Meeting on the Rights of Older Persons*, *supra* note 4, at 14.

[8]*See* I. Doron & I. Apter, *The Debate around the Need for an International Convention on the Rights of Older Persons*, 50 (5) The Gerontologist 590 (2010).

[9]*Report of the Special Rapporteur on the right to food, Mission to Cuba*: A/HRC/7/5/Add.3

[10]*See* O. Hathaway, *Why Do Nations Join Human Rights Treaties?*, 51 Journal of Conflict Resolution 590 (2007).

[11]David Hutton, Older People in Emergencies: Considerations for Action and Policy Development, WHO, 2008 World Health Organization.

should sensitize trainees to the rights and needs of older persons.[12] Insofar as this book will be used as an educational tool, translation of the rights of older persons into every sphere of the law can be worked into any curriculum. Unfortunately, aging and the situation of older persons continues to be addressed from a "humanitarian" perspective at best, and a "charity" perspective at worst, both of which ignore the fundamental equality of individuals, and deny societies of the potential of older persons to contribute to reaching national developmental objectives and goals.[13] Ignorance of the value of empowerment of older persons jeopardizes progress towards achieving global development goals.[14] Older persons' rights are an important part of their dignity, protection, and financial security.

Discrimination, violence and abuse, poverty and a lack of specific measures and services are all common threads running through the issues of older persons' human rights.[15] These issues have nothing to do with intrinsic vulnerability, and everything to do with disempowerment. The following chapters tell the story of the legal rights of this group in the international arena of the UN, the regional human rights systems, paying due attention to "soft law" mechanisms, and other international organizations which have begun to look at the critical challenges facing this diverse population group.

Soft and hard law continues to be a topic of discussion at the United Nations Open-Ended Working Group on Strengthening the Human Rights of Older Persons (OEWG), a working group that begun in 2011 and was tasked to explore main elements of a new legal instrument to protect the human rights of older persons, among other things. MIPAA, the group has discerned, has not had effects similar to the introduction of hard law.

Soft law and hard law are also part of the mandate of the new independent expert on the enjoyment of all human rights by older persons. On September 27, 2013, the Human Rights Council unanimously approved a resolution creating the new special procedure.[16] This position will assess the implementation of existing international instruments with regard to older persons, raise awareness of the challenges faced in the realization of all human rights by older persons, and ensure

---

[12]U.N. High Comm'r for Human Rights, *Consultation on the Focus of the Second Phase of the World Programme for Human Rights Education: Rep. of the High Comm'r for Human Rights*, U.N. Doc. A/HRC/12/36 (Aug. 20, 2009).

[13]*See Older Persons in Emergencies: An Active Ageing Perspective* World Health Org. (WHO), (2008), http://www.who.int/ageing/publications/EmergenciesEnglish13August.pdf (last visited Feb. 18, 2015).

[14]Id.

[15]U.N. Secretary-General, *Follow-up to the Second World Assembly on Ageing: Submitted Pursuant to G.A. Res. 65/182*, U.N. Doc. A/ 66/173 (Jul. 22, 2011).

[16]U.N. Human Rights Council, *Promotion and Protection of all Human Rights, Civil, Political, Economic, Social and Cultural Rights, Including the Right to Development, Twenty-Fourth Session, Agenda Item Three*, U.N. Doc. A/HRC/24/L.37/Rev.1 (Sep. 25, 2013), http://ap.ohchr.org/documents/dpage_e.aspx?si=A/HRC/24/L.37/Rev.1 (last visited Oct. 23, 2014).

that older persons receive information about those rights. It will work in cooperation with UN Member States in order to foster the implementation of measures that contribute to the promotion and protection of the rights of older persons; taking the views of many stakeholders into account, with a perspective inclusive of other marginalized group identities.[17]

These activities will be undertaken with the goal of producing annual reports for the next two years, and a comprehensive report of all findings at the Human Rights Council's thirty-third session. In the Independent Expert's preliminary considerations, she has given her views on the need to consider in her work both civil and political, and economic, social and cultural rights, as well as hard and soft law issues.[18]

B. Regional Efforts

1. The Americas

International law can inform regional laws, and vice versa. The potential for cross-pollination is present in the work undertaken by the Organization of American States (OAS) to draft an international convention on the human rights of older persons for the region. Its work has begun between the second and third sessions of the UN's OEWG.

The OAS process began years ago, thanks to the pressures of some state champions, including Argentina and Brazil. In 2010, the OAS requested the Permanent Council for convening a special meeting, in which different actors like national representatives, scholars, international organizations and civil society individuals. The meeting was held the 28th of October, and it was discussed the need of preparing an inter-American convention on the rights of older persons.[19] All states at that meeting, save the United States and Canada, positioned themselves supportively for embarking on a new program of work to prepare an inter-American convention on the rights of older persons.

In October 2011, a Work Plan of the Working Group on Protecting the Human Rights of Older Persons for 2010–2011 within the OAS was approved.[20] This plan outlined the way forward including contributions from civil society, and work

---

[17]See Id.

[18]U.N. Human Rights Council, *Report of the Independent Expert on the Enjoyment of All Human Rights by Older Persons*, U.N. Doc. A/ HRC/27/46 (Rosa Kornfeld-Matte Jul. 23, 2014), http://www.ohchr.org/EN/HRBodies/HRC/RegularSessions/Session27/Documents/A-HRC-27-46_en.doc (last visited Feb. 18, 2015).

[19]*Record of the Special Meeting of the Permanent Council*, OAS Doc. CP/SA. 1776/10 (Oct. 28, 2010).

[20]Permanent Council of the OAS, Committee on Juridical and Political Affairs, *Work Plan of the Working Group on Protecting the Human Rights of Older Persons*, OAS Doc. CAJP/GT/DHPM-1/11 rev. 2 (Oct. 21, 2011), http://www.oas.org/consejo/cajp/personas%20mayores.asp (last visited Feb. 18, 2015).

across the region with representatives from CARICOM, PAHO and ECLAC. In December 2011, the working group produced a plan of work that outlined its work on a Preliminary Draft Inter-American Convention on Protection of the Human Rights of Older Persons.[21] In June 2015 the OAS General Assembly finally adopted the Inter-American Convention on Protecting the Human Rights of Older Persons. This book was finalized before that date.

2. *Africa*

In 2011, the African Union began drafting of a treaty on the rights of older persons.[22] Also in 2011, the Network of African National Human Rights Institutions held its 8th Biennial conference in Cape Town, South Africa on the theme *Advancing the Human Rights of Older Persons and Persons with Disabilities – the Role of African National Human Rights Institutions*. At the opening of the 50th Session of the African Commission on Human and Peoples Rights, the situation of older persons were recognized as being "one of the most pressing human rights issues for states and public institutions."[23]

3. *Other Regions*

Other regional frameworks for the protection of the rights of older persons are not nearly so close at hand. The European Union opposed progress toward a dedicated human rights framework for older persons at each session of the UN's OEWG since 2011.[24] On February 14 2014, the Council of Europe (CoE)

[21]Permanent Council of the OAS, Committee on Juridical and Political Affairs Working Group on Protection of the Human Rights of Older Person, *Report on the Situation of Older Persons in the Hemisphere and the Effectiveness of Binding Universal and Regional Human Rights Instruments with Regard to Protection of the Human Rights of Older Persons*, OAS Doc. CAJP/GT/DHPM-14/11 rev. 1 (Dec. 21, 2011).

[22]African Comm'n on Human and Peoples' Rights [ACHPR]. *Statement of Reine Alapini-Gansou: The Legal Protection of the Rights of Older Persons in Africa* (Apr. 20, 2011), http://www.globalaging.org/agingwatch/convention/un/Alapini%20-PROTECTIONDPA%20AFRIQUE%5B1%5D.pdf.

[23]ACHPR, *Statement on Behalf of the Network of African National Human Rights Institutions at the Opening Session of the ACHPR 50th Ordinary Session* (Oct. 24, 2011–Nov. 7, 2011), http://www.achpr.org/sessions/50th (last visited Feb. 18, 2015).

[24]*See* U.N. Department of Economic and Social Affairs, Open-Ended Working Group on Ageing for the Purpose of Strengthening the Protection of the Human Rights of Older Persons, *EU Opening and Closing Statements* (July 30, 2014–August 1, 2014), http://social.un.org/ageing-working-group/govstatementfifth.shtml (last visited Oct. 23, 2014).

unanimously adopted a new soft law standard addressed to all 47 CoE member States.[25]

Asia has neither a human rights organ nor a dedicated field of legal inquiry for older persons. They do, however, have a roadmap toward promoting social justice for different groups, including older persons.[26] Such efforts mark progress that can be built upon as other international and regional efforts continue to develop and take shape in the international community.

Note on International Disaggregated Data Collection

Some UN member states have registered their opinions that a determination on how to proceed to best protect the human rights of older persons should be made only after data on the current situation of older persons is collected to verify that older persons are indeed facing challenges to exercise their rights. This line of reasoning sounds logical, until one attempt to find data on older people. UNAIDS, for example, collects HIV prevalence data only for persons under age 50. Not only is this sort of exclusion discriminatory, but disaggregated data collection on disease is necessary for a state to fulfill its human rights obligations to its citizens as well as its reporting to international human rights bodies. Where countries face no pressure to collect disaggregated data for their citizens, they will continue to draw a curtain over duties shirked and failures to protect their citizens. A normative change in the value placed on older persons could improve data collection. With the International Convention on the Rights of Persons with Disabilities (ICRPD), disaggregated data collection only really began once the ICRPD came into force.

Note on Nomenclature

Some may wonder why the term "older persons" is in use, rather than "the elderly" or "senior citizens," or "the aging." Self-definition of a group, reclamation of once-derogatory labels, and political positioning can all be reasons for certain identifications. The term "older persons" is used throughout these chapters for a few reasons. The paramount reason is that other successful rights movements have also used "person."[27] It is clearly understood -particularly for the purposes of

---

[25]Council of Europe Comm. of Ministers, *Recommendation to Member States on the Promotion of Human Rights of Older Persons*, CM/Rec(2014)2 (Feb. 14, 2014), https://wcd.coe.int/ViewDoc.jsp?id=2162283&Site=COE&BackColorInternet=C3C3C3&BackColorIntranet=EDB021&BackColorLogged=F5D383 (last visited Feb. 18, 2015).

[26]Association of Southeast Asian Nations [ASEAN], *Roadmap for an ASEAN Community 2009–2015 - Social Justice and Rights* (2009), http://www.asean.org/images/2012/publications/RoadmapASEANCommunity.pdf (last visited Feb. 18 2015).

[27]*See*, Convention on the Rights of Persons with Disabilities, May 3, 2008, 2515 U.N.T.S. 3.

human rights- that emphasis should always be on the person as an individual with humanity and dignity. Depending on the context, the idea of an "older person" may conjure images of the "golden years" of retirement, leisure, and golf courses. Or it may be seen as "nothing more than a euphemism used to mask the reality of old age, which is seen as a stigma, and to distance it from the associated idea of death."[28] "Older" can mean comparatively so, socially constructed as such, or experiencing natural physical processes after attaining adulthood.

[28]Sandra Huenchuan and Luis Rodríguez-Piñero, U.N. Economic Commission for Latin America and the Carribean, *Ageing and the Protection of Human Rights: Current Situation and Outlook*, 15, (Mar. 2011, Sandra Huenchuan & Luis Rodríguez-Piñero).

The United Nations and Older Persons: A Global Approach with Multiple Dimensions

A. A United Nations Introduction

Rich and poor countries alike are currently experiencing extraordinary demographic changes with respect to the rising number of older persons in their populations. Over the next four decades, the population over age 60 in developed countries is projected to double.[1] In developing countries, it is expected to triple, reaching a staggering 1.6 billion by 2050.[2] These changes will create a new set of challenges that should be dealt with coherently, including an acute need for redistribution of resources for access to justice, participation in political and social life, social security, and health care. Furthermore, states, as primary protectors of human rights, will face challenges related to upholding human rights in the new demographic landscape.

United Nations' Follow-Up to the Second World Assembly on Ageing[3] describes the problem this way: "[An] absence of nuanced, targeted, and comprehensive approaches toward the rights of older persons leave them with chronic poverty, untreated illness, homelessness or inadequate shelter, violence and abuse, lack of education, low paying jobs, ineffective legal protection, vulnerability,

[1]UN Secretary General, *Follow up to the Second World Assembly on Ageing*, 9 3, U.N. Doc. A/64/127 (Jul. 6, 2009).

[2]Peggy Kelly, *Intersection of Population and Human Rights: Rights of Older Persons in the International Context, Meeting of Experts Regarding Population, Inequality, and Human Rights* U.N. Econ. Comm. for Latin America and the Carribean (2006), *available at* http://www.eclac.org/celade/noticias/paginas/6/27116/KellyP.pdf. (last visited Oct. 23, 2014). (*hereinafter* Kelly Report).

[3]United Nations, General Assembly, *Follow-Up to the Second World Assembly on Ageing* G.A. Res. 63/151, U.N. Doc. A/63/424 (Feb. 11, 2009) http://www.worldlii.org/int/other/UNGARsn/2008/197.pdf (Follow-Up to the Second World Assembly on Ageing).

C. Martin et al., *Human Rights of Older People*, Ius Gentium: Comparative Perspectives on Law and Justice 45, DOI 10.1007/978-94-017-7185-6_2

exclusion, and isolation."[4] This diagnosis touches on almost every area of life - it is not just a problem of either economic, social and cultural rights, or civil and political rights.

This chapter focuses on international United Nations human rights frameworks and their impact on the status of the rights of older persons. The scope of analysis includes universal human rights instruments, as well as related work produced by treaty bodies including comments, recommendations, conclusions, and reports. This chapter will also endeavor to analyze older persons' human rights through the lens of the spectrum of the life course. Additionally, this chapter identifies a number of group rights which have further impact on the situation of older persons represented in particular groups.

Although international human rights standards have gained increasing recognition for the rights of individuals generally, the rights of older persons have not yet received the international legal treatment they require. The current international standards and recommendations on the rights of older persons derive from the fundamental premises established by United Nations (UN) human rights instruments starting with the Universal Declaration of Human Rights (UDHR). The UDHR, together with the International Covenant on Civil and Political Rights (ICCPR), its two Optional Protocols, and the International Covenant on Economic, Social and Cultural Rights (ICESCR) form the International Bill of Human Rights. Article 25, paragraph 1 of the UDHR states that:

> Everyone has the right to a standard of living adequate for the health and well-being of himself and of his family, including food, clothing, housing, and medical care and necessary social services, and the right to security in the event of unemployment, sickness, disability, widowhood, *old age* or other lack of livelihood in circumstances beyond his control.[5]

Since the adoption of the UDHR, more international and regional instruments addressing the protection and promotion of human rights have been created. The rights of older persons are theoretically addressed by virtue of their being human, even where they are not specifically mentioned, as above. Instruments have taken the form of treaties, agreements, conventions, and protocols. These instruments bind the contracting states to negotiated terms, agreed to by ratification or accession. However, none of these instruments explicitly refers to age as a prohibited ground for discrimination.[6]

The treaty bodies have developed protections that have expanded the scope of the respective treaties, including the rights of older persons, through General Comments (Economic, Social and Cultural Rights Committee) and through General Recommendations (Committee for the Elimination of all forms of Discrimination against Women). Standards for the rights of older persons are also enshrined in other instruments including plans of action, declarations, resolutions, and recommendations generated by the treaty bodies and a number of multilateral organizations.

---

[4]Id. at para. 49. *Id.* 49.

[5]Universal Declaration on Human Rights, Article 25. Universal Declaration of Human Rights, G.A. Res. 217 (III) A, Art. 25, U.N. Doc. A/RES/217 (III) (Dec. 10, 1948).

[6]U.N. Secretary-General, *Report on Ageing, 13* (2009).

In the context of an international human rights framework, the rights of older persons are universal and inalienable, indivisible, interdependent and interrelated,[7] just as they are for any human being. Additionally, the concepts of equality and non-discrimination apply to these rights.[8] This chapter will discuss the different types of rights as well as systems of protection in order to expand the understanding of the notion of older persons' rights, centering on specific international norms and their interpretation by treaty bodies and application in the Human Rights Council.

B. United Nations Human Rights Framework on Aging

a. The UN System

The UN aims to "bring all nations of the world together to work for peace and development, based on the principles of justice, human dignity and the well-being of all people."[9] It is structured such that states may "balance global interdependence and national interests when addressing international problems."[10]

Article 10 of the UN Charter empowers the General Assembly to discuss any questions or any matters within the scope of the Charter. Further, the General Assembly can initiate studies and make recommendations for the purpose of assisting in the realization of human rights.[11] The General Assembly is composed of all UN member states. Although its resolutions are not per se legally binding, they are an important reflection of the collective will of the world community, and in some cases reflect binding customary law. Much of the debate and drafting surrounding resolutions occurs in six main committees of the General Assembly. This chapter deals in particular with the Third Committee of the UN General Assembly (Third Committee), which focuses on social, humanitarian, human rights and cultural issues.[12]

The UN Charter also outlines the role of the organization with regard to respecting equal rights and promoting higher standards of living, employment and the economic conditions of social progress. It strives to promote universal

---

[7]*See Report of the Expert Group Meeting "Rights of Older Persons"* United Nations Dep't of Econ. and Soc. Affairs Division for Soc. Policy and Dev. Programme on Ageing (2009) http://www.ohchr.org/en/issues/Pages/WhatareHumanRights.aspx. Accessed 20 January 2014. *See* Expert Group Meeting on the Rights of Older Persons, Bonn, Ger., May 5–7, 2009, http://www.ohchr.org/en/issues/Pages/WhatareHumanRights.aspx.

[8]Id.

[9]U.N. Cyber School Bus, *An Introduction for Students* (Oct. 23, 2014), http://www.un.org/cyberschoolbus/unintro/unintro.asp.

[10]Id.

[11]U.N. Charter art. 13 (1945).

[12]Henry J. Steiner, Philip Alston, & Ryan Goodman, *Human Rights in Context 739 (3d ed. 2008).*

respect for human rights and fundamental freedoms for all, though it stops short of expressly recognizing prohibited distinctions other than race, sex, language, and religion.[13] It does not include age as a prohibited ground for discrimination.

Human rights scholars from different regions and cultures tend to agree that human rights stem principally from the liberal tradition of Western political and legal thought.[14] This tradition focuses on individuals rather than groups, and equality among individuals as paramount. The vital concept of equality informs the dignity of all human beings, the respect to which individuals are entitled, and the right for self-realization.[15] However, the notion of human rights has quickly evolved to include collective rights, among other contemporary forms of rights now recognized in many international instruments.

At the UN, human rights are monitored in two ways. Firstly, the organs created directly through a mandate of the UN Charter, such as the General Assembly and the Human Rights Council, monitor human rights. Secondly, treaty-based bodies created by a range of UN human rights treaties monitor states' compliance with their obligations under each treaty, respectively.[16]

UN Social Policy and Soft Law

The Third Committee is the main committee of the UN General Assembly that addresses agenda items relating to a range of social, humanitarian affairs and human rights issues by drafting resolutions for consideration by the Assembly.[17] The Commission for Social Development as a functional body of the Economic and Social Council has taken an increasingly active role in standard setting around social issues. For example, in 2001, a mandate for a draft of the Convention on the Rights of Persons with Disabilities stalled in the Commission on Human Rights. The Commission for Social Development established a special committee that successfully moved the CRPD forward. The Commission for Social Development has proven to be a body capable of developing new standards. This shows how bodies not belonging to the main UN human rights system can also be alternative fora to overcome stagnant processes. Its role in the

[13]United Nations, *Charter of the United Nations*, 1 UNTS XVI (October 24 1945), *available at* http://www.refworld.org/docid/3ae6b3930.html. Article 55, http://www.un.org/en/documents/charter/chapter9.shtml. (last visited Oct. 23, 2014).

[14]Philip Alston and Ryan Goodman, *International Human Rights*, 525 Oxford University Press (2013), at 525.

[15]Id.

[16]Id.

[17]U.N. General Assembly, *Social, Humanitarian, & Cultural* (Oct 23, 2014), *available at* http://www.un.org/ga/61/third/third.shtml.

Convention on the Rights of Persons with Disabilities may be considered evidence of this standard setting work.[18]

The UN secretariat, or agencies, also conduct social policy work, which can result in soft law. In 1982 the UN General Assembly held the first world conference on aging: the World Assembly on Ageing. This Assembly issued the Vienna International Plan of Action on Ageing (hereinafter "Vienna Plan").[19] This first world conference was followed in 2002 by the Second World Assembly on Ageing that adopted the Madrid International Plan of Action on Ageing (hereinafter "Madrid Plan" or "MIPAA").[20] The Second World Assembly was preceded a decade earlier by the 1991 United Nations Principles for Older Persons adopted by the General Assembly, and by the 1992 International Conference on Ageing as a follow up to the Vienna Plan. In this two-decade process, the UN refined its approach in identifying its main concerns regarding older persons.

In this chapter, the work of UN human rights mechanisms, other UN institutions, and UN specialized agencies are addressed. Older persons have been surprisingly absent from much of the ongoing work in each of these spheres, given the directives of the 1982 and 2002 world conferences and related UN resolutions, 2002 UN Madrid International Plan of Action on Ageing (MIPAA), which called for the mainstreaming of older persons in all UN programs of work.

Old age is a contextual, social concept with real implications for older persons. Older persons experience direct and indirect discrimination, and the cumulative effects of chronic poverty; violence and abuse; exclusion from social and political participation; and barriers to redress through access to justice.[21] In many developed countries, old age is sometimes thought to be commensurate with the age at which a person becomes eligible for a social pension. In those cases, society, through the government, determines that older persons reaching this age should be entitled to additional protections. In other societies, a person may become "old" when they have grandchildren, or great grandchildren. Whatever the societal construct, becoming "old" can mean facing discrimination, and invisibility.

Older persons' human rights have always been closely associated with the "economic and technical co-operation undertaken by United Nations bodies, agencies, funds, and regional commissions, particularly in the field."[22] Since 2002, the inter-

---

[18]The Commission is the Organ Implementing MIPAA (Madrid International Plan of Action on Ageing) (See Report of the Second World Assembly on Ageing, 1 43, U.N. Doc. A/CONF.197/9, U.N. Sales No. E.02.IV.4 (2002) (outlining the Madrid Plan)). (*See* U.N. World Assembly on Ageing, Apr. 8–12, 2002, *Report of the Second World Assembly*, U.N. Doc. A/CONF.197/9, 1–43.

[19]U.N. World Assembly on Ageing, Jul. 26-Aug. 6, 1982, *Vienna International Plan of Action on Ageing*, http://www.un.org/es/globalissues/ageing/docs/vipaa.pdf.

[20]U.N. World Assembly on Ageing, Apr. 8–12, 2002, *Political Declaration & Madrid International Plan of Action on Ageing*, http://www.un.org/en/events/pastevents/pdfs/Madrid_plan.pdf.

[21]*United Nations Mechanism for the Human Rights of the Older Person*, A/HRC/AC/4/CRP.1 (Chinsung Chun 2009).

[22]Sandra Huenchuan and Luis Rodríguez-Piñero, *Ageing and the Protection of Human Rights: Current Situation and Outlook*, 51 (Sandra Huenchuan & Luis Rodríguez-Piñero, 2011).

national community has engaged in a number of efforts geared toward the protection of older persons: the United Nations Department of Economic and Social Affairs has led 5-year and 10-year reviews of MIPAA, the Committee for the Elimination of Discrimination against Women came out with a General Recommendation on older women's rights[23]; and the UNFPA has turned its attention to the compilation of a report on the "State of the World's Older Persons," prepared in conjunction with the 10-year review of MIPAA.[24] MIPAA, as has been mentioned in the introduction, is a soft law instrument, and will be more fully addressed subsequently in this book. These activities did not all engage in the advancement of older persons' rights, but some may have heightened the international profile of the situation of older persons. UNFPA in particular has integrated a human-rights based approach to development,[25] which has been important in the discussion of the equal rights of older people.

The UN OEWG

Much of this thinking may be changing with the creation of the UN's Open-Ended Working Group on Strengthening the Human Rights of Older Persons (OEWG), a political meeting of the UN Member States. In September, 2010, the UN Secretary General released a report on the human rights of older persons.[26] This was released in response to a 2009 General Assembly resolution requesting more information on the full picture of the situation of older persons.[27] While it

[23]Committee on the Elimination of Discrimination Against Women, *General Recommendation No. 27: The Rights of Older Women*, U.N. Doc. CEDAW/C/GC/27 (Dec. 16, 2010). http://www2.ohchr.org/english/bodies/cedaw/comments.htm.

[24]Ann Pawliczko, Presentation at the U.N. Celebration of the International Day of Older Persons (Oct. 5, 2011).

[25]See, e.g. UNFPA, 2009. *Integrating Human Rights, Culture and Gender in Programming A Human Rights Based Approach to Programming* (2010). http://www.unfpa.org/webdav/site/global/shared/documents/publications/2010/hrba/module_1.pdf (last visited Feb. 18 2015); *See, e.g.* U.N. Population Fund, *Integrating Human Rights, Culture and Gender in Programming: A Human Rights Based Approach to Programming* (2009), http://www.unfpa.org/webdav/site/global/shared/documents/publications/2010/hrba/module_1.pdf. (last visited Oct. 23, 2014); *UNFPA at Work, Six Human Rights Case Studies*, http://www.unfpa.org/webdav/site/global/shared/documents/publications/2008/6human_rights_cases.pdf. (last visited Oct. 23, 2014). U.N. Population Fund, *Human Rights Based Programming: What It Is*, (2006). http://www.unfpa.org/webdav/site/global/shared/documents/publications/2007/human_rights.pdf (last visited Feb. 18 2015).

[26]U.N. Secretary-General, *"Follow-up to the Second World Assembly on Ageing" Submitted Pursuant to the General Assembly Res.* 65/182 U.N. Doc. A/66/173 (Dec. 2010) (July 22, 2011).

[27]*Follow-Up to the Second World Assembly on Ageing*, G.A. Res. 64/132, U.N. Doc. A/RES/64/132 (Feb. 5, 2010).

took note of policies and programs within the UN system targeting older persons, it also discussed the legal problems older persons face. The report outlined discrimination and ageism, poverty, violence and abuse, and the lack of specific measures and services for older persons today, and in the future. It sought to explore options to improve the situation of older persons worldwide, and created the OEWG.

In August 2011, the OEWG highlighted four main gaps: "norms, monitoring, implementation, and data surrounding older persons."[28] Likewise, is stressed that at an international level, the "existing rights have not been specifically applied, and thus proper implementation cannot be achieved."[29] Among the member states present, there was consensus for continued support of the work of the OEWG, and various expressions of the desire for the intensification of the inquiry.[30]

The OEWG's initial mandate allowed it to "consider the existing international framework of the human rights of older persons and identify possible gaps and how best to address them, including by considering, as appropriate, the feasibility of further instruments and measures."[31] The mandate of this working group changed with General Assembly resolution 67/139 of 20 December 2012. It is now charged to "Consider proposals for an international legal instrument to promote and protect the rights and dignity of older persons, based on the holistic approach in the work carried out in the fields of social development, human rights and non-discrimination, as well as gender equality and the empowerment of women..." This change in mandate was not reflected in the program of work at the fourth session of the OEWG in 2013. It was also not reflected in the program of work at the fifth session of the OEWG in 2014, although the lack of direct response to the changed mandate was discussed by Member States present.

A CESCR representative noted at the first OEWG meeting in April, 2011, that at this point, dedicated human rights committees *extend* human rights to older persons; before an argument or complaint can be made about the violation of an older person or group of older persons' rights, applicants must first explain how supposedly universal law applies to them.[32] As a matter of course human rights cannot be accessed by older persons in the current framework.

[28]U.N. Open-Ended Working Group on Ageing, *Daily Summary of the 2d Session* (Bethany Brown, Aug. 4, 2011), http://globalaging.org/agingwatch/convention/un/OEWG%20August%20Day%204.html.

[29]Id.

[30]NGO Committee on Ageing, *Recap Aug. 4: UN Open-Ended Working Group on Ageing* (Aug. 4, 2011), http://www.ngocoa-ny.org/the-second-un-open-ended/recap-aug-4-un-open-ended.html.

[31]http://social.un.org/ageing-working-group/. G.A. Res. 67/139, U.N. Doc. A/RES/67/139 (Feb. 13, 2013).

[32]*See* comments of Rocio Barahona Riera, April 19, 2011, OEWG, New York. *See* Rocio Barahona Riera, Comments at the U.N. Open-Ended Working Group on Ageing (April 19, 2011).

Civil Society and the UN OEWG

Through its mandate, the OEWG has given civil society the opportunity to play an active role in its work. In many sessions, equal time was devoted to interventions from states and civil society at the 2011 meetings of the OEWG. NGO's involved in the OEWG so far have ranged from groups of gerontologists, to retired civil servants, to the International Network for the Prevention of Elder Abuse. Through this exchange, organizations that serve or advocate for older persons directly have had a platform to share evidence of how the existing human rights framework does not adequately protect the rights of older persons.

Work is also being done in partnership with the UNFPA to strengthen civil society's support of ongoing work for older persons' rights. Professors Martin and Rodriguez-Pinzon, together with the UNFPA, have convened two training sessions for older persons' advocates from around the world in 2010 and 2011. Similar meetings should be convened to better train advocates at all levels in the understanding of human rights, and existing mechanisms.

A way to amplify civil society's voice is an important resource in an advocacy movement.[33] One such group was created among some of the largest NGO's after the first session of the OEWG. In 2013, it announced plans to accept membership of the broader community of organizations advocating for the rights of older persons. It aims to be a source of information, a communications resource, and to support the writing of strategic positioning papers.[34]

1. Overview of International Law

This chapter will review the instruments in greater depth, but this section provides an overview of basic human rights sources. The Universal Declaration of Human Rights (UDHR)[35] was adopted by the UN General Assembly. It is the world's common standard of respect and universal recognition and observance of basic rights and freedoms.[36] It is a statement of objectives to be pursued by governments, and therefore it is not per se part of binding international law. Nonetheless, it remains a potent way to apply moral and diplomatic pressure on states that

---

[33] J. Sciubba, *Explaining Campaign Timing and Support for a UN Convention on the Rights of Older People*, 18 Int'l J. Hum. Rts. 462, 474 (2014)., http://www.tandfonline.com/eprint/XuIpnacg7NKbEkTAFEb3/full. (last visited Oct. 23, 2014).

[34] *See* Global Alliance for the Rights of Older Persons, www.rightsofolderpeople.org. (last visited Oct. 23, 2014).

[35] Universal Declaration of Human Rights, G.A. Res. 217 (III) A, U.N. Doc. A/RES/217(III) (Dec. 10, 1948). http://www.un.org/en/documents/udhr/.

[36] Preamble, Universal Declaration of Human Rights http://www.un.org/en/documents/udhr/. (last visited Oct. 23, 2014).

violate the UDHR's principles, and many of its provisions are indeed a reflection of international customary law.[37]

The International Covenant on Civil and Political Rights (ICCPR),[38] is a multilateral treaty adopted by the General Assembly, which defends core civil and political rights including the right to life,[39] movement,[40] expression,[41] association[42] and stipulates that "no one shall be subjected to torture;"[43] slavery[44]; or forced or compulsory labor, among others.[45]

The Human Rights Committee (HRC),[46] a body of independent experts, monitors the implementation of the ICCPR by state parties. It accepts and examines reports and then addresses its concerns and recommendations to the state party in the form of "Concluding Observations." It also considers individual complaints against state parties to the ICCPR Optional Protocol, which provides for an individual petition system.[47]

The Human Rights Council (HR Council), which replaced the Commission on Human Rights, is responsible for strengthening the promotion and protection of human rights around the world.[48] The success of any human rights reporting "will frequently depend upon the composition of the supervisory body, its commitment to the case of human rights, its creativity and the larger political climate within which it exercises its functions."[49] Toward that end, the HR Council was created in 2006 as an organ of the General Assembly.[50] Its main purpose is to address situations of human rights violations and make recommendations, as well as administer a Universal Periodic Review ("UPR") mechanism, which assesses the human rights situations in all UN Member States over the course of a four-year

---

[37]U.N. Association in Canada, *Questions and Answers about the Universal Declaration of Human Rights*, http://www.unac.org/rights/question.html (last visited Oct. 23, 2014).

[38]International Covenant on Civil and Political Rights, 999 U.N.T.S. 171 (Dec. 16, 1966), http://www1.umn.edu/humanrts/instree/b3ccpr.htm. (*hereinafter* ICCPR).

[39]Id. at art. 6.

[40]ICCPR, art 12 *Id.* at art. 12.

[41]ICCPR, art 19 *Id.* at art. 19.

[42]ICCPR, art 22 *Id.* at art. 22.

[43]ICCPR, art 7 *Id.* at art. 7.

[44]ICCPR, art 8 *Id.* at art. 8.

[45]ICCPR art 8.3 literal c *Id.* at art. 8.3, lit. c.

[46]U.N. Human Rights Committee, *Monitoring Civil and Political Rights*, http://www2.ohchr.org/english/bodies/hrc/ (last visited Oct. 23, 2014).

[47]Id.

[48]U.N. Human Rights Council, *About the Human Rights Council*, http://www.ohchr.org/EN/HRBodies/HRC/Pages/AboutCouncil.aspx (last visited Oct. 23, 2014).

[49]Henry J. Steiner, Philip Alston & Ryan Goodman, *International Human Rights in Context* 347 (3d ed. 2008) (quoting Thomas Buergenthal, *The U.N. Human Rights Committee*, 2001 U.N.Y.B.341, 347).

[50]G.A.Res. 60/251.U.N.Doc.A/60/L.48 (Apr. 3, 2006) (creating the HRC).

schedule.[51] The most contentious issues in its creation were related to its deliberations, composition and functions, including the UPR, and its relationship to the special procedures and the mechanisms it 'inherited' from the Human Rights Commission. [52]

The International Covenant on Economic, Social and Cultural Rights (ICESCR)[53] commits state parties to the progressive realization of individual economic, social, and cultural rights within the global human rights framework.[54] The Committee on Economic, Social and Cultural Rights[55] (CESCR) is the body of independent experts that monitors implementation of the ICESCR by its state parties.[56] The United Nations has pointed out that "the process of reporting provides an opportunity for an individual state party to conduct a comprehensive review of the measures it has taken to bring its national law and policy into line with the provisions of the treaties to which it is a party. The preparation of reports provides a platform for national dialogue on human rights amongst the various stakeholders in a State party". Likewise, reporting "facilitates public scrutiny at the national level of government approaches to implementation, and stimulates constructive discussion with civil society of a way to advance enjoyment by all of the rights laid down in the various conventions."[57] The value of the reporting process is truly national and it can be assessed on how it helps improve the lives of the persons for which states are responsible.

After the submission of the State report on the implementation of the ICESCR, CESCR proceeds to its analysis and the issuing of its observations. This process can be described as follows: "Upon completion by the Committee of its analysis of reports by state parties, the committee concludes its consideration of the State parties' reports by issuing "Concluding Observations," which constitute the decision of the Committee regarding the status of adherence to the Covenant in a given

---

[51]U.N. Human Rights Council, *About the Human Rights Council*, http://www.ohchr.org/EN/HRBodies/HRC/Pages/AboutCouncil.aspx (last visited Oct. 23, 2014).

[52]Steiner &Alston, *supra* note 42, 791 and 803.

[53]International Covenant on Economic, Social & Cultural Rights, Dec. 16, 1966, 993 U.N.T.S. 3 (hereinafter ICESCR) http://www2.ohchr.org/english/law/pdf/cescr.pdf.

[54]*See Fact Sheet No. 16 (Rev. 1)*, Comm. on Econ., Soc., and Cultural Rights, (July 1 1991), http://www.ohchr.org/Documents/Publications/FactSheet16rev.1en.pdf.

[55]U.N. Committee on Economic, Social, and Cultural Rights, *Monitoring the Economic, Social, and Cultural Rights*, http://www2.ohchr.org/english/bodies/cescr/.

[56]Office of the High Commissioner for Human Rights, *Ratification o the International Covenant on Economic, Social, and Cultural Rights*, (Aug. 2014) http://www2.ohchr.org/english/bodies/cescr/.

[57]International Human Rights in Context, p. 851, quoting 'Concept Paper on the High Commissioner's proposal for a unified standing treaty body', UN doc. HRI/MC/2006/2 (2006) summarizing the approach of Committee on Economic, Social and Cultural Rights, General Comment 1 (1989): Reporting by States Parties, UN Doc. HRI/GEN/1/Rev.8, p.9, para.8. Henry J. Steiner, Philip Alston & Ryan Goodman, *International Human Rights in Context* 851 (3d ed. 2008) (*quoting* U.N. Secretariat, *Concept Paper on the High Commissioner's Proposal for a Unified Standing Treaty Body*, U.N. Doc. HRI/MC/2006/2 (2006)) (*summarizing the approach of* U.N. Committee on Economic, Social, and Cultural Rights, *General Comment 1: Reporting by States Parties*, 8, UN Doc. HRI/GEN/1/Rev.8 (1989)) http://www.iilj.org/courses/documents/ConceptPaperonUnifiedStandingTreatyBody.pdf. (last visited Oct. 23, 2014).

state party."[58] However, the CESCR has indicated that "while the Committee's concluding observations, in particular its suggestions and recommendations may not carry legally binding status, they are indicative of the opinion of the only expert body entrusted with and capable of making such pronouncements. Consequently, for state parties to ignore or not act on such views would be to show bad faith in implementing their Covenant-based obligations."[59]

The CESCR also publishes its interpretation of the provisions of the ICESCR as General Comments.[60] The most relevant General Comments for older persons so far are: General Comment No. 3 on The Nature of States Parties Obligations[61]; General Comment No. 4 on The Right to Adequate Housing[62]; General Comment No. 5 on the Rights of Persons with Disabilities[63]; General Comment No. 6 on The Economic, Social and Cultural Rights of Older Persons[64]; General Comment No. 12 on The Right to Adequate Food[65]; General Comment No. 13 on The Right to Education[66]; General Comment No. 14 on The Right to the Highest Attainable Standard of Health; and General Comment No. 19 on The Right to Social Security.[67]

---

[58]Committee on Economic, Social, and Cultural Rights. *Fact Sheet No. 16 (Rev. 1)*, 16 (July 1991), http://www.ohchr.org/Documents/Publications/FactSheet16rev.1en.pdf.

[59]Id. at 17.

[60]U.N. Committee on Economic, Social and Cultural Rights (CESCR), *General Comments*, http://tbinternet.ohchr.org/_layouts/treatybodyexternal/TBSearch.aspx?Lang=en&TreatyID=9&DocTypeID=11 (last visited Feb. 18 2015).

[61]U.N. Committee on Economic, Social and Cultural Rights (CESCR), *General Comment No. 3: The Nature of States Parties Obligations (Art. 2, Par. 1 of the Covenant)*, U.N. Doc. E/1991/23 (Jan. 1 1991), http://tbinternet.ohchr.org/_layouts/treatybodyexternal/Download.aspx?symbolno=E%2f1991%2f23(SUPP)&Lang=en.

[62]U.N. Committee on Economic, Social and Cultural Rights (CESCR), *General Comment No. 4: The Right to Adequate Housing (Art. 11.1 of the Covenant)*, U.N. Doc. E/1992/23 (Jan, 1, 1992), http://tbinternet.ohchr.org/_layouts/treatybodyexternal/Download.aspx?symbolno=E%2f1992%2f23&Lang=en.

[63]U.N. Committee on Economic, Social and Cultural Rights (CESCR), *General Comment No. 5: Persons with Disabilities*, U.N. Doc. E//1995/22 (Jan. 1, 1995), http://tbinternet.ohchr.org/_layouts/treatybodyexternal/Download.aspx?symbolno=E%2f1995%2f22(SUPP)&Lang=en.

[64]U.N. Committee on Economic, Social and Cultural Rights (CESCR), *General Comment No. 6: The Economic, Social and Cultural Rights of Older Persons*, U.N. Doc. E/1996/22 (Oct. 7, 1996), http://tbinternet.ohchr.org/_layouts/treatybodyexternal/Download.aspx?symbolno=E%2f1996%2f22(SUPP)&Lang=en.

[65]U.N. Committee on Economic, Social and Cultural Rights (CESCR), *General Comment No. 12: The Right to Adequate Food (Art. 11 of the Covenant)*, U.N. Doc. E/C.12/1999/5 (May 12, 1999), http://tbinternet.ohchr.org/_layouts/treatybodyexternal/Download.aspx?symbolno=E%2fC.12%2f1999%2f5&Lang=en.

[66]U.N. Committee on Economic, Social and Cultural Rights (CESCR), *General Comment No.13: The Right to Education (Art. 13 of the Covenant)*, U.N. Doc.E/C.12/1999/10 (Dec. 8, 1999), http://tbinternet.ohchr.org/_layouts/treatybodyexternal/Download.aspx?symbolno=E%2fC.12%2f1999%2f10&Lang=en.

[67]U.N. Committee on Economic, Social and Cultural Rights (CESCR), *General Comment No. 19: The Right to Social Security (Art. 9 of the Covenant)*, U.N. Doc. E/C.12/GC/19 (Aug. 11, 2000), http://tbinternet.ohchr.org/_layouts/treatybodyexternal/Download.aspx?symbolno=E%2fC.12%2fGC%2f19&Lang=en.

In General Comment No. 6 on the economic, social, and cultural rights of older persons, the CESCR developed a comprehensive legal analysis of the rights of older persons at the international level.[68] Adopted in 1995, it interprets the ICESCR from the perspective of older persons,[69] expands the scope of the ICESCR and provides insight into different mechanisms needed to protect the rights of older persons worldwide. It has structural divisions that can serve as a blueprint for further analysis, and elaboration of a comprehensive set of the rights of older persons. Additionally, the CESCR issued several other general comments that provide an authoritative guide to understanding the scope of the ICESCR.

There is no binding international instrument or even supervisory body dealing specifically with the rights of older persons. At the time it was drafting this General Comment, the CESCR gave additional weight to the importance of the Vienna Plan and the five UN Principles for Older Persons.[70] If it were making pronouncements today, it might further look to the Convention on the Rights of Persons with Disabilities, MIPAA, and the Convention on the International Protection of Adults.

In the International Convention on the Elimination of All Forms of Racial Discrimination ("ICRD"),[71] "state parties condemn racial discrimination and undertake to pursue by all appropriate means and without delay a policy of eliminating racial discrimination in all its forms and promoting understanding among all races."[72] They undertake "to engage in no act or practice of racial discrimination against persons, groups of persons or institutions and to ensure that all public authorities and public institutions, national and local, shall act in conformity with this obligation."[73]

The Committee on the Elimination of Racial Discrimination ("CERD") [74] is the body of independent experts that monitors implementation of the ICRD by state parties. The CERD receives regular reports from state parties and as it has stated it also "addresses its concerns and recommendations to the State party in the form of "Concluding Observations." Regarding other monitoring mechanisms, CERD's

---

[68]U.N. Committee on Economic, Social and Cultural Rights (CESCR), *General Comment No. 6: Economic Social and Cultural Rights of Older People.*, U.N. Doc.E/1996/22 (Oct. 7, 1996), http://tbinternet.ohchr.org/_layouts/treatybodyexternal/Download.aspx?symbolno=E%2f1996%2f22(SUPP)&Lang=en.

[69]Id. 9–15.

[70]U.N. Secretary-General. *Follow up to the Second World Assembly on Ageing*, U.N. Doc. A/64/127(July 6, 2009).

[71]U.N. General Assembly, *International Convention on the Elimination of All Forms of Racial Discrimination*, United Nations, Treaty Series, vol. 660, p. 195 (Dec. 21 1965) *available at* http://www.refworld.org/docid/3ae6b3940.html. (last visited Oct. 23, 2014). (hereinafter ICERD).

[72]ICERD, art 2. *Id.* at art. 2.

[73]ICERD, Article 2,1 literal a. *Id.* at art. 2.1, lit.a.

[74]U.N. Committee on the Elimination of Racial Discrimination, *Monitoring Racial Equality and Non-Discrimination*, http://www2.ohchr.org/english/bodies/cerd/. (last visited Oct. 23, 2014).

tools include an early-warning system, and the examination of inter-state complaints and individual complaints.[75]

Similarly, the Convention on the Elimination of All Forms of Discrimination against Women ("CEDAW")...[76] "defines what constitutes discrimination against women and sets up an agenda for national action to end such discrimination."[77] States that have ratified or acceded to the CEDAW have the legal obligation to implement it. They are required to submit national reports periodically (at least every four years) on the level of implementation of their treaty obligations.[78]

The Committee on the Elimination of Discrimination against Women[79] ("CEDAW Committee") is a body composed of experts on women's issues from around the world, which has the mandate to mark progress for women in countries party to CEDAW.[80] The CEDAW Committee makes "General Recommendations" on any issue affecting women to which it believes the states parties should devote more attention,[81] which are based on the examination of reports and information received from states parties.[82] Most recently, the CEDAW Committee adopted General Recommendation 27 on Older Women and Protection of their Human Rights.[83] Additionally, the Optional Protocol to the Convention on the Elimination of All Forms of Discrimination against Women[84] ("CEDAW Optional Protocol") establishes enforceability procedures.

---

[75]Id.

[76]UN General Assembly, *Convention on the Elimination of All Forms of Discrimination Against Women*, 18 December 1979, United Nations, Treaty Series, vol. 1249, p. 13, available at: http://www.refworld.org/docid/3ae6b3970.html (hereinafter CEDAW).

[77]U.N. Committee on the Elimination of Discrimination against Women. *Convention on the Elimination of All Forms of Discrimination Against Women: Overview of the Convention*, http://www.un.org/womenwatch/daw/cedaw/ (last visited Feb. 18 2015).

[78]U.N. Committee on the Elimination of Discrimination against Women. *Convention on the Elimination of All Forms of Discrimination Against Women: Country Reports*, www.un.org/womenwatch/daw/cedaw/reports.htm (last visited Feb. 18 2015).

[79]U.N. Committee on the Elimination of Discrimination against Women, *Overview of Current Working Methods of the Committee on the Elimination of Discrimination Against Women*, U.N. Doc. CEDAW/C/2007/I/4/Add.1(Oct. 25, 2006), http://www.un.org/womenwatch/daw/cedaw/committeerworkingmethods.html.

[80]U.N. Committee on the Elimination of Discrimination against Women, *Information Note 2*, (May 1999), http://www.un.org/womenwatch/daw/cedaw/cedaw20/committee.htm.

[81]Id.

[82]U.N. Committee on the Elimination of Discrimination against Women, *Convention on the Elimination of All Forms of Discrimination Against Women.: General Recommendations*. http://www.un.org/womenwatch/daw/cedaw/recommendations/index.html (last visited Feb. 18 2015).

[83]U.N. Committee on the Elimination of Discrimination against Women, *General Recommendation No 27 on older women and protection of their human rights,* U.N. Doc.CEDAW/C/GC/27 (Dec. 16, 2010).

[84]UN General Assembly, *Optional Protocol to the Convention on the Elimination of All Forms of Discrimination Against Women*, 2131 U.N.T.S. 83. p. 83, (Oct. 6 1999) *available at* http://www.refworld.org/docid/3ae6b3a7c.html.

The Convention Relating to the Status of Refugees[85] ("Refugees Convention") seeks "to assure refugees the widest possible exercise of fundamental rights and freedoms",[86] and facilitate co-operation between states and with the UN High Commissioner for refugees.[87]

The Convention on the Rights of Persons with Disabilities[88] ("CRPD") is a major step toward changing the perception of disability to ensure that societies recognize that all persons must be provided with the opportunities to live life to their fullest potential.[89] The CRPD encourages the right to live independently and be included in the community in its Article 19. It also promotes the habilitation and rehabilitation of the persons with disabilities in Article 26 and prohibits, in Article 27, discrimination on the basis on disability. It promotes opportunities for self-employment, entrepreneurship, the development of cooperatives and starting one's own business. Furthermore, an adequate standard of living and social protection are provided by the CRPD in its Article 28.

The Convention on the International Protection of Adults ("CIPA")[90] is not a UN convention; rather it was developed at the Hague Convention on Private International Law in response to the aging of the world's population.[91] This instrument provides "rules on jurisdiction, applicable law and international recognition and enforcement of protective measures and establishes a mechanism for co-operation between authorities of contracting states." While relatively few states have contracted to be bound by this treaty, it provides additional protections for adults of all ages should they require decision-making assistance, or experience incapacity. The related Explanation of Convention on the International Protection of Adults outlines the history and interprets the CIPA.[92]

Older persons' human rights should be included in all human rights monitoring and implementation. Member States should be encouraged by treaty bodies and other bodies such as the Commission for Social Development, to include the status

---

[85]UN General Assembly, *Convention Relating to the Status of Refugees*, United Nations, Treaty Series, vol. 189, p. 137, (July 28 1951) *available at* http://www.unhcr.org/protect/PROTECTION/3b66c2aa10.pdf. *Convention Relating to the Status of Refugees*, July 28, 1951, 189 U.N.T.S. 137.

[86]Id. at preamble.

[87]Id.

[88]UN General Assembly, *Convention on the Rights of Persons with Disabilities: resolution/ adopted by the General Assembly*, A/RES/61/106, (Jan. 24 2007) *available at* http://www.refworld.org/docid/45f973632.html.

[89]U.N. Convention on the Rights of Persons with Disabilities,*Why a Convention?*, http://www.un.org/disabilities/convention/questions.shtml. (last visited Oct. 23, 2014).

[90]Hague Conference on Private International Law. *Convention on the International Protection of Adults* (Jan. 13, 2000) http://www.hcch.net/index_en.php?act=conventions.text&cid=71.

[91]Id.

[92]Id.

of older persons' rights in their reporting.[93] Special procedures such as special rapporteurs and independent experts ought to closely observe and report on older persons' rights.[94] Cooperation among human rights-based organs including UNHCHR, the HR Council, the HRC Advisory committee, human rights NGO's and others is necessary to ensure full protection.[95] All these concerted efforts are crucial to ensure an integrated approach to the rights of older persons and to further develop more specific mechanisms that can appropriately undertake more specialized and permanent work in this area.

C. Civil and Political Rights of Older Persons

Basic civil and political human rights have been considered of utmost importance for the protection and empowerment of older persons. These rights are generally considered to be immediately enforceable, and older persons are able to have legal claims brought on their behalf to ensure full recognition of such rights.[96] Furthermore, older persons are considered to be - in certain circumstances - a vulnerable group.[97] As such, state plans of action on human rights should give special attention to them[98] and to issues of aging in general,[99] which enhances the direct and immediate effect of civil and political rights. Moreover, states are also required to implement positive obligations, which require programmatic measures, in order to secure adequate implementation of such rights. Civil and political rights are therefore reinforced by development cooperation, as recognized by the

[93]U.N. Human Rights Council, *The Necessity of a Human Rights Approach and Effective United Nations Mechanism for the Human Rights of the Older Person*, 62, A/HRC/AC/4/CRP.1 (Chinsung Chun 2009).

[94]Chung Working Paper, 2010 *Id.* 62.

[95]Chung Working Paper, 2010 *Id.* 62.

[96]U.N. Human Rights Council, *Report of the Independent Expert on the Question of Human Rights and Extreme Poverty*, U.N. Doc. A/HRC/11/9 (Magdalena Sepúlveda Carmona, Mar. 27, 2009), http://www.refworld.org/docid/49f846ff2.html.

[97]U.N. Human Rights Council, *Report of the Special Rapporteur on Violence against Women: Its Causes and Consequences on Intersections between Culture and Violence against Women*, U.N. Doc. A/HRC/4/34 (Jan. 17, 2007), http://www.refworld.org/docid/461e2c602.html.).

[98]U.N. Human Rights Council, *Report of the Special Representative of the Secretary-General on the Situation of Human Rights Defenders Addendum:Mission to Indonesia*, U.N. Doc. A/HRC/7/28/Add.2 (Hina Jilani, Jan. 28, 2008), http://www.refworld.org/docid/47baaeb62.html.

[99]UN Human Rights Council, *Report of the Special Rapporteur on contemporary forms of racism, racial discrimination, xenophobia and related intolerance and the independent expert on minority issuesaddendum: mission to Dominican Republic*, U.N. Doc. A/HRC/7/19/Add.5; A/HRC/7/23/Add.3 (Doudou Diène & Gay McDougall, Mar. 18, 2008), http://www.refworld.org/docid/47e23b252.html.

Economic, Social and Cultural Rights Committee in its General Comment No. 14 (2000), indicating that "it is incumbent on States parties and other actors in a position to assist, to provide international assistance and cooperation, especially economic and technical, which enable developing countries to fulfill their core obligations" to this group.[100] Additionally, Kelly Peggy, in her 2006 paper "Intersection of population and human rights: Rights of older persons in the international context," argues that "the first and most essential step to empowering older persons should include measures to protect their human rights."[101] These statements only confirm the evident interconnection between basic human rights, social and economic development, and empowerment of older persons.

In 2010, Chung stressed in one of her working papers that "the majority of existing legislation protecting elderly rights is limited to care and maintenance in family settings. These policy structures focus on health, nutrition, long-term care, social security, and welfare programs that may benefit older persons from a welfare-based approach."[102] This approach may not reflect the current requirements of rights protection, which incorporates independence and participation in society as important components. This next section examines the civil and political rights of older persons in the context of the full panoply of rights.

1. The Right to Recognition Before the Law

While it is important to recognize the need to protect older persons from an economic and social perspective, such efforts must be firmly grounded on the legal recognition that older persons have the inherent and inalienable right to a dignified life free from discrimination. Several human rights treaties have provisions on the right to be recognized before the law.[103] Such legal dimension includes equality under the law, equal access to justice and due process.

The issue of recognition under the law for older persons transcends national constitutional frameworks; it is reflected also in the international level where they are not specifically protected by a comprehensive thematic regime. The lack of an international convention on the human rights of the older persons also limits the

---

[100]UN Committee on Economic, Social and Cultural Rights (CESCR), *General Comment No. 14: The Right to the Highest Attainable Standard of Health (Art. 12 of the Covenant)*, 45, U.N. Doc. E/C.12/2000/4 (Aug. 11, 2000), http://www.refworld.org/docid/4538838d0.html.

[101]Peggy Kelly, *Intersection of Population and Human Rights: Rights of Older Persons in the International C ontext* (2006) http://www.eclac.org/celade/noticias/paginas/6/27116/KellyP.pdf (last visited Oct. 23, 2014).

[102]U.N. Human Rights Council, *The Necessity of a Human Rights Approach and Effective United Nations Mechanism for the Human Rights of the Older Person*, A/HRC/AC/4/CRP.1 (Chinsung Chun 2009).

[103]UDHR at art. 6; CEDAWat art. 15; CRPD at arts. 5, 12; ICCPR at art. 26.

legal recognition of this vulnerable group.[104] This situation simply compounds their vulnerable situation at both national and international levels.[105]

The difference between de jure and de facto status of State obligations to older persons is sometimes very great. Where states include national legal recognition to the rights of older persons, such recognition is not always upheld and may even be undermined by unacceptable national practices.[106] For example, despite having been born with these rights, a lack of identification papers such as a birth certificate to prove one's identity may prevent an older person from attaining full recognition and equality before the law. A lack of identification is just one of the barriers that may prevent older persons from fully realizing their rights.[107] Such situations can severely limit the exercise of many other rights by older persons, including autonomy and self-determination, and access to legal services as well as to health and other social services.

Some experts attribute discrepancies to the adoption of a welfare-oriented approach instead of a rights-based framework, combined with "a lack of international standards, effective implementation, and enforcement procedures."[108] In practice, the rights of older men and women are not being adequately protected or monitored. This lack of practical action is known as an "implementation gap." A study by HelpAge in 2008 showed that the treaty bodies tasked with monitoring how human rights conventions are implemented rarely ask countries to include older persons in their reporting. Governments rarely include older persons' rights in their reports to these treaty bodies.[109] What law there is that could apply to older persons is not being implemented. This situation could be related, at least in part, to the lack of comprehensive recognition of the rights of older persons as a distinct set of human rights recognized in international treaties and other instruments.

The general scope of the right to legal recognition is set out in Article 6 of the UDHR and Article 16 of ICCPR; everyone has the right to recognition everywhere as a person before the law. However, international human rights law lacks an explicit definition of the legal status of older persons that would reflect their particularly vulnerable position. Only the landmark General Comment No. 6of the CESCR, actually begins by defining older persons as those persons aged

---

[104]Chung Working Paper, U.N. Human Rights Council, *The Necessity of a Human Rights Approach and Effective United Nations Mechanism for the Human Rights of the Older Person*, 58, A/HRC/AC/4/CRP.1 (2010) (Chinsung Chun 2009).

[105]Expert Group Meeting on the Rights of Older Persons, Bonn, Ger., May 5–7, 2009, *Report of the Expert Group Meeting on the Rights of Older Persons*, p. 15.

[106]Bonn meeting, 2009, p. 14. *Id.* at 14.

[107]*Ageing and the Challenge of Non-Communicable Diseases in Low and Middle-Income Countries: A Position Paper*, HelpAge International, 3, (2009).

[108]Chung Working Paper, 2010. *The Necessity of a Human Rights Approach and Effective United Nations Mechanism for the Human Rights of the Older Person, supra* note 132, 48.

[109]HelpAge International, *supra* note 135, at 5.

sixty and above.[110] Since the ICESCR does not explicitly refer to older persons, the CESCR drew its authority first from interpreting Article 9 of the ICESCR, "the right of everyone to social security, including social insurance," to implicitly refer to the right to old-age benefits.[111] Secondly, the CESCR stated that since the ICESCR applies to all members of society, "it is clear that older persons are entitled to enjoy the full range of rights recognized in the Covenant." The CESCR then referred to the Vienna Plan for support.[112] Finally, the CESCR stated that special measures must be taken to protect the rights of older persons: "state parties are required by the Covenant to do so to the maximum of their available resources."[113]

With regard to the protection of adults, the preamble of the Convention on the International Protection of Adults ("CIPA")[114] stresses the need to provide for the protection in international situations of adults who, due to impairment or insufficiency of their personal faculties, are not in a position to protect their interests. CIPA's goals include also determining the law applicable to representation of the adult and provide for the recognition and enforcement of such measures of protection. This convention represents an attempt to fill a gap for adults of all ages who are not otherwise provided for in existing legal mechanisms. Treatment of this issue is necessary. Future articulations of older persons' rights will have to take into account the most recent and clearly articulated understanding of human rights and capacity in the International Convention on the Rights of Persons with Disabilities, and the impact it has in their full legal and practical recognition.

Article 3 of the CIPA lists the measures referred to in Article 1 that may deal in particular with issues such as (a) "the determination of incapacity and the institution of a protective regime," and (b) "the placing of the adult under the protection of a judicial or administrative authority." Institutionally within the United Nations, this has not been further articulated. However, future definitions of older persons' rights will have to take into account the most recent and clearly articulated understanding of human rights and capacity in the International Convention on the Rights of Persons with Disabilities.

It is also worth noting that the Statement on Aging and the Rights of Older Persons[115] of the 62nd Commission on Human Rights noted that the recognition

---

[110]UN Committee on Economic, Social and Cultural Rights (CESCR), *General Comment No. 6: The Economic, Social and Cultural Rights of Older Persons*, 9, U.N. Doc. E/1996/22 (Dec. 8, 1995), http://www.refworld.org/docid/4538838f11.html.

[111]Id. at 10.

[112]*See* id. 7 (describing other U.N.-sponsored activities related to ageing, including the Proclamation on Ageing and Vienna Plan).

[113]*See* id. 10 (noting that while neither of these instruments contain specific reference to discrimination, it is implicit that both of them apply regardless of age).

[114]Hague Conference on Private International Law. *Convention on the International Protection of Adults* (Jan. 13, 2000) http://www.hcch.net/index_en.php?act=conventions.text&cid=71.

[115]U.N. Economic and Social Council, *N.G.O. Statement to the 62nd Session of the Commission on Human Rights* (2006) http://www.un.org/esa/socdev/ageing/documents/StatementAgeingHRC06b.pdf. (last visited Oct. 23, 2014).

of the legal rights of older persons would be necessary to mainstream aging as a visible issue and to, therefore, ensure adequate recognition of their rights before the law, both in the national and international level.

In April 2012, the High Commissioner for Human Rights concluded that "the paradigm shift towards supported decision-making provided for by Article 12 of the CRPD places autonomy and independence of the individual at the centre of this notion."[116] In this report, one of the first issues related with this subject is that many older persons' testimonies repeatedly note that they are treated as incompetent and without due consideration to decades of independent work, productive lives and autonomy. Recurrently, they point to degrading attitudes by caregivers, civil servants or relatives in which they are "treated like children" purely because they cannot walk, talk or react as quickly."[117] These endemic social attitudes have a significant effect in preventing full recognition of their rights under the law. The 2009 Final Report of the Third Committee ("Final Report") encouraged member states "to consider how best the international framework of norms and standards can ensure the full enjoyment of the rights of older persons."[118] It is also recommended "that ongoing efforts to achieve the internationally agreed development goals, including those contained in the United Nations Millennium Declaration, take into account the situation of older persons."[119] The report further called "states to develop their national capacity for monitoring and enforcing the rights of older persons through, inter alia, national institutions for the promotion and protection of human rights where applicable."[120]

The Final Report also contained a request for the Secretary-General to submit a report to the General Assembly on "the status of the social situation, well-being, development and rights of older persons at the national and regional levels.[121] It attempts to highlight how older persons are affected by invisibility in law and "soft law" mechanisms, such as the Millennium Declaration mentioned above. The related draft resolution confirms the key principles of the Final Report, including a request for the governments to prevent age discrimination and to "consider how best to improve international norms and standards pertaining to older persons."[122] It called on "states to develop their national capacity for monitoring and enforcing

[116]U.N. Economic and Social Council, *Report of the United Nations High Commissioner for Human Rights*, 21, U.N. Doc. A/HRC/19/21/Add.3 (Apr. 20, 2012), http://www.un.org/ga/search/view_doc.asp?symbol=E/2012/51.

[117]Id. at 22.

[118]Follow-Up to the Second World Assembly on Ageing, G.A. Res. 64/132, U.N. Doc. A/RES/64/132 (Feb. 5, 2010), http://www.globalaging.org/agingwatch/GA/GAres64.pdf.

[119]Id. at 22.

[120]Id. at 17.

[121]Follow-Up to the Second World Assembly on Ageing, G.A. Res. 64/132, at 3, U.N. Doc. A/RES/64/132 (Feb. 5, 2010) http://www.globalaging.org/agingwatch/GA/GAres64.pdf.

[122]U.N General Assembly, A/RES/64/132; GA/SHC/3957 (Oct. 22 2009) *available at* http://www.un.org/News/Press/docs/2009/gashc3957.doc.htm; U.N. General Assembly, *Third Committee Approves Texts on Crime Congress, Cooperatives, Year of Family, Women's Anti-Discrimination Convention, African Crime Institute, Ageing*, U.N. Doc. GA/SHC/3957 (Oct. 22, 2009).

the rights of older persons in consultation with all sectors of society, including with organizations of older persons."[123] Likewise, it also included "the possibility of instituting new policies, instruments or measures to better improve the situation of older persons."[124]

The Second World Assembly on Ageing requested the UN Secretary General to submit to the General Assembly a report on the implementation of the promotion and protection of human rights as they pertain to older persons.[125] The Assembly also encouraged states to pursue efforts to promote the concerns of older persons, persons with disabilities, and their organizations in the planning, implementing and evaluating of all development programs and policies.

In a special session in 2009, the HR Council encouraged governments to increase national activity on the rights of older persons in cooperation with the UN Regional Commissions.[126] Finally, the Report of the Secretary General on the follow-up to the Second World Assembly on Ageing requested the General Assembly to "call upon governments, organizations and bodies of the United Nations system and the non-governmental community to reinforce their advocacy campaigns aimed at informing all major societal actors, including older persons and their organizations, about the decisions taken at the Second World Assembly on Ageing."[127]

2. The Right to Freedom from Violence

Several human rights treaties recognize the human rights to life, liberty, and security of person.[128] In this legal framework, older persons have the right to be free from all forms of abuse (physical, emotional, or financial, among others), as well as the right to enjoy the state's protection from all forms of exploitation and marginalization. Abuse of older persons can take the form of physical abuse, which refers to the infliction of physical or psychological harm or injury, or sexual abuse,

---

[123]Follow-Up to the Second World Assembly on Ageing, G.A. Res. 64/132, at 2, U.N. Doc. A/RES/64/132 (Feb. 5, 2010) *available at* http://daccess-dds-ny.un.org/doc/UNDOC/LTD/N09/547/37/PDF/N0954737.pdf?OpenElement.

[124]Id. at 2.

[125]U.N. General Assembly, *Social Development: Report of the Third Committee*, U.N. Doc. A/63/424, 11, (Dec. 2 2008) *available at* http://www.undemocracy.com/A-63-424.pdf.

[126]*Report of the Expert Group Meeting on the Rights of Older Persons*, 11 (May 5–7, 2009).

[127]U.N. General Assembly, *Social Development: Report of the Third Committee*, U.N. Doc. A/63/424 (Dec. 2, 2008); *Follow-up to the International Year of Older Persons: Second World Assembly on Ageing*, G.A. Res. 56/228, U.N. Doc. A/RES/56/228 (Feb. 28, 2002) http://www.undemocracy.com/A-63-424.pdf.

[128]UDHR, Art. 3; Convention on the Rights of the Child (CRC), Art. 2; CRPD, Art. 16.

in every country and every economic income bracket.[129] The United Nations General Assembly on the follow-up of the Second World Assembly on Ageing stressed that "Older women tend to be at most risk, as they live longer than men do and culturally are often more vulnerable to abuse."[130]

The Toronto Declaration on the Global Prevention of Elder Abuse defines 'elder abuse' as "a single or repeated act, or lack of appropriate action, occurring within any relationship where there is an expectation of trust, which causes harm or distress to an older person."[131] The UN High Commissioner for Human Rights in the report submitted before the UN during the first months of 2012 determined that "the abuse of older persons may be physical, psychological, sexual and emotional."[132]

Abuse committed against older persons is one of the most prevalent forms of abuse.[133] Elder abuse is "a single or repeated act or lack of appropriate action occurring within any relationship where there is an expectation of trust, which causes harm or distress to an older person."[134] Abuse, however, is not the only type of violence older persons face.

Many older men and women, especially older women, experience violence in their homes, in care settings, or in conflict or post-conflict situations.[135] On this issue, some authors have emphasized that elder abuse has only recently come to receive public attention. For instance, Peggy Kelly has stressed that "mistreatment of older persons violates their rights. The impact, consequences and costs of elder abuse, violence and neglect to societies and individuals may be considerable and have not been estimated, because of a lack of quantifiable data and information."[136]

Harmful practices targeted specifically at older women and men, including violence because of superstition or witchcraft accusations and elder abuse, have not

---

[129]U.N. Human Rights Council, *Agenda and Annual Programme of Work*, 11, U.N. Doc. A/HRC/4/121 (June 18, 2007).

[130]U.N. Secretary-General, *U.N. Department of Economic and Social Affairs Division for Social Policy and Development Programme on Ageing: Report of the Expert Group Meeting Rights of Older Persons*, 34 (2009).

[131]World Health Organization, *The Toronto Declaration on the Global Prevention of Elder Abuse*, 32 (2002), http://www.who.int/ageing/projects/elder_abuse/alc_toronto_declaration_en.pdf.

[132]http://www.un.org/ga/search/view_doc.asp?symbol=E/2012/51. (last visited Oct. 23, 2014). U.N. Economic and Social Council, *Substantive Session of 2012: Report of the High Commissioner for Human Rights*, U.N. Doc. E/2012/51 (Apr. 20, 2012).

[133]U.N. Human Rights Council, *UN Human Rights Council: Addendum to the Report of the Special Rapporteur on Violence against Women, Its Causes and Consequences, Mission to the Netherlands*, at nos. 29, 40, 41, U.N. Doc. A/HRC/4/34/Add.4 (Feb. 7, 2007).

[134]P.6Bonn meeting, 2009, referencing WHO World Report on Violence and Health, 2002: at 126. Expert Group Meeting on the Rights of Older Persons, Bonn, Ger., May 5–7, 2009, *Report of the Expert Group Meeting on the Rights of Older Persons*, at 6 (*referencing* World Health Organization, *Report on Violence and Health*, at 126 (2002)).

[135]HelpAge Position Paper at 2.

[136]Kelly Report, *supra* note 30, at 10.

been adequately addressed.[137] The victims of accusations of witchcraft who suffer the most serious violence are invariably older women; the international community should highlight the abuse of older women, and they should receive targeted funding.[138] The section of this chapter on older women, below, highlights state cases of witchcraft allegations and subsequent murders, monitored by the CEDAW Committee.

The 2012 report of the UN High Commissioner for Human Rights pointed out that "the consequences of physical violence for older persons can be serious and they are more likely to require longer periods of recuperation even from minor injuries. Apart from causing severe and lasting emotional distress, physical violence is also a cause of premature mortality among older persons."[139] Violence can cause irreparable harm.

In violent conflicts, older persons make up a disproportionate percentage of civilian casualties[140]; it is the recommendation of a group of special rapporteurs that the HR Council reiterate calls to end violence against civilians including older persons[141] in conflict areas.[142] One Special Rapporteur points out that in a conflict

---

[137]HelpAge International, *supra*, at 4.

[138]U.N. Office of the High Commissioner for Human Rights, *Report of the Special Rapporteur on violence against women, its causes and consequences, Mission to Ghana* U.N. Doc. A/HRC/7/6/Add.3, at no. 65 and conclusion (Feb. 21, 2008); U.N. Human Rights Council, *Report of the Special Rapporteur on Violence Against Women: Its Causes and Consequences,at no.123,* U.N. Doc. A/HRC/11/6/Add.5 (May 27, 2009); Human Rights Council, *Report of the Special Rapporteur on Extrajudicial, Summary or Arbitrary Executions, Philip Alston: Addendum: Mission to the Central African Republic*,at no. 49,, U.N. Doc. A/HRC/11/2/Add.3 (May 27, 2009).

[139]U.N. Economic and Social Council, *Substantive Session of 2012: Report of the High Commissioner for Human Rights*, 33, U.N. Doc. E/2012/51 (Apr. 20, 2012) http://www.un.org/ga/search/view_doc.asp?symbol=E/2012/51.

[140]U.N. Human Rights Council, *Report of the Independent Expert Appointed by the Secretary-General on the Situation of Human Rights in Somalia (Mr Ghanim Alnajjar)*, 18, U.N. Doc. A/HRC/7/26 (Mar. 17, 2008).

[141]U.N. Human Rights Council, *Report on the situation of human rights in Darfur prepared by a group composed of the special Representative of the Secretary-General for children and armed conflict, the Special Rapporteur on extrajudicial, summary on arbitrary executions, the Special Representative of the Secretary-General on the human rights defenders, the Representative of the Secretary-General on human rights internally displaced persons, the Special Rapporteur on the question of torture and the Special Rapporteur on violence against women: its causes and consequences*,at 11, U.N. Doc.A/HRC/5/6 (June 8, 2007), http://www2.ohchr.org/english/bodies/hrcouncil/5session/reports.htm.

[142]U.N. Human Rights Council, *Updated report on the situation of human rights in Darfur prepared by the United Nations Experts Group on Darfur, presided by the Special Rapporteur on the situation of human rights in Sudan and composed by the Special Representative of the Secretary-General for children and armed conflict, the Special Rapporteur on extrajudicial, summary on arbitrary executions, the Special Representative of the Secretary-General on the human rights defenders, the Representative of the Secretary-General on human rights internally displaced persons, the Special Rapporteur on the question of torture and the Special Rapporteur on violence against women: its causes and consequences*, U.N. Doc. A/HRC/6/7 (Sept. 22, 2007).

where half of all civilian casualties are children and older persons, the argument is strengthened that an occupying military operation, by its intrinsic nature, generates war crimes.[143]

Violence and abuse against elderly persons is a common issue for both developed and developing countries across the world.[144] The 2009 Final Report of the Third Committee specifically called upon the states to address "any cases of neglect, abuse and violence against older persons by designing more effective prevention strategies, stronger laws and policies to address these problems and their underlying factors."[145] Further, it reaffirmed that violence, includes domestic violence against older persons and other vulnerable groups and indicated that it is increasingly a threat to individuals and their families as well as communities. The report also stated that such threat "presents compelling and urgent reasons for action by governments and individually to foster social cohesion while recognizing, protecting and valuing diversity."[146]

The Final Report also called upon states to address "cases of neglect, abuse and violence against older persons, by designing more effective prevention strategies, and stronger laws and policies to address these problems and their underlying factors."[147] The findings of the Final Report on violence against older persons were also reaffirmed in the Follow-up to the International Year of Older Persons: Second World Assembly on Ageing report issued by the Secretary General.[148] Further, according to the Report of the Secretary-General on the follow-up to the tenth anniversary of the International Year of the Family, states are supposed to implement policies for early detection, reporting and intervention in cases of abuse of older persons.[149] For example, the CEDAW Committee has taken note that Brazil's Federal Constitution of 1988 establishes the duty of the State to suppress violence within the family, a common place where abuse of older persons takes place.[150]

In the Universal Periodical Review ("UPR") process, the Malta's Working Group Report reviewing states asked about the ongoing development of

---

[143]U.N. General Assembly, *Report of the Special Rapporteur on the Situation of human rights in the Palestinian territories occupied since 1967: note/by the Secretary-General*, at no. 23, U.N. Doc. A/HRC/10/20(Feb. 11, 2009).

[144]U.N. Human Rights Council, *The Necessity of a Human Rights Approach and Effective United Nations Mechanism for the Human Rights of the Older Person*, 11, U.N. Doc. A/HRC/AC/4/CRP.1 (Chinsung Chun 2009).

[145]U.N. General Assembly, *Social Development: Report of the Third Committee*, at 10, U.N. Doc. A/64/432 (Dec. 1, 2009) http://www.copac.coop/iyc/a-64-432-excerpt.pdf.

[146]Id. at17.

[147]Id. at.2.

[148]U.N. General Assembly, *Social Development: Report of the Third Committee*, U.N. Doc. A/63/424 (Dec. 2, 2008) http://www.undemocracy.com/A-63-424.pdf.

[149]U.N. Secretary-General, *Report on the Follow-Up to the Tenth Anniversary of the International Year of the Family*, U.N. Doc. A/64/134 (Feb. 1, 2010). http://daccess-dds-ny.un.org/doc/UNDOC/GEN/N09/402/16/PDF/N0940216.pdf?OpenElement.

[150]U.N. General Assembly, *Report of CEDAW on its 28th Session*, at 93 92, UN. Doc. A/58/38 (2003).

community-based services for older persons and the policy to increase awareness and understanding of what constitutes abuse and how to prevent it.[151] States often highlight the importance of combating violence, particularly against vulnerable groups. For example, Argentina, Kazakhstan, and the UK all noted the need for increased actions to solve the problems of violence against vulnerable groups in Slovakia.[152] Also, the HR Council noted that "the vulnerable situation of elderly persons placed in long-term care, which in some instances has resulted in degrading treatment and violated their right to human dignity." Therefore, it passed on to call upon the state "to pursue its efforts to improve the situation of elderly persons in nursing homes."[153]

3. The Right to Equality and Non-discrimination

Every human has inalienable rights regardless of their age.[154] However, the non-discrimination clauses in international law rarely specify age as a prohibited ground for discrimination. While human rights law prohibits discrimination on several grounds, age could be argued to be included in the phrase "and other status."[155] Some of this is a result of ageism. One definition of ageism is "a systematic stereotyping and discrimination against persons because they are old. Sometimes ageism is compared to sexism or racism because it discriminates against all members of a particular group."[156] The principles of equality and non-discrimination require taking special measures to protect segments of society with a history of discrimination as a matter of priority.[157] Clearly, as it was highlighted

[151]U.N. Human Rights Council, *Report of the Working Group on the Universal Periodic Review - Malta*, U.N. Doc. A/HRC/12/7, A/HRC/WG.6/5/L.6 (June 4, 2009).

[152]U.N. Human Rights Council, *Report of the Working Group on the Universal Periodic Review - Slovakia*, U.N. Doc. A/HRC/12/17, A/HRC/WG.6/5/L.16 (June 5, 2009).

[153]U.N. General Assembly, *Report of the Human Rights Committee-Germany*, at 39 68(17), U.N. Doc. A/59/40 vol. I (2004).

[154]*See e.g.*, the Universal Declaration of Human Rights (UDHR), http://www.un.org/en/documents/udhr/, Article 2; Convention on the Rights of Persons with Disabilities(CRPD), Article 5; International Covenant on Civil and Political Rights(I CCPR), Article 2.

[155]Lindsey Judge, *The Rights of Older People: International Law, Human Rights Mechanisms and the Case for New Normative Standards* 4 (2009).

[156]Kelly Report, *supra* note 30, at 8.

[157]U.N Human Rights Council, *Report of Independent Expert on human rights and extreme poverty, Rapporteur:. "To collect information to prepare this report the independent expert sent a questionnaire to Governments requesting information on their social pensions for older persons. The questionnaire addressed: (i) the legal and institutional framework; (ii) the schemes costs and coverage; (iii) implementation procedures; (iv) monitoring mechanisms and complaints procedures; and (v) studies and evaluations"*, 58 (Magdalena Sepulveda 2010)(hereinafter "Sepulveda Report").

in the report of Report of the Expert Group Meeting on the "Rights of Older Persons" "discrimination, neglect and abuse are a daily feature in the lives of many older persons."[158]

The right to equality before the law and the principle of non-discrimination are universal human rights foundations enshrined in Articles 1 and 2 of the UDHR that states that all human beings are "born equal in dignity and rights" and "everyone is entitled to the same rights and freedoms without distinction of any kind", including distinctions based on "other status." Article 26 of ICCPR further links these two principles, stating that "all persons are equal before the law and are entitled without any discrimination to the equal protection of the law. In this respect, the law shall prohibit any discrimination and guarantee to all persons equal and effective protection against discrimination on any ground." Theoretically, this should include "other status".

The ICCPR states that distinctions of "other status," such as age, are outlawed. It further requires in its Article 2 that the state parties "respect and ensure the rights recognized therein to all individuals within its territory and subject to its jurisdiction."[159] The ICCPR also lists non-derogable rights even in times of public emergency, which is of particular relevance to older persons, who are often more vulnerable in such situations. In Article 4, it establishes that even "in time of public emergency which threatens the life of the nation and the existence of which is officially proclaimed ... the state parties... may take measures derogating from their obligations under the ICCPR to the extent strictly required by the exigencies of the situation" and they cannot "involve discrimination solely on the ground of race, color, sex, language, religion or social origin."[160]

This group-based or "vulnerable group" approach is applicable in the context of civil and political, as well as economic, social, and cultural rights. Culture and tradition may influence the extent to which older persons are discriminated against in daily life.[161]

Equal treatment is important for all people.[162] In relation to government programs, older persons are often "denied equity in opportunities available to them and in resources allocated to them."[163] In its General Comment No. 6, the CESCR highlighted that when the ICESCR came to life "the problem of demographic aging

---

[158]Bonn meeting, 2009, at 15.

[159]ICCPR, Art. 1.

[160]ICCPR, No derogation from articles 6, 7, 8 (paragraphs 1 and 2), 11, 15, 16 and 18.

[161]Kelly Report, *supra* note 30, at 6.

[162]UN Human Rights Council, *Report of the Special Rapporteur on Contemporary Forms of Racism, Racial Discrimination, Xenophobia and Related Intolerance, Doudou Diène: addendum: mission to Lithuania*,, U.N. Doc. A/HRC/7/19/Add.4 (Feb. 7, 2008).

[163]Committee on Economic, Social and Cultural Rights, *General comment No. 6: The economic, social and cultural rights of older persons*, U.N. Doc. E/1996/22 (1996) http://dfasuomi.stakes.fi/NR/rdonlyres/ABF1AF26-5D33-458A-ABAD-3E4E284FD85D/0/Sidorenko.pdf. (last visited Oct. 23, 2014).

was not as evident or as pressing as it is now" being that the reason for the omission of age as a prohibited ground for discrimination. However, CESCR concluded that "the prohibition of discrimination on the grounds of 'other status' could be interpreted as applying to age." The Committee further stated "that while it may not yet be possible to conclude that discrimination on the grounds of age is comprehensively prohibited by the Covenant, the range of matters in relation to which such discrimination can be accepted is very limited."[164] Protection from age discrimination although not thoroughly espoused by human rights treaty texts can nonetheless be implicitly construed as a reality and a basic human right. It also has grounding in soft law documents such as the UN Principles for Older Persons and MIPAA.[165]

Additionally, the HRC stated in its General Comment No. 18 that the grounds for discrimination are not limited to those expressly mentioned in the ICCPR.[166] In litigation of the right to non-discrimination, the HRC has applied the right to equal protection under Article 26 of the ICCPR to economic, social and cultural rights. In *Brooks v. Netherlands*, the HRC stated that the scope of Article 26 extends to prohibit discrimination in law and in practice in any field regulated and protected by public authorities.[167] In this case, the petitioner claimed that her right to equality before the law under Article 26 was violated on the ground of sex and "other status" when her social security benefits, awarded on the basis of disability, were suspended.[168] According to domestic legislation, disability benefits were ensured only to married women who were "breadwinners" or separated from their husbands.[169] Since the petitioner did not fit either of these categories, her social security benefits were cut off after a period of time. The immediacy of the need for social security benefits in old age or with disability was not addressed. This case has served as an important precedent for subsequent complaints submitted to the HRC, arguing for the rights of older persons to social security protection without discrimination.[170]

Discrimination in the Netherlands was addressed in another HRC complaint, *Schmitz-de-Jong v. Netherlands*,[171] the HRC found "that the age limitation of allowing only partners who have reached the age of 60 years to obtain an entitlement to various rate reductions as a partner to a pensioner above the age of

---

[164]Id. at 4.

[165]Id. at 4.

[166]UN Human Rights Committee (HRC), *CCPR General Comment No. 18: Non-discrimination*, at 26, U.N. Doc. HRI/GEN/1/Rev.1 (Nov. 10, 1989).

[167]*Brooks v. Netherlands*, U.N. GAOR Hm. Rts. Comm., 39th Sess., Supp. No. 40, U.N. Doc. A/42/40 (1987), http://www1.umn.edu/humanrts/undocs/session42/172-1984.htm.

[168]*See id.* 2.3 (providing the basis for petitioner's claim).

[169]*See id.* 8.2.

[170]*See, e.g., S.B. v. New Zealand*, U.N. GAOR Hum. Rts. Comm., 50th Sess., U.N. Doc. CCPR/C/50/D/475/1991 (1994) (finding that New Zealand and the United Kingdom violated Article 26 of the covenant), http://www1.umn.edu/humanrts/undocs/html/dec475.htm.

[171]*Schmitz-de-Jong v. Netherlands*, Hm. Rts. Comm., U.N. Doc. CCPR/C/72/D/855/1999 (1999), http://www.bayefsky.com/pdf/170_netherlands855.pdf.

65 years is an objective criterion of differentiation... and not unreasonable."[172] Such decisions could pave the way for future discrimination against older persons based on age alone or in combination with other factors.

In the UPR process, India's Working Group Report commended its effective implementation of various policies to promote equality and justice as well as its affirmative action programs to address the situation of marginalized communities and support older persons. This support does not recognize older persons as rights holders. Peer review also praised the exemplary role of the National Human Rights Commission in India and asked the institution's experience to be shared with other countries as a best practice.[173] Such an institution could play a larger role in pushing for equal recognition of older persons' rights.

Likewise, the Philippines' Working Group Report positively commented on that country's constitutional provisions, which ensure the fundamental principle of equality before the law for all, including the rights of older persons, among others.[174] Poland's Working Group Report highlighted the need to raise awareness of various forms of discrimination and prejudice towards older persons, and to sensitize the society to manifestations of discrimination towards these groups, which include older persons.[175]

Regarding ageism, the United Kingdom mentioned in the UPR process that it has introduced regulations prohibiting age discrimination in the workplace and it is currently examining the possibility of prohibiting age discrimination in the provision of goods, facilities, and services and in the exercise of public functions. The United Kingdom also recognized that harmful age discrimination largely exists in a few sectors, including in the health and social sector and to a lesser extent the financial sector.[176]

The UPR report on Finland found a shortage of information and research on discrimination. It noted that effective anti-discrimination policy and sufficient support systems and prevention programs must be based on adequate information on discrimination in society. It further noted a particular shortage of information and research on the special needs of elderly lesbian, gay, bisexual, and transgender (LBGT) persons and the prevalence of multiple, intersecting forms of discrimination.[177]

The 2009 Final Report of the Third Committee also called upon states to "eliminate and address discrimination on the basis of age and gender," and

---

[172]Id.; ICCPR at165 2.1, 2.2, 3, 7.2.

[173]U.N. Human Rights Council, *Report of the Working Group on the Universal Periodic Review of India: Response of the Government of India to the recommendations made by delegations during the Universal Periodic Review of India, U.N. Doc. A/HRC/8/26 (Aug. 25, 2008).*

[174]UN Human Rights Council, *Report of the Working Group on the Universal Periodic Review - The Philippines*, U.N. Doc. A/HRC/8/28; A/HRC/WG.6/1/PHL/4 (May 23, 2008).

[175]A/HRC/8/30, 23 May 2008, Par. 8. *Id.* 8.

[176]Id. at 11.

[177]UN Human Rights Council, *Summary prepared by the Office of the High Commissioner for Human Rights, in accordance with paragraph 15(c) of the Annex to Human Rights Council Resolution 5/1 - Brunei Darussalam*, U.N. Doc. A/HRC/WG.6/1/FIN/3 (Mar. 6, 2008).

recommended to "engage with all sectors of society, including women's groups and organizations of older persons" in order to change "negative stereotypes about older persons, in particular older women, and promote positive images of older persons."[178] The report also reaffirmed that "social integration policies should seek to reduce inequalities, promote access to basic social services," increase the participation on equal terms and "integration of social groups, particularly… older persons and persons with disabilities."[179]

Finally, according to the Follow-Up to the Second World Assembly on Ageing Report of the Secretary General in 2012 (A/67/188), older persons face discrimination not just from society in general, but they may also face employers' negative perceptions. They are discriminated because of the "age limits, the penalties and denials of services imposed by insurance service providers and financial institutions; preconceived notions and negative attitudes on the part of medical staff and rationing of health care."[180] The aforementioned is a consequence of the general perception that "ageism rests on the assumption that neglect of and discrimination against older persons is the norm and acceptable."[181]

4. The Right to Freedom from Cruel, Inhuman or Degrading Treatment or Punishment

Article 5 of the UDHR states that "no one shall be subjected to torture or to cruel, inhuman or degrading treatment or punishment." Similarly, according to Article 7 of the ICCPR, "no one shall be subjected to torture or to cruel, inhuman or degrading treatment or punishment," including, "without his free consent, to medical or scientific experimentation." In General Comment No. 20 on Article 7 about the prohibition of torture, or other cruel, inhuman or degrading treatment or punishment, the Human Rights Committee stated that "special protection in regard to … experiments is necessary in the case of persons not capable of giving their consent." "More attention should be given to the possible need and means to ensure the observance of this provision."[182]

[178]U.N. General Assembly, *Social Development: Report of the Third Committee*, at 17, U.N. Doc. A/64/432 (Dec. 1, 2009) http://www.copac.coop/iyc/a-64-432-excerpt.pdf.

[179]Id. at 44.

[180]U.N. General Assembly, *Follow-Up to the Second World Assembly on Ageing: Report of the Secretary-General*, 57, U.N. Doc. A/67/188 (July, 26 2012) http://social.un.org/ageing-working-group/documents/A_67_188_E.pdf.

[181]Id. at 59.

[182]Torture or Cruel, Inhuman or Degrading Treatment or Punishment: General Comment 7, U.N. GAOR, Hum. Rts. Comm., 16th Sess., 7, U.N. Doc. HRI/GEN/1/Rev.1 (1994*); See also*, Javier Vasquez, Pan American Health Organization, Human Rights & Health: Older Persons at 1, (2008) available at http://www.paho.org/englisH/dd/Pub/10069_OlderPersons.pdf. (last visited Oct. 23, 2014).

In the Second Working Session of the UN Open-Ended Working Group on Ageing held on the August 2011 the Chair of the United Nations Committee against Torture emphasized that "the definition of torture takes into account the particular circumstances of an individual, including their health conditions and age. It is the duty of States to investigate and punish any conduct that can amount to inhuman or degrading treatment, conducing in places like psychiatric institutions or prisons where older persons may not be able to take action to protect themselves."[183]

The right to be free from torture enjoys in-depth protection thanks to the Convention against Torture and Other Cruel, Inhuman or Degrading Treatment or Punishment ("CAT"),[184] which is entirely dedicated to the topic. The CAT defines torture within the framework of the instrument as: "any act by which severe pain or suffering, whether physical or mental, is intentionally inflicted on a person, or for any reason based on discrimination of any kind, when such pain or suffering is inflicted by or at the instigation of or with the consent or acquiescence of a public official or other person acting in an official capacity."[185]

Likewise, Article 10 establishes the state's duty to "ensure that education and information regarding the prohibition against torture are fully included in the training of law enforcement personnel, civil or military, medical personnel, public officials and other persons who may be involved in the custody, interrogation or treatment of any individual subjected to any form of arrest, detention or imprisonment."[186] Article 16 further expands the scope by obliging the state parties to "undertake to prevent in any territory under its jurisdiction other acts of cruel, inhuman or degrading treatment or punishment which do not amount to torture as defined in Article 1." This may be particularly relevant for older persons in situations of vulnerability to inhuman or degrading treatment by non-state actors. Where the state has reason to know of the treatment, it has a duty to implement actions to prevent it. Its treaty body, the Committee Against Torture ("CAT Committee"), refines the meaning of the provisions through its general comments and individual complaint mechanism.

As dependency increases with age, many older persons become increasingly vulnerable to abuse and exploitation in residential care facilities and in families. The HRC, for example, noted "the vulnerable situation of elderly persons placed in long-term care homes, which in some instances has resulted in degrading

---

[183]Report of the Open-ended Working Group on Ageing for the Purpose of Strengthening the Protection of the Human Rights of Older Persons, G.A. Res. 65/182, U.N. Doc. A/Res/65/182 (Aug. 4, 2011) http://social.un.org/ageing-working-group/documents/Chair_summary_2nd_session_OEWG_final.pdf.

[184]*See* ICCPR, Art. 7 (commenting that individuals should not be subjected to torture or cruel punishment and stating that people should not be forced to participate in medical or scientific experiments without their consent).

[185]*Convention against Torture and Other Cruel, Inhuman or Degrading Treatment or Punishment*, G.A. Res. 39/46, art. 1, U.N. Doc. A/39/51 (June 26, 1987), *hereinafter* UNCAT.

[186]Id. Art. 10.

treatment and violated their right to human dignity and asked states to pursue its efforts to improve the situation of elderly persons in nursing homes."[187] Responsibility can be imputed to the state by its failure to hold private actors accountable and enacting domestic legislation that will preclude private actors from violating such rights of older persons, such as the responsibility Brazil has taken for preventing domestic violence.[188]

Where persons, including the more vulnerable older persons, are subjected to any form of arrest, detention or imprisonment, state parties "are to review interrogation rules, instructions, methods and practices" on a continual basis, "as well as arrangements for the custody and treatment" of them.[189] The CAT also sets basic procedural standards in its Article 13 when it calls upon state parties to "ensure that any individual who alleges he has been subjected to torture in any territory under its jurisdiction has the right to complain to, and to have his case promptly and impartially examined by, its competent authorities."[190]

The CAT and other human rights mechanisms have spoken forcefully against the abuse of older persons. The Special Adviser on the Human Rights situation in Myanmar requested the prioritization of the release of elderly political prisoners on humanitarian grounds.[191] The release of non-violent offenders from confinement in pretrial detention should also be expedited, he noted, beginning with older persons.[192] He further recommended that elderly prisoners be given priority release.[193]

In the case of Chile, the CAT further condemned the commission of abusive acts by police officers against members of indigenous peoples, especially members of the Mapuche tribe, and in particular when the victims of these acts include older persons.[194] The CAT asked Chile to carry out prompt investigations into these allegations.[195]

---

[187]U.N. General Assembly, *Report of the Human Rights Committee-Germany*, at 39 68(17), U.N. Doc. A/59/40 vol. I (2004).

[188]*See, e.g.* U.N. Committee on the Elimination of Discrimination Against Women, *Consideration of reports submitted by States parties under article 18 of the Convention on the Elimination of All Forms of Discrimination against Women: Brazil*, at 93 92, U.N. Doc., A/58/38 part II (Nov. 7, 2002). http://daccess-dds-ny.un.org/doc/UNDOC/GEN/N03/468/20/PDF/N0346820.pdf?OpenElement.

[189]UNCATArt. 11.

[190]Id., Art. 14.

[191]U.N. General Assembly, *Situation of Human Rights in Myanmar: Report of the Secretary-General*, at no. 31,, U.N. Doc. A/63/356 (Sept. 17, 2008).

[192]UN Human Rights Council, *Report of the Special Rapporteur on Torture and Other Cruel, Inhuman or Degrading Treatment or Punishment, Manfred Nowak: addendum: mission to Nigeria (4 to 10 March 2007)*, at 45 and 100,, U.N. Doc. A/HRC/7/3/Add.4 (Nov. 22, 2007).

[193]U.N. Human Rights Council, *Report of the Special Rapporteur on the Situation of Human Rights in Myanmar*, at no. 95, U.N. Doc. A/HRC/10/19 (Mar. 11, 2009).

[194]U.N. Committee Against Torture, *Concluding Observations on Chile*, U.N. Doc. CAT/C/CHL/CO/5 (June 23, 2009) *available at* http://acnudh.org/wp-content/uploads/2011/01/CAT-Chile-2009-ENGLISH.pdf.

[195]Id.

In the context of the UPR process, a non-governmental organization, INREDH, reported in Ecuador's Working Group Report that a lack of application of norms protecting older persons represents a serious human rights violation in the context of the criminal justice system in that country.[196] In the Russian Federation's Working Group Report, the delegation recognized that domestic violence is a problem that affects older persons.[197]

The 2009 Final Report of the Third Committee called upon Member States to address "any cases of neglect, abuse and violence against older persons, by designing more effective prevention strategies, stronger laws."[198] In the Note by the Secretary-General transmitting the interim report of the Special Rapporteur on the situation of human rights in Myanmar, the Special Rapporteur positively noted that in 2005, the International Committee of the Red Cross continued to monitor the treatment and living conditions of persons in need of particular protection, such as the elderly.[199]

Finally, the Report of the High Commissioner for Human Rights of the UN, concluded that "the realization by older persons of the right to liberty and security of person, the right to privacy, freedom of movement, freedom of expression, freedom from torture and other cruel, inhuman and degrading treatment, the right to personal integrity, the right to an adequate standard of living and the right to the highest attainable standard of physical and mental health, among others, are common but have often not been adequately regulated."[200]

5. *The Right to Life*

According to Article 3 of the UDHR, "everyone has the right to life, liberty and security of person."[201] Similarly, Article 6 of the ICCPR states that "every person has the inherent right to life"... "protected by law" and "shall not be arbitrarily deprived of life." The ICCPR does not restrict capital punishment with respect to older persons.

---

[196]U.N. Human Rights Council, *Report of the Working Group on the Universal Periodic Review–Ecuador*, 22, U.N. Doc. A/HRC/WG.6/1/ECU/3 (Mar. 6, 2008) http://daccess-dds-ny.un.org/doc/UNDOC/GEN/G08/113/10/PDF/G0811310.pdf?OpenElement.

[197]U.N. Human Rights Council, *Report of the Working Group on the Universal Periodic Review - Russian Federation*, U.N. Doc. A/HRC/11/19 (Mar. 3, 2009).

[198]U.N. General Assembly, *Social Development: Report of the Third Committee*, at 10, U.N. Doc. A/64/432 (Dec. 1, 2009) http://www.copac.coop/iyc/a-64-432-excerpt.pdf.

[199]U.N. General Assembly, *Situation of Human Rights in Myanmar*, U.N. Doc. A/60/221 (Aug. 12, 2005).

[200]U.N. Economic and Social Council, *Substantive Session of 2012: Report of the High Commissioner for Human Rights*, 23, U.N. Doc. E/2012/51 (Apr. 20, 2012), http://www.un.org/ga/search/view_doc.asp?symbol=E/2012/51.

[201]UDHR at art. 3.

According to HRC General Comment No. 6, the right to life is a right that should be broadly interpreted, and includes all possible measures to increase life expectancy, giving older persons an enforceable right to government protection and positive measures to prolong lives. It reminded that "the HRC has noted that the right to life has been too often narrowly interpreted." In relation with the expression "inherent right to life," it sustained that is to be interpreted broadly, and the protection of this right requires that States adopt positive measures to protect lives in all stages of the life course. In another broader interpretation of the right to life, the HRC has considered that it would be desirable for state parties to take all possible measures to increase life expectancy.[202]

The HRC noted the U.S. Government's failure to protect the right to life during the Hurricane Katrina emergency, establishing that residents trapped in nursing homes and hospitals were especially endangered.[203] The ongoing violence and murder surrounding witchcraft allegations, detailed above in the Right to Freedom from Violence, is situation where states have a duty to intervene to protect the right to life of older women.

The August 2012 third working session of the UN Open-Ended Working Group on Ageing for the purpose of strengthening the protection of the human rights of older persons, recognized that the situation of older persons has clearly degraded from older persons being respected and recognized by their communities to increased abuse and violence, especially for older women.[204]

According to the fiftieth session of the Commission for Social Development related with second review and appraisal of the Madrid International Plan of Action on Ageing, held in February 2012, there are numerous cases of older women being accused of witchcraft and subjected to violence and murder reported by civil society and the media across sub-Saharan Africa.[205] Ghana, which currently houses six refugee camps for older women fleeing such violence, is currently undertaking efforts to reintegrate older women into communities.[206] The government has a duty to ensure that its actions do not endanger the lives of the older women it is seeking to disperse.

---

[202]U.N. Human Rights Committee, *General Comment No. 6: The Right to Life*, U.N. Doc. HRI/GEN/1/Rev.9 (Vol. I) (Apr. 30, 1982).

[203]Report Submitted to the 87th Session of the Human Rights Committee: U.N. Human Rights Committee, *Hurricane Katrina and Violations of ICCPR Articles 6 and 26: A Response to the Third Periodic Report of the United States of America (Submitted by the U.S. Human Rights Network to the 87th Session)* (Jean M. Carmault, 2006), http://www2.ohchr.org/english/bodies/hrc/docs/ngos/USHRN.pdf.

[204]U.N. Open-Ended Working Group on Ageing, *Summary of the 3rd Working Session*, 27 (Aug. 21–24, 2012), http://social.un.org/ageing-working-group/documents/Chairssummary3rdsessionOEWGfinal.pdf.

[205]Id. at 13.

[206]http://www.theafricareport.com/index.php/society-and-culture/witches-camps-in-ghana-to-be-disbanded-50176642.html. (last visited Oct. 23, 2014). Dasmani Laary, *Ghana Shuts Down Witches' Camp*, The Africa Report, (Dec. 19, 2014) http://www.theafricareport.com/Society-and-Culture/ghana-shuts-down-witches-camp.html.

6. The Right to Liberty and Security

Pursuant to Article 9 of the ICCPR, everyone has the right to liberty and security; "no one shall be deprived of his liberty except on such grounds and in accordance with such procedure as are established by law", and arbitrary arrest or detention is prohibited.

The ICCPR establishes that "persons deprived of their liberty should be treated with humanity and with respect for the inherent dignity of the human person."[207] For older persons, this is an important protection for those in state and private long-term care facilities, and living with other forms of institutionalization. One's liberty is directly affected by legal capacity, defined as "the right to access the civil and juridical system and the legal independence to speak on one's own behalf."[208] The Convention on the Rights of Persons with Disabilities ("CRPD") outlines that the right to make choices freely is not contingent on legal capacity; it influences how much support is needed to exercise one's rights.[209]

The ICCPR also states in its Article 12 that "everyone lawfully within the territory of a state shall, within that territory, have the right to liberty of movement and freedom to choose his residence."[210] A further right related to the protection of liberty and security is the right to fair trial as enshrined in Article 11 of the UDHR. This requires procedural due process for restrictions of liberty in advance of a criminal trial, regardless of age or other status. For older persons, procedural due process may include a balancing of the detriment of detention on an older person's health, versus the risk of flight pending a trial, for example.

In the UPR process, a reviewing state in Ecuador's Working Group Report noted that older persons represent an important segment of the population and asked what kind of measures Ecuador was taking to ensure their enjoyment of basic rights.[211] Concerning the social and economic inclusion of older persons, the report noted that Ecuador has developed a new system and training programs for community-based services, in particular together with private sector institutions, carrying out comprehensive care programs through homes for older persons deprived of liberty.[212] Such questioning, and such initiatives are necessary both for the well-being of older persons and their enjoyment of the full panoply of rights.

---

[207]ICCPR at Art. 10

[208]*Explanatory Note on Legal Capacity and Forced Interventions*, http://www.un.org/esa/socdev/enable/rights/ahc8docs/ahc8idc1218ex.doc. (last visited Oct. 23, 2014).International Disability Caucus, *Explanatory Note on Legal Capacity and Forced Interventions* (2014), http://www.un.org/esa/socdev/enable/rights/ahc8docs/ahc8idc1218ex.doc.

[209]Convention on the Rights of Persons with Disabilities, at Art. 12.2.

[210]Article 12, ICCPR. ICCPR at Art. 12.

[211]UN Human Rights Council, *Report of the Working Group on the Universal Periodic Review - Ecuador*, A/HRC/8/20; A/HRC/WG.6/1/ECU/4, 35 (May 13 2008) *available at*: http://www.refworld.org/docid/4857aa1d0.html. U.N. Human Rights Council, Report of the Working Group on the Universal Periodic Review - Ecuador, 35, U.N. Doc. A/HRC/WG.6/1/ECU/3 (Mar. 6, 2008).

[212]Id. at 53.

Another interesting example is the position of Finland during the UPR process in 2012. In its report, Finland indicated that "the Parliamentary Ombudsman of that nation has repeatedly raised the issue of the rights and treatment of persons deprived of their freedom, and of persons who are subject to restraining measures." Likewise, it affirmed that:

> legislation on the requirements for deprivation of liberty or on the use of force or restraint in this connection is either inadequate or there are no legal norms on the issue at all in certain situations. There is no legislative basis for the use of coercive measures in caring for the elderly. The use of force in caring for persons with disabilities is based on a single provision that does not provide for judicial control of restraining measures. It has been observed that the instructions of psychiatric hospital wards do not always respect a principle known as the ban on institutional authority. This ban forbids any restriction of the rights of patients based on internal ward norms, and restrictions must instead be based in law and applied on the basis of individual discretion.[213]

7. The Right to Respect for Home and Family

General human rights law includes a number of provisions for the environment of the home and family. Article 23 of the ICCPR establishes that "the family is the natural and fundamental group unit of society and is entitled to protection," including that "the right of men and women of marriageable age to marry and to found a family shall be recognized." States are expected "to take appropriate steps to ensure equality of rights and responsibilities of spouses as to marriage, during marriage and at its dissolution."[214]

Article 10 of the ICESCR provides that with regard to the family, states are to "take appropriate steps to ensure equality of rights and responsibilities of spouses as to marriage, during marriage and at its dissolution. The CESCR encouraged states to support, protect, and strengthen the family because many cultures believe that the family unit must provide care for older persons.[215] It also suggested that those elderly persons who prefer to stay at home should not be discriminated against, and

[213]UN Human Rights Council, *Summary: [Universal Periodic Review]: Finland/ prepared by the Office of the High Commissioner for Human Rights in accordance with paragraph 5 of the annex to Human Rights Council resolution 16/21*, A/HRC/WG.6/13/FIN/3 (Mar. 9 2012), *available at* http://www.refworld.org/docid/5007e1b72.html. U.N. Human Rights Council, Report of the Working Group on the Universal Periodic Review–Finland, 149, U.N. Doc. A/HRC/WG.6/13/FIN/3 (Mar. 9, 2012).

[214]ICCPR at Art. 23.

[215]*See* General Comment 6, para. 31 (stating that Recommendation 29 encourages governments and non-governmental organizations to implement social service programs to support the whole family). U.N. Human Rights Committee, *General Comment No. 6: The Right to Life*, 31, U.N. Doc. HRI/GEN/1/Rev.9 (Vol. I) (Apr. 30, 1982)(stating that Recommendation 29 encourages governments and non-governmental organizations to implement social service programs to support the whole family).

should be provided for through services from both government and non-governmental agencies. The CESCR further suggested that states provide financial support to family members caring for older persons at home, particularly poor families. States should also provide financial and logistical assistance to older persons living alone and to elderly couples who prefer to stay in their own home. The assistance should include transportation, food delivery, nursing care, and doctors' visits. In sum, the state should encourage the cultural decision to keep the family unit together. Resources that might be spent building nursing homes should instead be spent on compensating families for caring for their elderly relatives at home.

In the UPR process, Malaysia's Working Group Report noted the government's pledge that it would also be undertaking greater efforts to ensure that the role of the family as the basic stabilizing and care agent for young and older persons is maintained.[216]

Further, the 2009 Final Report of the Third Committee recognized the importance of strengthening intergenerational partnerships and solidarity among generations and called upon the member states to promote opportunities for interaction between young persons and older generations in the family, the workplace and society.[217]

8. The Right to Information, Association and Participation

Many international human rights instruments embody the guarantee to information, association and participation by preserving rights to expression, association and participation in cultural life as Article 15 of the ICESCR. Information is often not available in a form that is suitable and accessible to older people,[218] which can make meaningful participation in society impossible.

The scope and status of the right is also outlined in general terms in the primary instruments of the International Bill of Rights. Article 27 of the UDHR elaborates that "everyone has the right freely to participate in the cultural life of the community, to enjoy the arts and to share in scientific advancement and its benefits."[219] The International Bill of Rights includes a number of provisions that establish the rights to association and participation. Article 20 of the UDHR guarantees the right of everyone to freedom of peaceful assembly and association. Furthermore, Article 21 of the ICCPR states that "no restrictions may be placed on the exercise of this right other than those imposed in conformity with the law."

---

[216]UN Human Rights Council, *Report of the Working Group on the Universal Periodic Review - Malaysia*, A/HRC/WG.6/4/L.16; A/HRC/11/30, 102 (Mar. 3 2009), *available at* http://www.refworld.org/docid/49f964f10.html. U.N. Human Rights Council, *Report of the Working Group on the Universal Periodic Review - Malaysia*, 102, U.N. Doc. A/HRC/11/30 (Mar. 3, 2009).

[217]U.N. General Assembly, *Social Development: Report of the Third Committee*, at 4, U.N. Doc. A/64/432 (Dec. 1, 2009) http://www.copac.coop/iyc/a-64-432-excerpt.pdf.

[218]HelpAge International, *supra*, at 3.

[219]Cross-reference to the ICESCR Art. 15.

Additionally, Article 8 of the ICESCR establishes the right to labor organization without state interference as the "state parties undertake to ensure the right of everyone to form trade unions and join the trade union of his choice, subject only to the rules of the organization concerned, for the promotion and protection of his economic and social interests." So far, no such organization of labor has taken up the issue of the human rights of its aging members.

Furthermore Article 25 of the ICCPR stipulates that "every citizen shall have the right and the opportunity, without any of the distinctions mentioned in Article 2 and without unreasonable restrictions: to take part in the conduct of public affairs either directly or through freely chosen representatives." General Comment No. 25 supports the principle that each person's vote should be equal to each other person's vote.[220] An older person's vote should in no way be diminished from a younger person's vote. The General Comment also mentioned that, "the drawing of electoral boundaries and the method of allocating votes should not distort the distribution of voters or discriminate against any group and should not exclude or restrict unreasonably the right of citizens to choose their representatives freely."[221] Where older persons are evenly dispersed within different electoral boundaries, this is not as much of a concern. In contrast, if older persons lived in a retirement community or institution, this would concentrate their numbers.

Older persons often lack avenues to participate in society, access information, and some may face barriers to free association. All of these rights require conscious efforts for inclusion.[222] A clear example of how access to information affects one's legal rights is in the case of class action settlements.[223] Without access to information, a person might not know he or she is even eligible for a complaint. Exclusion often results from the negative stereotypes of older people.[224]

Exclusion can be exacerbated by physical barriers to participation. Article 2 of the CRPD defines, inter alia, "universal design" as "the design of products, environments, programs and services to be usable by all people, to the greatest extent possible, without the need for adaptation or specialized design."[225] States must

---

[220]Human Rights Committee, General Comment 25 UN Human Rights Committee (HRC), *CCPR General Comment No. 25: Article 25 (Participation in Public Affairs and the Right to Vote), The Right to Participate in Public Affairs, Voting Rights and the Right of Equal Access to Public Service*, CCPR/C/21/Rev.1/Add.7, (July 12 1996), *available at* http://www.refworld.org/docid/453883fc22.html.

[221]Id. 21.

[222]U.N. Human Rights Council, *Addendum to the Report of the Special Rapporteur on the Promotion and Protection of Human Rights and Fundamental Freedoms While Countering Terrorism: Mission to Turkey*, at no. 45, U.N. Doc. A/HRC/4/26/Add.2 (Nov. 16, 2006).

[223]Id. at 20, 47.

[224]Id. pg. 7.

[225]General Assembly resolution 61/106, annex I. UN General Assembly, *Convention on the Rights of Persons with Disabilities: resolution/ adopted by the General Assembly*, A/RES/61/106, (Jan. 24 2007) *available at* http://www.refworld.org/docid/45f973632.html. *Convention on the Rights of Persons with Disabilities*, 2515 U.N.T.S. 3. (May 3, 2008).

create sanitation facilities designed to facilitate access to water and sanitation for everyone.[226]

One way that is appropriate for combatting exclusion of some older persons is through volunteering. The follow-up to the Second World Assembly on Ageing Report of the Secretary-General affirmed that, "at the individual level, volunteerism helps to keep older persons active and engaged."[227] The Report of the Secretary-General on follow-up to the International Year of Volunteers[228] mentions older persons' volunteer activities. According with the follow-up to the Second World Assembly on Ageing Report of the Secretary-General, "macro-level factors and general social conditions influence the readiness of older persons to engage in voluntary activities."[229] Citing the case of Germany, the report showed that "the volunteerism rate of persons with an intermediate level of education was nine percent, while the rate rose to seventeen percent among persons with higher education."[230]

One key aspect to ensure the fulfilment of many of their rights is the ability of older persons to adequately participate in the political system, and especially in issues that can directly affect them. This includes the possibility for them to participate in political movements or other social organizations through which they can be involved in the discussion of public affairs in their country.[231] Many older persons encounter problems while attempting to establish their legal identities due to the absence of identification documentation. As a result, they are unable to access services or speak out against objectionable public policies.[232]

The 2009 Final Report of the Third Committee also encouraged states to build national networks, including civil society organizations that can enhance the local capacity to confront challenges regarding the rights of older persons. The related draft resolution called upon member states "to develop their national capacity for monitoring and enforcing the rights of older persons, in consultation with, organizations of older persons through, inter alia, national institutions for the promotion

---

[226]U.N. Human Rights Council, *Report of the Independent Expert on the Issue of Human Rights Obligations related to Access to Safe Drinking Water and Sanitation, Catarina de Albuquerque*, at no. 76, U.N. Doc. A/HRC/12/24 (July 1, 2009).

[227]U.N. General Assembly, Report of the Secretary General. 2012. A/67/188. Follow-up to the Second World Assembly on Ageing, 16, U.N. Doc. A/67/188 (July, 26 2012) *available at* http://social.un.org/ageing-working-group/documents/A_67_188_E.pdf.

[228]U.N. General Assembly, *Report of the Secretary-General on follow-up to the International Year of Volunteers*. U.N. Doc. A/60/128 (Aug. 30, 2005), http://www.unv.org/fileadmin/docs/old/pdf/2005/050830INT_res_iyvffup.pdf.

[229]U.N. General Assembly, *Follow-Up to the Second World Assembly on Ageing: Report of the Secretary-General*, 18, U.N. Doc. A/67/188 (July, 26 2012) *available at* http://social.un.org/ageing-working-group/documents/A_67_188_E.pdf.

[230]Id.

[231]Id. 21(The General Assembly has recognizing that "in some countries, the high rate of voter turnout among older persons has helped to ensure that their concerns are addressed.").

[232]Id. pg. 7–8 *Id.* at 7–8.

and protection of human rights where applicable."[233] Furthermore, the report "encourages the international community to support national efforts to forge stronger partnerships with civil society, including organizations of older persons, in an effort to help build capacity on aging issues."[234]

In the follow-up to the Second World Assembly on Ageing[235] and follow-up to the Tenth Anniversary of the International Year of the Family and Beyond,[236] participating states highlighted the ability of cooperatives in social development to contribute to poverty eradication by promoting the participation of older persons in the economy and in society.[237] The Assembly recognized that cooperatives, promoted the fullest possible participation of all persons in their economic and social development, including, inter alia, older people, and that they contributed to the eradication of poverty.

The Report of the Secretary-General on cooperatives in social development pointed out that "in many developed countries, retired older persons have organized to advocate for their own interests in public and private programs that affect them, as well as to provide personal advice and counsel for those within their own age group."[238] It also referred to Switzerland indicating that, "cooperatives have programs to help members to assist older persons maintain and make use of their abilities.[239] To address the demand created by the rapid aging of the rural population, agricultural cooperatives in Japan have expanded services for older members."[240] Cooperatives can have a deleterious effect where some election techniques, including verification requirements, disproportionately disenfranchise older persons.[241]

---

[233]Follow-Up to the Second World Assembly on Ageing, G.A. Res. 64/132, 11, U.N. Doc. A/RES/64/132 (Feb. 5, 2010) http://daccess-dds-ny.un.org/doc/UNDOC/LTD/N09/547/37/PDF/N0954737.pdf?OpenElement.

[234]U.N. General Assembly, *Social Development: Report of the Third Committee*, 18, U.N. Doc. A/64/432 (Dec. 1, 2009) http://www.copac.coop/iyc/a-64-432-excerpt.pdf.

[235]Follow-Up to the Second World Assembly on Ageing, G.A. Res. 64/132, 11, U.N. Doc. A/RES/64/132 (Feb. 5, 2010). Press Release, U.N. General Assembly, Third Committee Approves Texts on Crime Congress, Cooperatives, Year of Family, Women's Anti-Discrimination Convention, African Crime Institute and Ageing, U.N. Press Release GA/SHC/3957 (Oct. 22, 2009) *available at* http://www.un.org/News/Press/docs/2009/gashc3957.doc.htm.

[236]*Follow-up to the Tenth Anniversary of the International Year of the Family and Beyond*, G.A. Res. 64/133, U.N. Doc. A/RES/64/133 (Feb. 5, 2010) http://www.un.org/News/Press/docs/2009/gashc3957.doc.htm.

[237]U.N. General Assembly, *Argentina, Bangladesh, Guatemala, Mongolia, Myanmar, Nepal, Panama and Sudan: draft resolution: Cooperations in Social Development*, U.N. Doc. A/C.3/64/L.10 (Oct. 7, 2009) http://daccess-dds-ny.un.org/doc/UNDOC/LTD/N09/547/92/PDF/N0954792.pdf?OpenElement.

[238]U.N. General Assembly, *Cooperatives in Social Development: Report of the Secretary-General*, 34, U.N. Doc. A/60/138 (Oct. 7, 2009) http://www.copac.coop/publications/un/a60138e.pdf.

[239]Id.

[240]Id.

[241]U.N. General Assembly, *Report of the Third Committee. 59th Session: Human rights questions: human rights situations and reports of special rapporteurs and representatives*. U.N. Doc. A/59/503/Add.3 (Dec. 9, 2004).

D. Economic, Social and Cultural Rights of Older Persons

The Rights of Progressive Realization

The rights of older persons are recognized in isolated provisions of international treaties that primarily protect economic, social and cultural rights (ESCR). Many of these rights are regarded as the rights of progressive realization whereby states, as primary duty-bearers, are bound to provide international economic and technical assistance - especially to developing countries - in order to enable and cooperate with states in fulfilling their human rights obligations.[242] This can create a perception that many rights of older persons are only programmatic aspirations without hard-and-fast standards. Even though many of the rights of older persons' in international human rights law have been identified as ESCR, they cannot be treated as a separate category of standards. Moreover, these rights can have obligations that require progressive realization and in many instances immediate enforceability.

Additionally, because older persons face obstacles to the full enjoyment of their rights due to age discrimination, they are often considered a disadvantaged or vulnerable group – a group that can and should be encompassed in the often repeated "any other status." This can enhance the protection standards that states are required to ensure in order to comply with international human rights obligations. This doctrine imposes the obligations upon duty-bearers to treat older persons as a group that deserves special protection, which carries with it a higher duty. International human rights law has established the doctrine that vulnerable groups are entitled to special protection to ensure their enjoyment of human rights guaranteed to all persons. This doctrine is based on the principle of equality before the law as enshrined in international law, which can also provide legitimacy for 'positive discrimination' measures and policies based on affirmative actions.

Interrelatedness of the Civil and Political Rights and Economic, Social and Cultural Rights

The well-being of older persons is not only protected by the provisions and instruments that guarantee ECSR, but also by the norms whose primary purpose and definition is to guarantee civil and political rights. The right to a fair trial and the right to non-discrimination, for instance, are fundamental rights that can also be used to protect older persons from certain practices such as forced retirement

[242]U.N. Committee on Economic, Social and Cultural Rights (CESCR), *General Comment No. 14: The Right to the Highest Attainable Standard of Health (Art. 12 of the Covenant)*, 45, U.N. Doc. E/C.12/2000/4 (Aug. 11, 2000).

or slow judicial proceedings regarding immediately necessary health or social benefits. Similarly, provisions prohibiting cruel and degrading treatment provide for specific protection of older persons who are confined in detention centres and prisons as well as health care facilities. In other instances, the pensions and social benefits that older persons receive from a state can be protected by the standards of the right to property and due process of law. Furthermore, all ESCR and civil and political rights are indivisible and interdependent from each other. Thus the protection of one right often allows for the implementation and protection of other types of rights. Civil and political rights and ESCR can also be immediately applicable and subject to judicial adjudication.

Normative Framework of the General Comment No. 6 – the Economic, Social and Cultural Rights of Older Persons

The CESCR summarized the responsibilities of member states toward the protection of older persons in its landmark *General Comment No. 6,* in which it declared that state parties to the ICESCR are obligated to pay particular attention to promoting and protecting the economic, social and cultural rights of older persons.[243] The CESCR further established its own mandate to monitor the implementation of international protection for older persons, particularly since no international convention exists for their specific rights.[244]

The CESCR further reasoned the necessity for this approach in state reports issuing little information on the status of the rights of older persons.[245] The CESCR emphatically concluded that future state reports ought to address situations and circumstances of older persons in relation to each of the rights recognized in the ICESCR.[246]

Furthermore, *General Comment No. 6* notes that neither the ICESCR nor the UDHR explicitly prohibits age discrimination, but this should not be seen as an intentional normative omission by the international community with regard to the legal status of older persons.[247] In that regard, the CESCR stated that the prohibition of discrimination based on other status could include age.[248] The CESCR rationalized its conclusion by summarizing the numerous policy documents in international forums as well as domestic legislation that support this principle, as well as the need to eliminate age discrimination.

[243]UN Committee on Economic, Social and Cultural Rights (CESCR), *General Comment No. 6: The Economic, Social and Cultural Rights of Older Persons*, 13, U.N. Doc. E/1996/22 (Dec. 8, 1995).

[244]Id.

[245]Id. at paras. 14–15.

[246]Id.

[247]Id.

[248]Id. at para. 12.

1. The Right to an Adequate Standard of Living

Older persons are a particularly vulnerable group that is specially exposed to poverty and its consequences. Even in the wealthiest countries on earth that pride themselves on social protection in old age, older women have a poverty risk rate of about 22 percent and older men have a poverty risk rate of 16 percent.[249] Precise information about poverty in old age is difficult to ascertain. As the Independent Expert on Human Rights in Extreme Poverty has suggested, "weaker knowledge of the specificities of old-age poverty…most likely leads to policy choices that ignore and possibly exclude older persons living in extreme poverty."[250]

The CESCR linked this right to the inherent dignity of all persons and pronounced its indispensability to the fulfillment of virtually every other right in the International Bill of Rights, including the ICCPR and ICESCR, among others.[251] Moreover, it has attached great importance to the right to an adequate standard of living, as it includes the right to "adequate food, water, shelter, clothing and health care through the provisions of income, family and community support and self-help."[252] In addition, Article 25 of the UDHR sets the standard that requires states to ensure the "right to security in the event of unemployment, sickness, disability, widowhood, old age or other lack of livelihood in circumstances beyond his control."[253]

An adequate standard of living encompasses many different aspects of life. As already mentioned, the CESCR has interpreted this to include several social and economic rights.[254] Article 10 of the ICESCR requires states to accord "the widest possible protection and assistance …to the family, which is the natural and fundamental group unit society."[255] As Lindsay Judge puts it "CESCR General Comment No. 6 also points to the need to support families in responding to the needs of older relatives. Specifically, it stresses the need for measures to support low-income families in caring for older persons as well as single persons or elderly couples who wish to remain in their homes."[256]

---

[249]"Sepulveda Report", *supra*, 20.

[250]Id. 14.

[251]U.N. Committee on Economic, Social and Cultural Rights (CESCR), *General Comment No. 6: The Economic, Social and Cultural Rights of Older Persons*, 32, U.N. Doc. E/1996/22 (Dec. 8, 1995)(describing how the right to adequate food is linked to human dignity and social justice).

[252]*See id.* 32 (referring to Principle 1 of the United Nations Principles for Older Persons).

[253]Universal Declaration of Human Rights, art. 25(1) (emphasizing that everyone has the right to "a standard of living adequate for the health and well-being of himself and of his family").

[254]Diego Rodriguez-Pinzon and Claudia Martin, *The International Human Rights Status of Elderly Persons*, 18 Am. U. Int'l L. Rev. 915, 960 (2003).

[255]Lindsey Judge, *The Rights of Older People: International Law, Human Rights Mechanisms and the Case for New Normative Standards* 7 (2009).

[256]Id. at pg. 7.

Access to the right to housing, for example, shows the complex interrelation that such a right has with other rights. According to the CESCR, the focus should not be on providing just shelter, but providing adequate housing, as the right to housing is intimately linked with other human rights.[257] For example, the CESCR linked the right to housing with freedom of expression and association, traditionally held as civil and political rights, and stated that the full implementation of these other rights is "indispensable if the right to adequate housing is to be realized."[258]

On the other hand, the ICESCR has raised concerns regarding the standard of living of older persons in relation to the "extent of poverty among older persons and about the fact that home care is still insufficiently developed."[259] In this vein, it recommended that Serbia and Montenegro "take special measures to alleviate the extent of poverty among older persons and that priority be given to home care rather than institutionalization of older persons in need of care, including to allocate sufficient funds to that effect and strengthen the role of non-profit organizations in the provision of home care and other social services."[260] The right to adequate housing can be negatively impacted when older persons are involuntarily excluded from their communities.

The 2009 *Final Report of the Third Committee* encourages governments "to pay greater attention to building capacity to eradicate poverty among older persons, in particular older women, by mainstreaming aging issues into poverty eradication strategies and to include both age-specific policies and age-mainstreaming efforts in their national strategies."[261] The report further called upon governments "to ensure conditions that enable families and communities to provide care and protection to persons as they age."[262] Further, the draft resolution encouraged "the international community to enhance international cooperation to support national efforts to eradicate poverty, in order to achieve sustainable social and economic support for older persons."[263]

---

[257]Id. at para. 7 (stating that, in the Committee's view, "the right to housing should not be interpreted in a narrow or restrictive sense … rather it should be seen as a right to live somewhere in security, peace and dignity"). General Comment 6, *supra note 279*, 7 (stating that, in the Committee's view, "the right to housing should not be interpreted in a narrow or restrictive sense … rather it should be seen as a right to live somewhere in security, peace and dignity").

[258]Id. at para. 9.

[259]U.N. Econ. and Soc. Council Official Record, 2006. Supp. No. 2. Comm. on Econ., Soc., and Cultural Rights. Report on the 34th and 35th Sess., E/2006/22-E/C.12/2005/5. 288, http://bayefsky.com/general/e_2006_22.pdf (last visited Oct. 23, 2014).

[260]Serbia and Montenegro, Int'l Covenant on Econ. And Soc. and Cultural Rights, Rep. on its 34th and 35th Sess., Supp. No. 2 E/2006/22 (2005) 41 at. 288 and 315. 288 http://www.refworld.org/pdfid/474d35ca2.pdf (last visited Oct. 23, 2014).

[261]G.A. Res. A/64/432; http://www.copac.coop/iyc/a-64-432-excerpt.pdf. (last visited Oct. 23, 2014).

[262]Id. para. 8.

[263]U.N. G.A, 64th Sess. at 3A/C.3/64/L.6 http://daccess-dds-ny.un.org/doc/UNDOC/LTD/N09/547/37/PDF/N0954737.pdf?OpenElement, (last visited Oct. 23, 2014). *See also* 13 and 2, of Follow-up to the International Year of Older Persons: Second World Assembly on Ageing G.A. Res. 63/151, U.N. Doc. A/63/424 http://www.un.org/en/ga/search/view_doc.asp?symbol=A/RES/63/151&Lang=E. (last visited Oct. 23, 2014).

A notion closely related to the concept of 'adequate standard of living' is that of sustainable livelihood. Article 11 of the ICESCR recognizes "the right of everyone to an adequate standard of living for himself and his family, including adequate food, clothing and housing, and to the continuous improvement of living conditions." Similarly, it indicates that "states will take appropriate steps to ensure the realization of this right."[264]

In the development context, the concept of sustainable livelihood focuses on "reducing poverty by empowering the poor to build on their opportunities."[265] Among older persons, as highlighted by the Independent Expert on the Question of Human Rights and Extreme Poverty, the factors for poverty "include: lack of access to regular income, work, and health care, declining physical and mental capacities, and dependency within the household."[266] These are the main tools required to maintain a sustainable livelihood, building on one's own resources.

An important part of sustainable livelihoods is the right to own property. Every person has the right to own property, and "[n]o one shall be arbitrarily deprived of his property."[267] Regarding the right of property, it has been underscored that "in many parts of the world, inheritance laws, both statutory and customary, deny women of all ages the right to own or inherit property when their husband dies. Family members sometimes force widows off their land or seize their property."[268] Their ability to maintain a sustainable livelihood is severely restricted when their right to own property is not protected.

The scope of the right to a sustainable livelihood includes pensions and other related benefits, as well as access to services. In *Love*, et al. *v. Australia*, the HRC concluded that mandatory retirement in the interest of safety was based on reasonable and objective considerations, yet noted that age discrimination, while not enumerated, is a ground for violation of Article 26 of the ICCPR.[269] The CESCR expressed concern in 2001 about the widespread and unacceptable incidence of poverty in the Hong Kong Special Administrative Region, particularly the large

---

[264]Int'l Covenant on Econ., Soc. and Cultural Rights (ICESCR) *supra* note 288 at Article 11.

[265]Diana Carney, *Sustainable Livelihoods Approaches, Progress, and Possibilities for Change*, DEPARTMENT FOR INTERNATIONAL DEVELOPMENT http://www.eldis.org/vfile/upload/1/document/0812/SLA_Progress.pdf (last visited Oct. 23, 2014).

[266]Magdalena Sepulveda ,*Report of Independent Expert on Human Rights and Extreme Poverty, Rapporteur*, Para.15, (2010)("To collect information to prepare this report the independent expert sent a questionnaire to Governments requesting information on their social pensions for older persons. The questionnaire addressed: (i) the legal and institutional framework; (ii) the schemes costs and coverage; (iii) implementation procedures; (iv) monitoring mechanisms and complaints procedures; and (v) studies and evaluations").

[267]Universal Declaration of Human Rights, G.A. Res. 217 (111), A, UN. Doc A/Res/217(111) Article 17 (Dec. 10, 1948).

[268]HelpAge Position Paper (Convention) at 3.

[269]*Love v Australia*, Int'l Covenant on Civil and Political Rights, No. 983/2001, A/58/40 vol. II 286 at 2.1, 8.2, 8.3, and Individual Opinion of Mr. Prafullachandra Natwarlal Bhagwati, 302 (Mar. 25, 2003) http://www1.umn.edu/humanrts/undocs/983-2001.html.

number of older persons continuing to live in poverty without effective access to social services.[270] In the cases of United Kingdom of Great Britain and Northern Ireland, the CESCR reiterated "its concern about the persistence of considerable levels of poverty, especially in certain parts of the country", and among certain sections of the population, including older persons.[271] Finally, in Korea, the CESCR expressed its concern about "the consequences of the widespread famine suffered by the country from the mid-1990s, and that certain groups," including "older persons, have been more severely affected than others and have not received proper assistance in order to alleviate their plight."[272]

In the UPR process, Monaco's Working Group Report noted that older persons enjoy a long-standing policy of guaranteeing an income enabling them to live decently. Geriatric care arrangements were in place to allow older persons to remain living in their home as long as they wished, with the necessary financial assistance. When they had to leave their home, there were graduated establishments according to their needs. Notably, they could accede to all establishments, whatever their level of income.[273] In the Republic of Korea's Working Group Report a reviewing state praised the government for its awareness of challenges, such as the increasing older population, though it did not offer solutions.[274]

2. *The Right to Food*

According to the General Comment No. 12 on the right to food, issued by the Committee of the ICESCR, "the human right to adequate food is of crucial importance for the enjoyment of all rights," as "the right to adequate food is indivisibly linked to the inherent dignity of the human person and is indispensable for the fulfillment of other human rights," such as the right to life.[275] The comment underlines that the right to adequate food "means physical and economic access at all

[270]China, 2001, Comm. On Econ. Soc. and Cultural Rights, Report on the 25th, 26th and 27th Sess. E/2002/22 E/C.12/2001/17 42 http://www.bayefsky.com/general/e_2002_22_2002.php. (last visited Oct. 23, 2014).

[271]United Kingdom of Great Britain and Northern Ireland, ICESCR, E/2003/22 39 at 221 (2002) http://www.unhchr.ch/tbs/doc.nsf/0/8c3da14278eb6162c1256d4f00561c38/$FILE/G0341455.pdf. (last visited Oct. 23, 2014).

[272]Democratic People's Republic of Korea, ICESCR, E/2004/22 (2003) 71 at 530.

[273]UN Human Rights Council, *Report of the Working Group on the Universal Periodic Review - Monaco*, Part. 75, A/HRC/12/3, A/HRC/12/3, A/HRC/WG.6/5/L.2, (Jun. 4, 2009) *available at*: http://www.refworld.org/docid/4a9cdb950.html.

[274]UN Human Rights Council, *Report of the Working Group on the Universal Periodic Review - Republic of Korea*, 14, A/HRC/8/40, A/HRC/8/40; A/HRC/WG.6/2/L.6 (May 29, 2008), *available at*: http://www.refworld.org/docid/485b96303.html.

[275]*General Comment 12 Right to Adequate Food*, U.N. Doc. E/C.12/1999/5 CESCR, 20th Sess. 4, (1999) *available at* http://www1.umn.edu/humanrts/gencomm/escgencom12.htm.

times to adequate food or means for its procurement."[276] This is a right recognized by many states as a fundamental right.[277] A state's economic vulnerability and widespread poverty affect vulnerable groups, including older persons,[278] who can be prone to food insecurity.[279]

The CESCR interprets the scope of the right to adequate food in General Comment No 12, stating that pursuant to Article 11 of the ICESCR, states recognize "the right of everyone to an adequate standard of living for himself and his family, including adequate food, clothing and housing, and to the continuous improvement of living conditions," which necessarily implies the right of everyone to be free from hunger.[280] The core aspects of the right to adequate food include "the availability of food in a quantity and quality sufficient to satisfy the dietary needs of individuals, free from adverse substances, and acceptable within a given culture; and the accessibility of such food in ways that are sustainable and that do not interfere with the enjoyment of other human rights."[281]

The right to food encompasses economic as well as physical access. The General Comment indicates that "socially vulnerable groups such as landless persons and other particularly impoverished segments of the population may need attention through special programs."[282] Likewise, it refers that "physical accessibility implies that adequate food must be accessible to everyone, in particular physically vulnerable individuals including elderly people."[283] The General Comment also states that "any discrimination in access to food, as well as to means and entitlements for its procurement, including on the grounds of … age… with the purpose or effect of nullifying or impairing the equal enjoyment or exercise of economic, social and cultural rights, constitutes a violation of the Covenant."[284] A violation of this right would occur if a state refused access to ade-

[276] Id. at 14.

[277] UN Human Rights Council, *Report of the Special Rapporteur on the Right to Food, Jean Ziegler: addendum: mission to Bolivia*, A/HRC/7/5/Add.2, (Jan. 30 2008) *available at* http://www.refworld.org/docid/47c6c3332.html:.

[278] UN Human Rights Council, *Report of the High Commissioner for Human Rights on the Situation of Human Rights in Afghanistan and on the Achievements of Technical Assistance in the Field of Human Rights*, A/HRC/7/27, at no. 10. (Feb. 21 2008) *available at* http://www.refworld.org/docid/47cfb4b22.html .

[279] UN Human Rights Council, *Report of the Special Rapporteur on the Situation of Human Rights in the Democratic People's Republic of Korea, Vitit Muntarbhorn*, no. 41 15, A/HRC/7/20, (Feb. 2008) *available at* http://www.refworld.org/docid/47ce78b22.html.

[280] General Comment 12 *supra* note 303 at para. 1.

[281] Id. at 8 (explaining in more detail the core aspects of the right to adequate food named in paragraphs 9–13).

[282] Id. at 13.

[283] Id.

[284] Id. at 18.

quate food on the grounds of age.[285] When it comes to food, states have a positive duty to protect persons in their jurisdiction from actions they undertake that may indirectly violate the right to an adequate standard of living. According to the Special Rapporteur on the right to food, privatization of state resources can lead to increased prices for the necessities of an adequate standard of living, which disproportionately affect poor and older persons.[286]

The CESCR has explained that this also encompasses protection of productive lands as well as a functioning distribution of the market systems that move foodstuffs.[287]

Economic accessibility refers to maintaining costs at a level that is attainable and does not infringe on the exercise of other basic rights.[288] Physical accessibility refers to the ability of those who are physically impaired or disabled, which includes many older persons, to have access to food. States should guarantee that, during times of economic and national emergencies, older persons are given special protection[289] and priority if possible.[290] For example in natural disasters, priority access to humanitarian aid goods and services should be afforded to older persons.[291] The obligation to guarantee the right to food also requires states to legislate and provide judicial review to protect and enforce the enjoyment of the right to food.[292] States must also refrain from using access to food as a political weapon.[293]

States should reduce environmental pollution that affects food production and should provide older persons with adequate access to potable water to reduce the

---

[285]Id. (stating that any discrimination in access to food on the grounds of age would result in a violation of the Covenant).

[286]*Report of the Special Rapporteur on the Right to Food, Communications to and from Governments*: A/HRC/4/30/Add.1, no. 38. (May 18 2007), http://daccess-dds-ny.un.org/doc/UNDOC/GEN/G07/126/30/PDF/G0712630.pdf?OpenElement.

[287]Id. at 12 (stating that "availability refers to the possibilities either for feeding oneself directly from productive land or other natural resources, or for well-functioning distribution, processing and market systems that can move food from the site of production to where it is needed in accordance with demand").

[288]General Comment 12 *supra* note 303 13.

[289]Id. at 28. (stating that even where a state faces "severe resource constraints, whether caused by a process of economic adjustment, economic recession, climatic conditions or other factors, measures should be undertaken to ensure that the right to adequate food is especially fulfilled for vulnerable population groups and individuals").

[290]Id. 38 (stating that "priority in food aid should be given to the most vulnerable populations").

[291]*Annual Report of the Representative of the Secretary-General on Human Rights of Internally Displaced Persons*: A/HRC/4/38/Add.1, at B.1.3. (Jan 23, 2006) *available at* http://daccess-dds-ny.un.org/doc/UNDOC/GEN/G07/105/12/PDF/G0710512.pdf?OpenElement.

[292]General Comment 12 *supra* note 303 at 32 (stating that "any person or group who is a victim of a violation of the right to adequate food should have access to effective judicial or other appropriate remedies" and that such victims are entitled to "adequate reparation, which may take the form of restitution, compensation, satisfaction or guarantees of non-repetition").

[293]Id. at 37 (stating that "food should never be used as an instrument of political and economic pressure").

spread of disease.[294] Furthermore, they should provide older persons with access to up-to-date information and education regarding food and nutrition.[295] Additionally, states should refrain from inhibiting persons' access to adequate food,[296] and should guarantee that third parties do not inhibit individuals' right to food. This requires states to formulate national food programs that are culturally appropriate.[297]

The 2005 Report on the World Social Situation expressed the concern about added dimensions to the patterns of inequality characterizing food crises. Food may not be divided equitably among household members, with women, children, and older persons receiving proportionately less than adult men.[298] The topic was also addressed by the Report of the Secretary General on the follow-up to the Second World Assembly on Ageing, noting that the Food and Agriculture Organization mainstreamed its work on rural aging by focusing on enhancing income opportunities for rural older persons in agriculture. This included providing older rural residents with the agro-technical and business management skills they need and preparing agricultural extension services to address extension and training needs of older farmers.[299]

According to the Report of the United Nations High Commissioner for Human Rights from April, 2012, "many older persons are capable of maintaining food security for themselves and those in their care through production or procurement of food by their own means." However, older persons may be more likely of losing access to resources, like "the situation of older widows, who in some cases may not control access to agricultural land." Furthermore, the report indicates that "older persons may require support for procuring, accessing and preparing food adequate for their nutrition."[300] Overall, the Report appears to suggest that states have a key obligation under international human rights law of providing special support to older persons to allow them to exercise their right to seek food security.

---

[294]Eide, at 91–92 (stating that disease and malnutrition can be avoided by removing the dangers and risks associated with environmental pollution).

[295]Id. (stating that all segments of society should have access to education and information about essential needs and associations with nutrition and health care).

[296]General Comment 12 *supra* note 303 15 (noting that states have an obligation to protect vulnerable individuals by taking measures to "ensure that enterprises or individuals do not deprive [them] of their access to adequate food").

[297]Id. at 39 (stating that "products included in international food trade or aid programs must be. culturally acceptable to the recipient population").

[298]Press Release, General Assembly, Equality Should be at the Forefront of Socio-economic Policymaking, Third Committee Told at Beginning of Debate on Social Issues, A/60/117 210, http://www.un.org/press/en/2005/gashc3814.doc.htm (last visited Oct. 23, 2014).

[299]U.N. G.A., 60th Sess. 34, *available at* http://www.un.org/ga/60/third/docslist.htm. (last visited Oct. 23, 2014).

[300]*Report of the United Nations High Commissioner for Human Rights*, U.N. Econ. And Soc. Council, 41, 42, (Apr. 20, 2012) http://www.un.org/ga/search/view_doc.asp?symbol=E/2012/51.

3. The Right to Education and Culture

The right to education is essential to ensure the fulfillment of the civil, political, economic, social and cultural rights.[301] This right clearly articulates the notions of indivisibility and interdependence of all human rights.[302] Older persons generally experience severe limitations in the enjoyment of the right to education, including access to literacy training as well as other holistic educational activities (cultural, spiritual, recreational, etc.),[303] due to, among others, the social preconception that education is to be delivered to younger generations. The right of everyone to education is stipulated in Article 26 of UDHR, which states "that technical and professional education shall be made generally available and higher education shall be equally accessible to all on the basis of merit." Further, Article 26 elaborates upon the scope of the right stating that "education shall be directed to the full development of the human personality and to the strengthening of respect for human rights and fundamental freedoms."

The right to education includes a variety of dimensions, such as access to educational institutions, freedom to seek education and deliver education, academic freedom (which is closely related to freedom of expression and thought) and non-discrimination in any of the mention dimensions.[304] Article 13 of the ICESCR states that "higher education shall be made equally accessible to all, including that fundamental education shall be encouraged or intensified as far as possible for those persons who have not received or completed the whole period of their primary education." The CESCR has acknowledged that this provision "is the most wide-ranging and comprehensive article on the right to education in international human rights law."[305] According to the CESCR, this right, in all its forms and levels, should be available, accessible, acceptable, and adaptable.[306]

---

[301]Manfred Nowak, *The Right to Education, in Economic, Social and Cultural Rights: A Textbook* 189 (Asbjorn Eide et al. eds., 1995) (explaining that people cannot exercise various civil, political, economic, social, and cultural rights without attaining a minimum level of education, such as literacy).

[302]*Plans of Action for Primary Education* (Art. 14): General Comment 11, U.N. Econ., Soc., & Cultural Rights. Comm., 20th Sess., 2, UN Doc. E/C.12/1999/4 (1999).

[303]CRC, Article 28; CRPD, Article 24; ICESCR, Article 13.

[304]*See generally* Nowak, supra note 222, at 198–211 (listing and discussing his delineation of secondary rights stemming from the core right to education).

[305]*Right to Education* (Art. 13): General Comment 13, U.N. Econ., Soc., & Cultural Rights. Comm., 21st Sess., 2, U.N. Doc. E/C.12/1999/10 (1999) *available at* http://www.unhchr.ch/tbs/doc.nsf/0/ae1a0b126d068e868025683c003c8b3b?Opendocument. (last visited Oct. 23, 2014).

[306]General Comment 13, 6(a)-(d) (explaining that although each state party contends with differing conditions in providing education to its people, the right to education depends on these "interrelated and essential features").

Furthermore, the CESCR stated in its General Comment 13 that the right to fundamental education is not limited by age and should be extended to older persons.[307] It considered fundamental education an important aspect of adult education and required states to provide all aspects of the right to education in a suitable manner for all different age groups.[308] It also indicated that states should establish educational institutions that specialize in gerontology, geriatrics, and geriatric psychology, in order to ensure that the education system is supporting the needs of the whole population.[309] The CESCR has also established fellowship systems that enhance educational access for disadvantaged groups, which could include older persons.[310] Additionally, in General Comment 6 the CESCR stressed the need to interpret these rights from two different perspectives, including the right of older persons to benefit from educational programs, and the right to have the knowledge and experience of older persons passed on to persons of younger generations.[311]

Education permeates every social group in society. Its implementation must always take into account the diverse audience it addresses, based on age, among many other factors.[312] To ensure these rights, the CESCR has also requested that states provide community centers, both educational and recreational, to assist the development of life-long education for older persons.[313] The United Nations Educational, Scientific and Cultural Organization (UNESCO) further recommends "informal, community-based and recreation-oriented programs for older persons to help to develop their sense of self-reliance."[314]

One example of a culture-preserving program is an intergenerational center, which may have both educational and inter-personal benefits. Furthermore, such

---

[307] Id. 24.

[308] Id. (clarifying that states must develop educational "curricula and delivery systems" that are suitable for students of any age).

[309] Id. 42 (providing various recommendations to States that ensure that older persons benefit from scientific progress).

[310] Id. 26 (incorporating the non-discrimination and equality clauses of the Covenant into the requirement of establishing a fellowship system).

[311] *See, e.g.*, General Comment 6, 36. (outlining two views on how states should approach the education rights of older persons).

[312] *See* id. 37 (asserting that access to these educational programs fulfills the recommendations provided in Principle 16 of the United Nations Principles for Older Persons); *See e.g.*, *Concluding Observations of the Committee on Economic, Social and Cultural Rights: Germany*, U.N. Econ., Soc., & Cultural Rights. Comm., 9th Sess., 46th mtg., 11, U.N. Doc. E/C.12/1993/17 (1994) (recommending that Germany provide information "to facilitate the access and the participation of older persons in educational programs")(this recommendation stems from the Committee's review of Germany's second periodic report regarding the status of rights protected by Articles 13 to 15 of the ICESCR. Id. para 1).

[313] General Comment 6, 37 (drawing from the Vienna International Plan of Action and its recommendations in developing a sense of independence for older persons and the community's responsibility for ensuring the rights of older persons to education).

[314] Report of the Expert Group Meeting "Rights of Older Persons," 29 (2009), *available at* http://www.un.org/esa/socdev/ageing/documents/egm/bonn09/report.pdf. (last visited Oct. 23, 2014) (hereinafter "U.N. SG Report on Aging").

centers would provide older persons with a forum in which to teach others about their experiences and insights. To respect the right to culture, the CESCR has urged states to refrain from segregating older persons in society. States should actively integrate older persons into society acknowledging their right to education as well as the education they provide with a wealth of information and skills.[315]

Literacy is an important outcome of education. The 2009 report of the expert group meeting on the rights of older persons showed that "older persons, especially older women and those residing in rural areas of developing countries, are more likely to be illiterate than the other parts of the population."[316] In Western Asia, for example, illiteracy among older women is high because many were denied access to educational opportunities throughout their lives.[317] Such isolation has long-lasting effects on social integration, access to information, and even access to the law. A knowledge-based and service-based society, as many developed countries have become and many developing countries aspire to become, requires policies that ensure lifelong access to education and training.[318]

According to the second review and appraisal of the Madrid International Plan of Action on Ageing in 2002, "lifelong education and training are critical for the participation of older persons in the labor force. Yet the (current) level of education among persons aged 60 years and above remains much lower than among (current) younger generations. As a result, older persons face challenges in adjusting to economic and technological changes."[319] Likewise, it was mentioned that the "lifelong education for older persons has typically taken the form of "third age" universities, in countries such as Argentina, Canada, China, Indonesia, Hungary, the Russian Federation and Serbia. For instance, in Argentina there is a specific program, the University Extension for Older Persons (*Extensión Universitaria para Adultos Mayores*), which provides courses for older persons at national universities."[320]

Furthermore, the second review indicated that "while there have been reports of small-scale, local-level initiatives that offer older persons training in information and communications technology, there are no such national programmes on record."[321] It further stated that it is "worth noting … the case of Hungary, where

---

[315]*See* id. 38 (acknowledging the significance of preserving the role of older persons as sources of "information, knowledge, traditions and spiritual values").

[316]*See, e.g.*, UNDESA, World Report on Ageing 2009.

[317]Bonn Meeting 2009, p. 5.

[318]Lindsay Judge, *The Rights of Older People: International Law, Human Rights Mechanisms and the Case for New Normative Standards* 16 (Int'l Symp. on the Rts. of Older People, Background Briefing Paper, Jan. 2009), at 4–5.

[319]*Second Review and Appraisal of the Madrid International Plan of Action on Ageing*, 47, (2002) *available at* http://daccess-dds-ny.un.org/doc/UNDOC/GEN/N11/579/60/PDF/N1157960.pdf?OpenElement. (last visited Oct. 23, 2014).

[320]Id. at para. 48.

[321]Id. at para. 49.

recently implemented programmes aimed at improving ICT skills among the general population, such as the net ready programme launched in 2007 and the e-Hungary programme in 2004,"[322] which are good examples of programs targeting older persons.

4. The Right to Work

States should protect older workers' right to work in just and favorable work conditions through regulation of employment conditions. Access to education throughout the life course is necessary to ensure this right. Adequate work and employment are important components of the right to access to an adequate income. The right to work encompasses the right to earn a livelihood in decent work conditions and the right to individual and personal freedom and dignity.[323] Older persons have the right to be free from discrimination in hiring, promotion, wages, and discharge, and they have equal rights to form and to join trade unions.[324] For most of the world's older people, retirement is not an option. Those who are poor have no choice but to work, and policy makers often ignore the needs and contributions of older workers.[325]

The scope of the right to work for everyone according to Article 23 of UDHR includes a "free choice of employment, just and favorable conditions of work and to protection against unemployment", as well as "the right to equal pay for equal work." The provision also includes "the right to just and favorable remuneration ensuring for himself and his family an existence worthy of human dignity, and supplemented, if necessary, by other means of social protection." Older persons' access to these rights has not been universally defended in national laws and regulations.

Article 6 of the ICESCR is a landmark provision establishing the scope of the right to work. This provision recognizes the right to work as the right of everyone to have the opportunity to gain his or her living by work that he or she freely chooses or accepts. Further, under that provision, states agree "to achieve the full realization of this right", including "technical and vocational guidance and

[322]Id.

[323]*See* Krzysztof Drzewicki, *The Right to Work and Rights in Work, in Economic, Social and Cultural Rights: A Textbook* 169 (Asbjorn Eide et al., eds., 1995) (stating that "the right to work and rights in work constitute a core of not only socio-economic rights, but also fundamental human rights").

[324]As stated in many human rights declarations including the UDHR, Article 23; CRPD, Article 27.

See also, Kelly Report, *supra* note 30, at 12. Committee on Economic, Social and Cultural Rights General Comment 20, 29, 25, UN SG Report on Aging, *supra* note 342.

[325]*Working for Life: Making Decent Work and Pensions a Reality for Older People*, Aging and Development, HelpAge International, p. 2.

training programs, policies and techniques to achieve steady economic, social and cultural development and full and productive employment under conditions safeguarding fundamental political and economic freedoms to the individual."[326]

Also, CESCR General Comment No. 6 specifically recognizes the difficulty older persons face in finding and sustaining jobs, and advocates measures that will prevent age discrimination. In developed states, mandatory retirement at a certain age is a contentious issue. In certain circumstances it can be discriminatory, although it is often embraced by persons who would rather receive retirement benefits than continue working. In developing countries, where no such retirement benefits are generally available, conditions of work and fair income, lack of access to credit, and training are particularly salient to older workers.[327] Aware of the problems that older persons may encounter to find and keep a job, the General Comment "stresses the need for measures to prevent discrimination on grounds of age in employment and occupation."[328]

According to the Department of Economic and Social Affairs of the United Nations (DESA), "some form of legislation against age discrimination in employment exists in approximately 50 countries around the world", but "it is difficult to assess the effectiveness of such legislation in combating age discrimination."[329] Furthermore, DESA underscored that "effective means of publicizing relevant legal provisions and of monitoring and enforcing compliance are needed if laws are to have an impact" and it indicated that "efforts to combat negative stereotypes held by employers may also have an effect; in some countries informational campaigns have been launched to address this issue."[330] Age discrimination in employment affects many other rights, and should be a priority for any country seeking to ensure the rights of older persons are protected.

Working older persons face age and gender discrimination. A report of HelpAge indicated that older men and women are often denied "access to social protection, income security, legal protection for their rights as workers, or formal support networks."[331] States should look to empower older persons to participate actively in their societies, including opportunities for older persons to work as long as they wish, in productive and satisfying jobs, without being forced into retirement. The CESCR recommends strengthened unemployment programs for groups with high rates of long-term unemployment, particularly persons over 40.[332] The follow-up to the International Year of Older Persons: Second World

---

[326]Article 6 thorough 8, ICESCR.

[327]Lindsey Judge, *The Rights of Older People: International Law, Human Rights Mechanisms and the Case for New Normative Standards* pg. 4 (Jan. 2009).

[328]Id.

[329]*Current Status of the Social Situation, Wellbeing, Participation in Development and Rights of Older Persons Worldwide,*United Nations, 43 (2011) *available at* http://www.un.org/esa/socdev/ageing/documents/publications/current-status-older-persons.pdf. (last visited Oct. 23, 2014).

[330]Id.

[331]Chung Working Paper at para. 16, 2010.

[332]Spain, ICESCR, E/2005/22 (2004) 34 at 235 and 252.

Assembly on Ageing also recognized "that specific programmes should be developed aimed at overcoming discrimination in employment and at integrating into the labour market women... and older persons."[333] The goal of promoting full employment as outlined in Commitment 3 of the Copenhagen Declaration includes older persons as long as they are able and willing to work.[334]

Older workers make up an increasing proportion of the labor force, though often in informal labor rather than formal labor. Formal labor force participation after age 65 has changed little in the last thirty years, remaining constant at 10 percent in developed countries, and 27 percent in developing countries.[335] It is projected to remain unchanged despite the drastic increase in numbers of older persons,[336] signaling a great increase in the numbers of older persons in the workforce.

The report on the follow-up to the Second World Assembly on Ageing expressed that "the differences in levels of labour force participation between developed and developing regions reflect the typical inverse relationship between labour force participation at older ages and the proportion of the older population receiving a pension."[337] Furthermore, the report indicated that 'in developing countries, employment opportunities for older persons often are only available in the informal sector, where jobs typically offer relatively low pay, insecure tenure and limited opportunities for advancement, and lack retirement benefits.[338]

The right to work is one of the most clearly connected to economic, social, and cultural rights[339] because these rights relate to not only protection from labor-related violations such as slavery, forced and compulsory labor, and discrimination, but also incorporate such positive aspects as the freedom to work, the choice of occupation, and the choice of place of work.[340] These rights also encompass the right to free employment services and guidance and assistance for job seekers and access to information.[341]

---

[333]G.A. Res 63/51 23, U.N. Doc. A/63/424 (Feb. 11, 2009) *available at* http://www.un.org/en/ga/search/view_doc.asp?symbol=A/RES/63/151&Lang=E.

[334]Robert Venne Position Paper at 8.

[335]*World Population Ageing*, United Nations Dept. of Economic and Social Affairs, Population Division, 41. (2009) *available at* http://www.un.org/esa/population/publications/WPA2009/WPA2009_WorkingPaper.pdf. (last visited Oct. 23, 2014).

[336]Id.

[337]U.N. Secretary-General, *Follow-up to the Second World Assembly on Aging*, U.N Doc. A/67/188 6 (Jul. 26, 2012), http://daccess-dds-ny.un.org/doc/UNDOC/GEN/N12/442/82/PDF/N1244282.pdf?OpenElement.

[338]Id. at 11.

[339]Id. at 171 (explaining that the "relatively developed stage of juridization of labour rights has, to a larger extent, been a contribution of the objective law (international labor legislation) than of the subjective law (international law of human rights)").

[340]Id. at 178 (mentioning that human rights and labor law include freedoms such as the choice of occupation and freedom of choice for a place of their performance).

[341]Id. at 180 (stating that before this right emerged in the international human rights field, the International Labour Organization Conventions primarily regulated the issue).

Regarding the challenges faced by the older workers, the DESA has indicated that there is "discrimination in hiring, promotion and access to job-related training" and that when older persons lose their jobs they "tend to remain unemployed for longer periods than their younger counterparts."[342] The CESCR expressed a concern about older persons under the retirement age who had trouble finding or keeping a job; the working conditions of older persons and the lack of preparation they receive prior to actually retiring.[343] To respect these rights, the CESCR again maintained that states refrain from discriminating against elderly persons based on their age, both during the hiring process, and during the evaluation process. The CESCR suggested that there should not be age requirements for all jobs, nor mandatory retirement provisions. In addition, to respect these rights, the CESCR stated that rights categorized in Article 8, namely trade union rights, should apply to elderly workers.[344]

In the UPR process, Switzerland emphasized older persons' greater difficulty than the population on average in entering or returning to the labor market. Similarly, reentry into the labor market for workers over age 50 is much more difficult than workers under 50 years of age.[345] Of course, while older persons may be those carrying this burden of discrimination, the problem points to discrimination in hiring practices. This is a little-addressed issue in the international human rights framework. CESCR expressed deep concern about discrimination in employment on the basis of age in Croatia in 2001.[346] Some states have instituted special measures to identify and protect older workers. Canada's UPR Working Group Report listed single adults as especially vulnerable, and therefore entitled to special changes in the federal budget regarding the older workers.[347] Ecuador highlighted its employment policies covering various groups, including older

---

[342]*Current Status of the Social Situation, Well-Being, Participation in Development and Rights of Older Persons Worldwide* U.N. Dep't Econ. Soc. Affairs *available at* http://www.un.org/esa/socdev/ageing/documents/publications/current-status-older-persons.pdf. (last visited Oct. 23, 2014).

[343]*See* General Comment 6, 22–24 (stating that older people often have difficulties finding and keeping jobs after they reach a retirement age).

[344]*See* Id. at 25 (implying that this appears to be a duty that states should undertake immediately).

[345]*Compilation Prepared by the Office of the High Commissioner for Human Rights, in accordance with paragraph 15(b) of the Annex to Human Rights Council Resolution 5/1- Switzerland* 66, A/HRC/WG.6/2/CHE/1 (2008) UN Human Rights Council, (Apr. 7, 2008) *available at* http://www.refworld.org/docid/4858ef1b2.html.

[346]Croatia, ICESCR, E/2002/22 125 at 895. (2001) http://tbinternet.ohchr.org/_layouts/j63treatybodyexternal/Download.aspx?symbolno=E%2f2002%2f22%28SUPP%29&Lang=en. (last visited Oct. 23, 2014).

[347]Art 39, A/HRC/11/17 http://www.ohchr.org/EN/HRBodies/UPR%5CPAGES%5CCASession4.aspx; *Report of the Working Group on the Universal Periodic Review - Canada*, UN Human Rights Council, A/HRC/WG.6/4/L.3; A/HRC/11/17, (Mar. 3 2009) *available at* http://www.refworld.org/docid/49f964ec2.html.

persons.[348] Japan's Working Group Report requested additional information on measures to ensure the rights of elderly workers.[349] Motivated by the need for equal access and training, Poland's Working Group Report asked the responding government to educate older workers.[350]

5. *The Right to Adequate Housing*

The right to housing is delineated in Article 11.1 of the ICESCR, as a part of an adequate standard of living.[351] Applying the norm requires states to respect, ensure, and fulfill this right in a manner that creates a living environment suited to the culture, skills, needs, and wishes of the people.[352] This right is complemented by the right to be protected against "arbitrary or unlawful interference" within one's home.[353]

According to CESCR General Comment No. 4, everyone is entitled to adequate housing, and individuals as well as families are entitled to this right "regardless of age, economic status, group or other affiliation or status and other such factors."[354]

CESCR General Comment No. 6 affords protection to older persons with respect to housing, expanding on the need for family support; housing "must be viewed as more than mere shelter as it clearly contains a psychological and social

[348]Id. at 48.

[349]*Report of the Working Group on the Universal Periodic Review - Japan*, UN Human Rights Council, A/HRC/8/44; A/HRC/WG.6/2/L.10, (May 30 2008) *available at* http://www.refworld.org/docid/485b962fd.html.

[350]*Report of the Working Group on the Universal Periodic Review - Poland*, UN Human Rights Council, 24 A/HRC/8/30; A/HRC/WG.6/1/POL/4, (May 23, 2008) *available at* http://www.refworld.org/docid/4857aa220.html.

[351]Eide, *supra* note 312, at 89–105 (discussing the right to housing as one of many factors associated with adequate living standards: UDHR and ICESCR both espouse the view that shelter is part of the right to an adequate standard of living). Cf. Revised European Social Charter, supra note 10, art. 23 (addressing housing within the context of the rights of older persons).

[352]Scott Leckie, *The Right to Housing, in Economic, Social and Cultural Rights: A Textbook* 113–14 (Asbjorn Eide et al., eds., 1995) (States are required to: "place sufficient legal and policy emphasis on the full realization of housing rights, through a series of active measure including national and/or local legislative recognition of the right, the incorporation of housing rights imperatives into housing and related policies, and the identification of discernable "bench marks' towards the full enjoyment of housing rights by all sectors of society").

[353]ICCPR article 17.1.

[354]*General Comment No. 4: The Right to Adequate Housing (Art. 11 (1) of the Covenant)*, CESCR, E/1992/23 6th Sess., at 53, (Dec. 13, 19911) *available at* http://www.refworld.org/docid/47a7079a1.html.; *Compilation of General Comments and General Recommendations Adopted by Human Rights Treat Bodies*, U.N. Doc. HRI/Gen/1/Rev.1 (1994), http://www1.umn.edu/humanrts/gencomm/hrcom22.htm. (last visited Oct. 23, 2014).

dimension."[355] Likewise, a report of the Secretary General of the UN pointed out that "to the extent possible, older persons should be enabled to continue to live in their homes for as long as possible, to avoid both the physical and psychological disruptions involved in moving to an unfamiliar location," and stated that "appropriate and adequate means of transportation are also needed so that older persons can remain mobile and remain integrated into the larger community."[356] Approximately 14 per cent of the world's elderly live alone, a situation that can lead to greater vulnerability. Older persons who live alone are more likely to need assistance in the case of illness or disability and are at greater risk of social isolation and poverty.[357]

Furthermore, according to CESCR's General Comment No. 4, everyone is entitled to adequate housing, and individuals as well as families are entitled to this right regardless of age.[358] The CESCR has further indicated that an initial step toward the realization of those rights is to identify the sectors of the population that should be the focus of positive state action.

Throughout General Comment No. 4, the CESCR reiterated that the right to housing is an immediate right. "In essence, the obligation is to demonstrate that, in aggregate, the measures being taken are sufficient to realize the right for every individual in the shortest possible time in accordance with the maximum of available resources."[359] The definition of available resources in relation to this right might include international assistance; such as the importance of the right to housing. Rights enforcement is a key to the immediacy of the right. According to the CESCR, "all persons should possess a degree of security of tenure which guarantees legal protection against forced eviction, harassment and other threats."[360]

The right to housing includes the right to be free from arbitrary or forced evictions.[361] Demolitions of homes can have disproportionate effects on older persons.[362] The CESCR notes that efficient legal appeals to prevent evictions and

---

[355] *The Economic, Social, and Cultural Rights of Older Persons General Comment 6*, CESCR E/1996/22 Article 11, 33 (1996). http://www.unhchr.ch/tbs/doc.nsf/0/482a0aced8049067c12563ed005acf9e?Opendocument. (last visited Oct. 23, 2014).

[356] U.N. SG Report on Aging, *supra* note 342, para. 27.

[357] Chung Working Paper, 2010 at para. 12.

[358] *General Comment No. 4: The Right to Adequate Housing (Art. 11 (1) of the Covenant)*, CESCR, E/1992/23, (Dec. 13, 1991) *available at* http://www.refworld.org/docid/47a7079a1.html.

[359] Id. at para. 14.

[360] General Comment 4, supra note 243, para. 8(a).

[361] Id. para. 18 (quoting the Committee"s statement that "instances of forced eviction are prima facie incompatible with the requirements of the Covenant and can only be justified in the most exceptional circumstances, and in accordance with the relevant principles of international law"); *See also* General Comment 7, (noting that older persons "suffer disproportionately from the practice of forced evictions").

[362] *Report of the Special Rapporteur on the Situation of Human Rights in the Palestinian Territories Occupied since 1967, Richard Falk*, A/HRC/10/20 at no. 78. UN Human Rights Council, (Feb. 11 2009) *available at* http://www.refworld.org/docid/49d33a8a2.html.

demolitions, legal procedures for compensation, complaints against landlords for inadequate housing conditions, and class action suits addressing discrimination claims are appropriate remedial measures.[363] States should monitor the housing market for discriminatory practices, excessive rent, inflated housing costs and taxes, availability of adequate public housing, and access to basic services such as heating.[364] Discriminatory practices in housing should be addressed by ensuring that victims have access to legal representation and, where a quick settlement is not reached, prompt access to hearings and remedies.[365]

The CESCR stated that "disadvantaged groups must be accorded full and sustainable access to adequate housing resources."[366] Sustainable access for older persons is unlikely to be achieved through voucher programs alone.[367] The CESCR listed older persons as a disadvantaged group, and stated that disadvantaged persons are entitled to priority consideration in national housing schemes.[368] Homes should be located near employment options, schools, and other social facilities, but should not be located near polluted areas that threaten the right to health.[369] Finally, the CESCR recognized the need for cultural adequacy.[370] Building materials and policies, as examples, should enable cultural expression and allow for technological advances.

The United Nations High Commissioner for Human Rights published a report in April 2012 in which it referred to some forms of contemporary practices that severely impair the full enjoyment of the right to housing. These practices appear to exist both in developing countries and developed countries. The High Commissioner stated in this regard that "older women face obstacles to inheriting housing, land and property. In other regions, long-term older residents may be harassed by landowners or real estate agents to force them to evacuate their housing so that it can be redeveloped or sold at a higher price.[371]

---

[363]Id. at para. 17 (listing potential domestic legal remedies for violations of the right to adequate housing).

[364]Leckie at 115 (explaining that in order to fulfill the right to adequate housing, governments must guarantee persons access to those housing rights otherwise unattainable by their own efforts).

[365]UN Human Rights Council, *Report of the Special Rapporteur on Adequate Housing as a Component of the Right to an Adequate Standard of Living, and on the Right to Non-Discrimination in This Context, Miloon Kothari: addendum: mission to Canada* A/HRC/10/7/Add.3, *(9 to 22 Oct. 2007)*, (Feb. 17 2009) *available at* http://www.refworld.org/docid/49b7af2c2.html.

[366]Id. at para. 8(e).

[367]UN Human Rights Council, *Report of the Representative of the Secretary-General on the Human Rights of Internally Displaced Persons, Walter Kälin: addendum: mission to Georgia*, A/HRC/10/13/Add.2, (Feb. 13 2009) *available at* http://www.refworld.org/docid/49b7c0172.html.

[368]Id. (stating that "such disadvantaged groups as older persons … should be ensured some degree of priority consideration in the housing sphere").

[369]Id. at para. 8(f).

[370]Id. at para. 8(g).

[371]*Report of the United Nations High Commissioner for Human Rights*, U.N. Econ. and Soc. Council, E/2012/51, para. 45 (Apr. 20, 2012), http://www.un.org/ga/search/view_doc.asp?symbol=E/2012/51.

Likewise, the High Commissioner emphasized that

architectural barriers can greatly affect older persons. For instance, older persons may be unable to leave their apartments for long periods in buildings lacking functioning elevators. This phenomenon, described as "prison flats", may isolate older persons, and may also pose important obstacles to basic activities of daily life, like buying food or attending medical treatment. The systematic inclusion of specific accessibility criteria and the universal design in housing and building codes, and city planning, would serve various parts of the population, including older persons, to enjoy adequate housing.[372]

States should not evict an older person where the action would expose older persons to other human rights violations; homelessness and exposure to the elements are particularly dangerous for this population.[373] Older persons who are evicted from their homes and lacking family support should be provided with humanitarian assistance.[374] States should provide suitable facilities for elderly persons who cannot afford them.[375] Additionally, states should protect older persons from third party abuse, such as arbitrary evictions, discrimination, and harassment, [376] and heavily penalize "mobbing."[377] There is a critical lack of housing in some

[372]Id. at para. 47.

[373]*Report of the Special Rapporteur on Adequate Housing as a Component of the Right to an Adequate Standard of Living, Government Communications, Bangladesh*: A/HRC/7/16/Add.1, (Mar. 4 2008) *available at* http://www.internal-displacement.org/8025708F004CE90B/%28httpDocuments%29/63F06A78949814D6C125770A0048CA1E/$file/G0811338+4+March+2008.pdf.

[374]*Report of the Representative of the Secretary-General on the Human Rights of Internally Displaced Persons, Walter Kälin*: *addendum*: *mission to Georgia*, A/HRC/10/13/Add.2, (Feb. 13 2009) *available at* http://www.refworld.org/docid/49b7c0172.html.

[375]Protocol of San Salvador, *supra* note 9, art. 17 (declaring that everyone has a "right to special protection in old age" and therefore requiring States Parties to provide food and specialized medical care for elderly individuals); *Additional Protocol to the American Convention on Human Rights in the Area of Economic, Social and Cultural Rights ("Protocol of San Salvador")*, Organization of American States (OAS), A-52, (Nov. 16 1999) *available at* http://www.refworld.org/docid/3ae6b3b90.html.

[376]Leckie, *supra* note 241, at 114–15 (specifying that where third parties such as landlords, property developers, and land owners infringe on the rights of housing rights beneficiaries, public authorities should act to prevent future deprivations).

[377]*UN Human Rights Council: Addendum to the Report of the Special Rapporteur on Adequate Housing as a Component of the Right to an Adequate Standard of Living, Mission to Australia (31 July to 15 August 2006)*, A/HRC/4/18/Add.2, (May 11 2007), *available at* http://www.refworld.org/docid/46652ba52.html.; *Preliminary note on the mission to Spain* A/HRC/4/18/Add.3, no. 11 (June 5 2007) http://daccess-dds-ny.un.org/doc/UNDOC/GEN/G07/128/14/PDF/G0712814.pdf?OpenElement; *Report of the Special Rapporteur on Adequate Housing as a Component of the Right to an Adequate Standard of Living, Mission to Spain*: A/HRC/7/16/Add.2 at 52. (Feb. 7 2008) *available at* http://daccess-dds-ny.un.org/doc/UNDOC/GEN/G08/105/31/PDF/G0810531.pdf?OpenElement ("Mobbing" is illegal eviction for speculative purposes).

states.[378] The CESCR, for example, has urged Italy "to take effective measures to ensure that forced evictions of Roma and tenants who cannot pay their rents comply with the guidelines established by the Committee in its General Comment No. 7 on the right to adequate housing" and the prohibition of "forced evictions, and to provide more housing units to cater for the needs of older persons."[379] Going a step further, the 2009 Final Report of the Third Committee encouraged states to "promote the concerns of older persons and persons with disabilities in the planning, implementing and evaluating of all development programmes and policies."[380]

The rights of older persons should be integrated into all national housing sector policies.[381] Specific funding should be directed to older persons due to increased housing discrimination, and plans to make available various forms of housing should be nationally adopted.[382] To guarantee these rights, states should acknowledge the particular circumstances of older persons when establishing urban development plans and should include community centers and other facilities that account for the needs of all. This would include providing transportation to older residents.[383] Other considerations, such as the fact that older persons are disproportionately affected by electricity outages when dependent on lifts in multi-story buildings, should also be considered.[384] Local authorities are obligated to take into

[378] *Report of the Special Rapporteur on Adequate Housing as a Component of the Right to an Adequate Standard of Living, and on the Right to Non-Discrimination in this Context, Miloon Kothari: Addendum: Mission to South Africa*, UN Human Rights Council, A/HRC/7/16/Add.3, (Feb. 29 2008) *available at* http://www.refworld.org/docid/47d55d3f2.html.

[379] U.N. Report on the 32nd and 33rd Sess. Sup. No. 2, ICESCR, E/2005/22 (2004) 54 at 458. http://daccess-dds-ny.un.org/doc/UNDOC/GEN/G05/409/40/PDF/G0540940.pdf?OpenElement. (last visited Oct. 23, 2014).

[380] U.N. Res A/64/432 28 http://www.copac.coop/iyc/a-64-432-excerpt.pdf. (last visited Oct. 23, 2014).

[381] *Report of the Special Rapporteur on Adequate Housing as a Component of the Right to an Adequate Standard of Living, and on the Right to Non-Discrimination in This Context, Raquel Rolnik: addendum: follow-up to country recommendations: Afghanistan, Mexico, Peru and Romania*, A/HRC/10/7/Add.2, UN Human Rights Council, (Feb. 24 2009), *available at* http://www.refworld.org/docid/49c78d342.html.

[382] *Report of the Special Rapporteur on adequate housing as a component of the right to an adequate standard of living, and on the right to nondiscrimination in this context, Miloon Kothari: addendum: Preliminary note on the Mission to Canada (9 to 22 October 2007)*, A/HRC/7/16/Add.4, UN Human Rights Council, (Feb. 28, 2008) *available at* http://www.refworld.org/docid/47d549a62.html.

[383] General Comment 6, art. 11, 33 (emphasizing the need to facilitate the "mobility and communication" of elderly people through national policies).

[384] *Report of the Special Rapporteur on Adequate Housing as a Component of the Right to an Adequate Standard of Living, Government Communications, Tajikistan*: A/HRC/7/16/Add.1, (March 4 2008) *available at* http://daccess-dds-ny.un.org/doc/UNDOC/GEN/G08/113/38/PDF/G0811338.pdf?OpenElement.

account needs of older persons in housing schemes.[385] States should encourage older persons to participate in the development of legislation and policies committed to achieving the full implementation of housing rights.[386]

6. The Right to Social Protection

Article 22 of the UDHR states that everyone "has the right to social security and is entitled to its realization, through national effort and international co-operation in accordance with the organization and resources of each State, of the economic, social and cultural rights indispensable for his dignity."[387] One of the most substantial contributions to the understanding of the human rights dimension of social protection for older persons was made by the UN independent expert on the question of human rights and extreme poverty. In 2008, the independent expert indicated that "extreme poverty and social exclusion constitute a violation of human dignity", and can be prevented it by adopting social security measures.[388] In 2010, the report of the independent expert stated that "establishing or extending social security systems is more than a policy option or a means for reducing poverty; it is first-and-foremost a duty of States stemming directly from human rights norms and standards, in particular the right to social security and the right to an adequate standard of living."[389]

The 2010 report also noted that "when establishing or extending social pensions", the states must "recognize social security as a human right and establish appropriate legal frameworks in domestic law"... "ensure universal access to social security", and "integrate human rights principles and standards throughout the design, implementation and evaluation of social pensions."[390] Likewise, the independent expert highlighted that "as a necessary part of a rights-based approach to social pensions, the right to social security must be included in

[385]*Report of the Special Rapporteur on Adequate Housing as a Component of the Right to an Adequate Standard of Living, and on the Right to Non-Discrimination in This Context, Raquel Rolnik: addendum: follow-up to country recommendations: Afghanistan, Mexico, Peru and Romania*, UN Human Rights Council, A/HRC/10/7/Add.2, (Feb. 24, 2009) *available at* http://www.refworld.org/docid/49c78d342.html.

[386]Leckie, *supra* note 241, at 114 (stressing the need to consult with those most affected by existing housing policies, such as the homeless and the inadequately housed, in developing national housing strategies).

[387]UDHR, Article 22; ICESCR, Article 9.

[388]*Report of the independent expert on the question of human rights and extreme poverty, Arjun Sengupta*, UN Human Rights Council, A/HRC/7/15, (February 28 2008), *available at* http://www.refworld.org/docid/47d163902.html.

[389]Sepulveda Report, 2010 at para. 51.

[390]Conclusions and Recommendations, Sepulveda Report 2010.

national legal frameworks."[391] Furthermore, the social security systems should have a complaint system. According to the expert, "[i]ndividuals and organizations should have the right to seek, receive and impart information in a clear and transparent manner."[392]

During the second working session of the UN Open-Ended Working Group on Ageing for the purpose of strengthening the protection of the human rights of older persons, the independent expert on the question of human rights and extreme poverty sustained that "the right to social security is firmly grounded in international human rights law and that several treaties contain specific references to old age protection through social security schemes." However, in practice, states are not complying with the obligation to offer the protections that older persons deserve as right-bearers under those human rights instruments. One possible solution is to establish a universal pension structure, which is in line with human rights obligations as it adequately responds to the claim of universality of human rights norms and the principle of equality and non-discrimination, as well as addressing the issue of corruption and the problem of stigma attached, as it could be made available to all who comply with the age requirement.[393] This, of course, does not exclude other options, as long as the system in place ensures equal access to all persons without discrimination.

UDHR Article 25, states "that everyone has the right to a standard of living adequate for the health and well-being of himself and of his family, including necessary social services and the right to security in the event of unemployment, sickness, disability, widowhood, old age or other lack of livelihood in circumstances beyond his control."[394] Further to articles 22 and 25 of the UDHR, articles 9, 10 and 11 of the ICESCR, the CERD,[395] the CEDAW,[396] article 26 of the CRC,[397] the International Convention for the Protection of Migrant Workers and Their Families,[398] and the CRPD set out a right to social protection.[399]

---

[391]Sepulveda Report, 2010 at para. 53.

[392]Sepulveda Report, 2010 at para 77.

[393]G.A. Res. 65/182 (Aug. 1–4 2011) *available at* http://social.un.org/ageing-working-group/documents/Chair_summary_2nd_session_OEWG_final.pdf.

[394]Universal Declaration of Human Rights, Art. 25(1).

[395]CERD article 5 (iv).

[396]CEDAW Art 11.

[397]Art 26, CRC.

[398]*International Convention for the Protection of Migrant Workers and Their Families*, UN General Assembly, A/RES/45/158, art. 27 (Dec. 18 1990) available at: http://www.refworld.org/docid/3ae6b3980.html.

[399]CPRD Art. 28.

Article 9 of the ICESCR describes "the right of everyone to social security, including social insurance." Regarding this, Lindsay Judged mentions that "it is clear from that Article 9 envisages non-contributory schemes for those who fall outside the ambit of contributory pensions, as well as a system of compulsory old-age insurance."[400] Article 9 outlines an entitlement and not just a benefit. Because social security payments allow older persons to procure basic needs like food and shelter, social security is an entitlement that the state must provide and protect due to its connection to the acquisition of basic needs for survival.[401] This provision has been further elaborated in CESCR General Comment 6 and in General Comment 19 to produce a number of clearer obligations on states vis-à-vis older people.

As with all rights, states have a duty to uphold equality and non-discrimination, and should endeavor to eliminate sex discrimination.[402] Older women already face higher risk of poverty. Older women may be disproportionately barred from receiving old-age contributory pensions because of years spent outside of the formal labor market, often caring for family.[403] Eligibility criteria are important for transparency of government actions, as well as to ensure equality and non-discrimination.[404] With regard to Iceland, for example, the CEDAW Committee recommended that the state take action to avoid poverty among older women.[405] Current standards for social security[406] enable a decent living and are encouraged.[407] Successful plans are currently more likely to be found in medium-income countries than low-income countries.[408]

UN independent expert on human rights and extreme poverty also underscored that 'the principle of equality and non-discrimination also requires States to ensure that social pensions are physically... accessible to all older people". In that sense,

---

[400]Lindsey Judge, *The Rights of Older People: International Law, Human Rights Mechanisms and the Case for New Normative Standards* pg. 5 (Jan. 2009).

[401]Eide, *The Right to an Adequate Standard of Living Including the Right to Food, in Economic, Social and Cultural Rights: A Textbook* 94–96 (1995) (emphasizing that the right to social security is particularly important for older persons, the unemployed, and the disabled).

[402]Sepulveda Report, 2010, at para. 36.

[403]UN SG Report on Aging, *supra* note 342, para. 26.

[404]Sepulveda Report, 2010, at para. 77.

[405]Iceland, CEDAW Committee, at para. 250.

[406]*Report of the Special Rapporteur on Adequate Housing* A/HRC/10/7/Add.2, at No. 331 (Feb. 24 2009).

[407]*Report of the Working Group on the Universal Periodic Review - Monaco*, UN Human Rights Council, A/HRC/12/3; A/HRC/WG.6/5/L.2, (Jun. 4 2009) *available at* http://www.refworld.org/docid/4a9cdb950.html.

[408]*Report of the Special Rapporteur on Violence against Women, Its Causes and Consequences, Yakin Ertürk: addendum: mission to Tajikistan*, UN Human Rights Council, A/HRC/11/6/Add.2, (Apr. 29, 2009) *available at* http://www.refworld.org/docid/4a0bda662.html.

special measures should be taken to ensure beneficiaries have access to benefits.[409] In addition, the independent expert addressed the accessibility criterion and underscored the states duty of "ensuring that benefits are distributed … at a reasonable geographic location is part of ensuring accessibility. Limited physical strength and mobility can be a major obstacle for older persons, thus particular attention should be paid to older persons' opportunity cost in term of transport, loss of labor, or caretaker time."[410]

A 2012 report of the United Nations High Commissioner for Human Rights (UNHCHR) estimated that "80 percent of the world's population, many of them older persons, lacks access to any kind of social security."[411] Also, the independent expert on human rights and poverty stated that "less than 20 percent of older persons are covered by pensions today, and only about 25 percent of the labour force is currently contributing or accruing pension rights."[412] As Eide has stated, "the right to social security is essential, particularly when a person does not have the necessary property available, or is not able to secure an adequate standard of living through work, due either to unemployment, old age or disability."[413]

It is noteworthy that the UNHCHR has also pointed out that social protection includes social insurance and social assistance, adjusting to the true cost of living.[414] The independent expert Sepulveda also indicated that "contributory and non-contributory schemes should be regarded as complementary and mutually reinforce aspects of social protection,"[415] and "social pensions should be seen… as only one component of a broad social protection system." According to the expert, "in the absence of contributory social security, social assistance and, in particular, the provisions of non-contributory pensions to older persons play a central role in bridging the existing protection gap."[416] The UNHCHR has also mentioned that

---

[409] *Report of the Independent Expert on the Question of Human Rights and Extreme Poverty, Magdalena Sepúlveda Carmona*, UN Human Rights Council, A/HRC/11/9, (Mar. 27, 2009) *available at* http://www.refworld.org/docid/49f846ff2.html.

[410] Sepulveda Report, 2010 at para. 74.

[411] *Report to the Economic and Social Council,* U.N. Econ. and Soc. Council, E/2012/5, 48, (Nov. 14 2011) http://www.un.org/ga/search/view_doc.asp?symbol=E/2012/5.

[412] Sepulveda Report, 2010, at para 29.

[413] Eide, *Economic, Social and Cultural Rights as Human Rights, in Economic, Social and Cultural Rights: A Textbook* 32 (Asbjorn Eide et al., eds., 1995) (noting that everyone is "the beneficiary of human rights").

[414] U.N. Econ. and Soc. Council, *Report of the United Nations High Commissioner for Human Rights,* 51, http://www.un.org/ga/search/view_doc.asp?symbol=E/2012/51. (last visited Oct. 23, 2014).

[415] Sepulveda Report, 2010, at para. 28.

[416] Sepulveda Report, 2010, at para 32. (The UN independent expert on the question of human rights and extreme poverty underlined that: "in its general comment No.19 on the right to social security, the CESCR spells out the key features of this right and the content of respective State obligations. It notes that states parties are obliged to ensure progressively the right to social security to all individuals within their territories, providing specific protection for disadvantaged and marginalized individuals and groups.").

"the right to social security includes contributory and non-contributory benefits, and both cash and in-kind benefits."[417] For example, in Ecuador, the NGO INREDH emphasized the achievement of the government decision to the government to pay a 'solidarity bonus' to elderly persons.[418] The solidarity bonus, while a step in the right direction, has not had as far-reaching results as were hoped.[419]

The CESCR definition of social security includes coverage for all "risks involved in the loss of means of subsistence for reasons beyond a person's control."[420] To ensure this right, states must immediately establish old-age insurance programs that would start at an age prescribed by law, but maintain flexibility.[421] Additionally, states must protect the benefits of survivors and orphans upon the death of a pension recipient or person entitled to social security.[422]

The CESCR has highlighted many situations where older persons are not entitled to enjoy their rights under the ICESCR.[423] The CESCR expressed concern with the "lack of the exact percentage of GDP spent on social security", especially, since "that comprehensive social protection is not available to the vast majority of the population, in particular workers over 55 years of age", and urged the state party to extend the protection.[424] The CESCR has expressed concern about decreases in state expenditures while the number of older persons in the system is increasing.[425] Since the 1980's, a number of governments have undertaken

---

[417] *Report of the United Nations High Commissioner for Human Rights*, U.N. Econ. and Soc. Council, E/2012/51, 50, (Apr. 20 2012) *available at* http://www.un.org/ga/search/view_doc.asp?symbol=E/2012/51.

[418] U.N. General Assembly, *Sumary Prepared by the Office of the High Commissioner for Human Rights, in Accordance with Paragraph 15(c) of the Annex to Human Rights Council Resolution, Ecuador*, A/HRC/WG.6/1/ECU/3, 48, (Mar. 6 2008) *available at* http://daccess-dds-ny.un.org/doc/UNDOC/GEN/G08/113/10/PDF/G0811310.pdf?OpenElement. (last visited Oct. 23, 2014).

[419] UN Human Rights Council, *Summary prepared by the Office of the High Commissioner for Human Rights, in accordance with paragraph 15(c) of the Annex to Human Rights Council Resolution 5/1 - Ecuador*, A/HRC/WG.6/1/ECU/3, (Mar. 6 2008) *available at* http://www.refworld.org/docid/4857a6ef0.html.

[420] General Comment 6, *supra* note 1, para. 26 (stating that the Committee does not specify the type and level of social security protection).

[421] Id. at para. 27 (mentioning that establishing some sort of pension or insurance for older persons is in agreement with Article 9 of the Covenant and International Labour Organization social security conventions).

[422] Id. at para. 29 (enacting those requirements is necessary in order to place into effect the provisions of Article 9 of the Covenant).

[423] United Kingdom of Great Britain and Northern Ireland, ICESCR, E/1995/22 (1994) 52 at. 294 http://tbinternet.ohchr.org/_layouts/treatybodyexternal/Download.aspx?symbolno=E%2f1995%2f22%28SUPP%29&Lang=en. (last visited Oct. 23, 2014).

[424] Zambia, ICESCR, E/2006/22 (2005) 19 at 90 and 112. http://www.refworld.org/pdfid/474d35ca2.pdf. (last visited Oct. 23, 2014).

[425] *Report on the Twenty-Fifth, Twenty-Sixth, and Twenty-Seventh Sessions*, CESCR E/2002/22, E/C.12/2001/17 http://www.bayefsky.com/general/e_2002_22_2002.php (voicing concern that the Jamaican social security scheme excludes a considerable portion of the disadvantaged and marginalized groups in society, including older persons). (last visited Oct. 23, 2014).

measures to reduce spending on social services, and some of the member countries of the Organization for Economic Cooperation and Development, have pursued policies to reduce expenditures on old-age pensions.[426]

In specific examples of its analysis of situations of social security, the CESCR regretted that "the Hong Kong Special Administrative Region has not implemented a number of the recommendations contained in its concluding observations of 2001."[427] It reiterated its concern for the exclusion of many individuals, including "older persons, from the Mandatory Provident Fund Scheme, due to their inability to make sufficient voluntary contributions."[428] Further, the CESCR expressed the concern that "under the existing social security system… the levels of benefit are not sufficient to guarantee a decent standard of living and that many low-income persons, in particular older persons, are not covered by the scheme."[429]

The HR Council has also taken note of some improvements. For example, in the UPR process for Malaysia, reviewing states noted the impressive progress in poverty reduction and in providing social safety nets to older persons.[430] In Korea's Working Group Report, the reviewing state raised the issue of how the Korea, in addition to providing a social safety net to older persons, addresses the concern expressed by CESCR in 2001 in relation to the hardship caused by family breakdown.[431] The responding state informed that the persons above the age of 65 have a basic old age pension, long-term protection and medical treatment and other job opportunities. Furthermore, subsidies and services are provided to for single-parent families and families in which children are living with their grandparents as well.[432] Ecuador also emphasized their new system for managing comprehensive care programs and creating conditions for comprehensive care in 64 geriatric units as well as training programs for community-based services, providing homes for older persons deprived of liberty.[433]

[426]Report on the World Social Situation, A/60/117, (2005) *available at* http://www.un.org/esa/socdev/rwss/docs/2005/rwss05.pdf (last visited Oct. 23, 2014).

[427]Hong Kong Special Administrative Region (China), ICESCR, E/2002/22 39 at 183, 198. (2001) http://www.bayefsky.com/general/e_2002_22_2002.php. (last visited Oct. 23, 2014).

[428]China (Hong Kong Special Administrative Region), ICESCR, E/2006/22 34 at 207, 213, 215, 219, 225 and 227 (2005) http://www.refworld.org/pdfid/474d35ca2.pdf. (last visited Oct. 23, 2014).

[429]Id. at para. 213.

[430]U.N. General Assembly, *Report of the Working Group on the Universal Periodic Review, Malaysia*, A/HRC/11/30, (Oct. 5 2009), *available at* http://daccess-dds-ny.un.org/doc/UNDOC/GEN/G09/163/60/PDF/G0916360.pdf?OpenElement.

[431]U.N. General Assembly, *Universal Periodic Review, Republic of Korea*, A/HRC/8/40 53 (May 28 2008) *available at* http://lib.ohchr.org/HRBodies/UPR/Documents/Session2/KR/A_HRC_8_40_RoK_E.pdf .

[432]Id. at para. 55.

[433]U.N. General Assembly, *Universal Periodic Review, Ecuador*, A/HRC/20 (May 13 2008) *available at* http://daccess-dds-ny.un.org/doc/UNDOC/GEN/G08/133/24/PDF/G0813324.pdf?OpenElement.

The Report of the Secretary-General on the implementation of the outcome of the World Summit for Social Development highlighted the needs of aging populations with regard to the necessary changes "in current systems for the financing of pensions and long-term care" or, the "future generations of older persons may be left without adequate social protection."[434] Furthermore, the report stressed that "the social integration policies should seek to reduce inequalities, to promote access to education and health care and to increase the participation and integration of older persons."[435] Moreover, it mentioned that "the private sector should be encouraged to make investments in social sectors and provide civil society organizations with support to carry out programs," which "should widen public participation in decision-making by ensuring access of information to citizens and establishing mechanisms for reviewing Government policies", as well as "the collection of statistics to better identify the most vulnerable groups."[436]

In the *Note Transmitting the interim Report of the Special Rapporteur on the Situation of Human Rights in Myanmar*, the Secretary General suggested that the government should consider the establishment of social safety nets for the most vulnerable groups, including poor persons, women, youth, older persons and persons with disabilities.[437]

7. The Right to Health

Everyone has the right to the health and wellbeing of himself and of his family.[438] According to HelpAge International, the area of health care is a domain in which "older persons are frequently subject to discrimination, either by being denied access to services, or receiving substandard or insufficient care on account of their age."[439] States should give particular attention to securing the right to health,

---

[434]U.N. Secretary General, *Follow-up to the implementation of the outcome of the World Summit for Social Development and of the Twenty-Fourth Special Session of the General Assembly*, A/60/80, (May 23 2009) *available at* http://www.iom.int/jahia/webdav/shared/shared/mainsite/policy_and_research/un/60/A_60_80_en.pdf.

[435]Id. at para. 62.

[436]Id.

[437]United Nations General Assembly, *Situation of Human Rights in Myanmar*, A/60/221, (Aug. 12 2005) *available at* http://daccess-dds-ny.un.org/doc/UNDOC/GEN/N05/459/65/PDF/N0545965.pdf?OpenElement; UN Commission on Human Rights, *Situation of Human Rights in Myanmar: Report of the Special Rapporteur, Paulo Sérgio Pinheiro*, E/CN.4/2005/36, 62, (Dec. 2, 2004) *available at* http://www.refworld.org/docid/42d66e359.html.

[438]UDHR, Article 25; CRPD, Article 25; ICESCR, Article 12.

[439]HelpAge Position Paper (Convention), 2009, p. 3.

ensuring equality and non-discrimination.[440] Aging can present a significant challenge to achieving effective health care.[441] Health systems should be responsive to the particular needs of older persons.[442] As stated by the United Nations High Commissioner for Human Rights "in the context of health and end-of-life care, dignity and respect for all human rights is key for the well-being of older persons."[443]

The General Assembly on the follow-up to the Second World Assembly on Ageing, pointed out that "older persons are entitled to the enjoyment of a satisfactory standard of physical and mental health, with due attention to preventive, curative, rehabilitative and long-term care."[444] Likewise, the independent expert on human rights and extreme poverty ascertained that without this, "many may fall into a vicious cycle where poor health engenders poverty and poverty engenders poor health."[445] The independent expert Sepulveda also underlined that "social pensions should be planned and implemented in close coordination with other initiatives aimed at ensuring the enjoyment of the right the highest attainable standard of health by older persons." And in fact, providing pensions grants an opportunity for better articulation of health services to other government programs.[446]

---

[440] UN Human Rights Council, *Report of the Special Rapporteur on the Right of Everyone to the Enjoyment of the Highest Attainable Standard of Physical and Mental Health, Anand Grover*, A/HRC/11/12, (Mar. 31 2009) *available at* http://www.refworld.org/docid/49faf7652.html; *Missions to the World Bank and the International Monetary Fund, and to Uganda* A/HRC/7/11/ Add.2 (Mar. 5 2008); UN Human Rights Council, *Report of the Special Rapporteur on the Right of Everyone to the Enjoyment of the Highest Attainable Standard of Physical and Mental Health, Paul Hunt: addendum: Missions to the World Bank and the International Monetary Fund in Washington, D.C. (20 October 2006) and Uganda (4–7 February 2007)*, A/HRC/7/11/ Add.2, (Mar. 5 2008) available at http://www.refworld.org/docid/47da904c2.html; (Noting older persons affected by leprosy face discrimination in health systems: A/HRC/10/62, 23 February 2009 at (i)); UN Human Rights Council, *Elimination of Discrimination Against Persons Affected by Leprosy and their Family Members: Report of the Office of the United Nations High Commissioner for Human Rights*, A/HRC/10/62, (Feb. 23 2009) available at http://www.refworld.org/docid/49c78f842.html. (last visited Oct. 23, 2014).

[441] U.N. General Assembly, *Implementation of General Assembly Resolution 60/251* (May 31 2007) *available at* http://daccess-dds-ny.un.org/doc/UNDOC/GEN/G07/126/94/PDF/G0712694.pdf?OpenElementReport.

[442] *Report of the Special Rapporteur on the Right of Everyone to the Enjoyment of the Highest Attainable Standard of Physical and Mental Health*: A/HRC/7/11, 31 at 42 (Jan. 2008) http://www.who.int/medicines/areas/human_rights/A_HRC_7_11.pdf.

[443] *Report of the United Nations High Commissioner for Human Rights*, U.N. Econ. and Soc. Council, E/2012/51 53 (Apr. 20 2012) http://www.un.org/ga/search/view_doc.asp?symbol=E/2012/51.

[444] General Comment 14, para. 34 (noting that exception to these prohibitions should only be taken for the treatment of either mental illness or the prevention of communicable diseases), UN SG Report on Aging, *supra* note 342.

[445] Sepulveda Report, 2010 at para. 17.

[446] Sepulveda Report, 2010 at paras. 90, 91.

According to the Report of the United Nations High Commissioner for Human Rights of April 2012, older persons may face severe health conditions, often over extended periods of time, and are confronted with profound psychosocial and spiritual questions as they face illness or death and physical pain. Persons who live with chronic pain are more likely to suffer from depression or anxiety, according to certain studies.[447] Similarly, the General Assembly has also indicated that "depression is known to be common among older persons, although in developing countries, precise data are scarce.... [a]nd depression, loneliness and anxiety may arise as a result of major life changes such as the death of a spouse or a sudden decline in health" and it "is also considered to be a main factor in explaining the surge in suicide rates among persons aged 75 and over."[448]

Article 12 of the ICESCR recognizes "the right of everyone to the highest attainable standard of physical and mental health."[449] It goes on to highlight particular priorities states should consider, including "the prevention, treatment and control of epidemic, endemic, occupational and other diseases" and "the creation of conditions which would assure to all medical service and medical attention in the event of sickness."[450] States should have review mechanisms in place to address violations of the right to health.[451] In addition, states should maintain adequate training standards for health care professionals.[452] For older persons, this might encourage the mainstreaming of geriatric medicine in medical training and professional programs.

In CESCR General Comment No. 14, which addresses health, there are four core aspects to the right to health. *Availability* refers to the sufficient quantity and sustainable quality of health care facilities, goods, services and programs. *Accessibility* includes the notions of non-discrimination, and physical, economical, and informational accessibility. *Acceptability* refers to cultural appropriateness, including gender sensitivities and data confidentiality. Finally, *quality* refers to the

[447]*Report of the United Nations High Commissioner for Human Rights,* U.N. Econ. and Soc. Council, E/2012/51 54 (Apr. 20 2012) *available at* http://www.un.org/ga/search/view_doc.asp?symbol=E/2012/51.

[448]U.N. General Assembly, *Follow-up to the Second World Assembly on Ageing* A/67/188, (July 26 2012) 33 *available at* http://daccess-dds-ny.un.org/doc/UNDOC/GEN/N12/442/82/PDF/N1244282.pdf?OpenElement.

[449]*Report of the Special Rapporteur on the Right of Everyone to the Enjoyment of the Highest Attainable Standard of Physical and Mental Health*: A/HRC/7/11, 31 at 42 (Jan. 2008) http://www.who.int/medicines/areas/human_rights/A_HRC_7_11.pdf. (last visited Oct. 23, 2014).

[450]Lindsey Judge, *The Rights of Older People: International Law, Human Rights Mechanisms and the Case for New Normative Standards*, pg. 6 (Jan. 2009).

[451]General Comment 14, para. 59 (emphasizing the importance of providing adequate judicial reparations to victims of right to health violations, including restitution and compensation, as well as guarantees that the violations will not reoccur) (The Committee also encourages formation of patients' rights associations, consumer forums, and other relevant institutions to address remedy and accountability issues). Id.

[452]Id. at para. 36 (indicating that in order for the States to fulfill their legal obligations to the ICESCR, they must ensure that medical personnel and doctors are adequately trained).

scientifically and medically appropriate production of medical goods, services, facilities, and drugs.[453] The CESCR reaffirms paragraphs 34 and 35 of General Comment No. 6, which reiterates the need for preventative, curative and rehabilitative health treatment for older persons, including rehabilitative measures. Likewise, the CESCR reiterates the recommendations of the Vienna International Plan of Action on Ageing, which took a similar comprehensive view of all stages of health.

Further, in General Comment No. 14, the CESCR describes how accessibility includes health facilities, goods, and services that shall be within safe physical reach for all sections of the population, especially vulnerable or marginalized groups, such as, *inter alia*, older persons, as well as adequate access to buildings for older persons.[454] Furthermore, Lindsay Judge highlighted that "the CRPD may provide some additional protection for older persons who fall into the category of living with a disability," as well as the "article 25 states that disabled persons have "the right to the highest attainable standard of health without discrimination on the basis of disability."[455] Access to the right to health should not be based on any type of discrimination, and older persons should enjoy access on an equal basis to others.

Health care data must be disaggregated by age in order to realize a progressive right to health.[456] Addressing these issues will create an overall better life for the aging. In order to respect the right to health, the CESCR noted that states should refrain from denying elderly persons access to health services, and from prohibiting or impeding traditional preventative care.

Development of home care services for older persons should be a priority.[457] In this regard, the General Assembly has pointed out that "the health-care systems of developing countries were established to focus primarily on combating communicable diseases, and they remain poorly adapted to, and have insufficient resources for, the care and prevention of chronic diseases.[458] Family members should be allowed to accompany injured elderly persons in their receipt of health care.[459]

---

[453]General Comment 14, para. 12(a)-(d) (defining the essential elements of the right to health).

[454]Id. at 128 paras. 12(b)(ii) and 25.

[455]Id. at pg. 6.

[456]*Report of the Special Rapporteur on the Right of Everyone to the Enjoyment of the Highest Attainable Standard of Physical and Mental Health*: A/HRC/7/11, 31 at 42 (Jan. 2008). http://www.who.int/medicines/areas/human_rights/A_HRC_7_11.pdf. (last visited Oct. 23, 2014).

[457]*Report of the Special Rapporteur on Adequate Housing* A/HRC/10/7/Add.2, at no. 331 (February 24 2009) (last visited Oct. 23, 2014).

[458]U.N. Secretary General, *Follow-up to the Second World Assembly on Ageing*, A/67/188, 35, (July 26 2012) *available at* http://daccess-dds-ny.un.org/doc/UNDOC/GEN/N12/442/82/PDF/N1244282.pdf?OpenElement.

[459]UN Human Rights Council, *Report of the High-Level Fact-Finding Mission to Beit Hanoun: established under Council resolution S-3/1*, A/HRC/9/26, at (a) (Sept. 1 2008) *available at*: http://www.refworld.org/docid/48cfa3a22.html.

Both patent holders and pharmaceutical companies have a right-to-health responsibility to ensure that medicines are developed for, among others, older persons.[460] Pharmaceutical companies should give proper attention to the needs of older persons in clinical trials and with informed consent.[461] Furthermore, the CESCR notes that states should provide older persons with up-to-date health information, maintain the confidentiality of personal health data[462] and strive to ensure "facilities, goods, and services are physically accessible, near in proximity, and affordable to older persons."[463] States should refrain from interfering with an elderly person's decision regarding his or her medical treatment. "Information should be tailored to suit their needs and communicated to them in an appropriate, comprehensible manner allowing them to make fully informed decisions about their health condition and treatment."[464] Where disability prevents unassisted decision-making, the CRPD requires support.[465]

Responsibility for the enforcement of health rights lies with the state. Japan, for example, has been received positively for high legal protections for vulnerable groups, including, expressly, older persons, and delineating rights including the quality of life of older persons, affording a right to health.[466] Health rights are also related to the right to adequate food, housing, water, working conditions, and environment, since the reduction of malnutrition and other inadequacies in these areas increase an individual's chance to enjoy a healthy life.[467] Toward this end, national

---

[460]UN Human Rights Council, *Report of the Special Rapporteur on the Right of Everyone to the Enjoyment of the Highest Attainable Standard of Physical and Mental Health, Paul Hunt: annex: mission to GlaxoSmithKline*, A/HRC/11/12/Add.2, (May 18 2009), *available at* http://www.refworld.org/docid/4a1d0af22.html.

[461]Id. at 23.

[462]General Comment 14, para. 12(b) (explaining that accessibility of information and ideas concerning health issues should not encroach upon the right to confidentiality).

[463]Id. (stating that potable water and sanitation facilities should be physically accessible, even in rural areas). Health services should also be affordable for all, including socially disadvantaged groups. Id.

[464]UN Human Rights Council, *Thematic study on the realization of the right to health of older persons by the Special Rapporteur on the right of everyone to the enjoyment of the highest attainable standard of physical and mental health*, A/HRC/18/37, **25** (July 4 2011) *available at* http://www.refworld.org/docid/4e7322ac2.html.

[465]Convention on the Rights of Persons with Disabilities, Article 12.2.

[466]A/HRC/8/44 Japan's Working Group Report UN Human Rights Council, *Report of the Working Group on the Universal Periodic Review - Japan*, A/HRC/8/44; A/HRC/WG.6/2/L.10, (May 30 2008), available at http://www.refworld.org/docid/485b962fd.html.

[467]The Right to the Highest Attainable Standard of Health: General Comment 14, U.N. ESCOR, Econ., Soc., & Cultural Rts. Comm., U.N. Doc. E/C.12/2000/4 4 (2000) (explaining that the reference to the "highest attainable standard of physical and mental health" in the International Covenant on Economic, Social and Cultural Rights expands beyond the right to health care). [hereinafter General Comment 14].

health plans should address older women's' specific health issues as equal to other, predictable health issues throughout the life course.[468]

In order to guarantee these rights adequately, the CESCR declares that states provide access to facilities, goods, and services without any discrimination based on age, sex or financial ability.[469] Discrimination in this area involves not only discrimination based on impaired health, which particularly affects older persons, but also discrimination in the diagnosis (or failed diagnosis) of disease or mental illness.[470] States need to enforce non-discrimination and the right to health, ensuring the right to health encompasses states' duties to:

> implement preventative measures to reduce occupational accidents and diseases; to provide an adequate supply of safe, potable water as well as basic sanitation; to reduce and prevent the exposure of the population to dangerous substances such as radiation, harmful chemicals, or other detrimental environmental conditions; to minimize health hazards in industrial workplaces; to administer checkups for older persons on a regular basis, giving special attention to chronically and terminally ill persons.[471]

During the UPR process, a reviewing state recommended in Cuba's Working Group Report that Cuba share its experiences and good practices with regard to the right to health, in particular in the view of the growing aging population.[472] Also, in regard to Malaysia, the reviewing state emphasized that in order to meet the healthcare demands of an increasingly aging population, the state t worked to provide geriatric services in all government hospitals.[473] The United Arab Emirates Working Group Report noted that rapid advancement in health-care facilities had led to increased life expectancies for men and women,[474] though they were requested to provide more information on healthcare systems for the aged.[475]

---

[468]*See* General Comment 14, paras. 20–21 (advocating the integration of a gendered perspective in health related policies).

[469]ICESCR, Article 12, The Right to Health; ICESCR Article 2.2, (noting that the right to non-discrimination applies to all rights enumerated in the ICESCR).

[470]*See* Katarina Tomaschevski, *Health Rights, in Economic, Social and Cultural Rights: A Textbook* 140–41 (Asbjorn Eide et al., eds., 1995) (recounting how, in the past, non-conformity with moral, social, cultural, or political values had been a factor in diagnosing mental illness).

[471]Diego Rodriguez-Pinzon and Claudia Martin, *supra* note 254 (discussing General Comment 14 paras. 15, 25 providing general commentary on Article 12 of the ICESCR).

[472]UN Human Rights Council, *Report of the Working Group on the Universal Periodic Review - Cuba*, A/HRC/WG.6/4/L.8; A/HRC/11/22, 26 (Mar. 3 2009), *available at* http://www.refworld.org/docid/49f964ee0.html.

[473]UN Human Rights Council, *Report of the Working Group on the Universal Periodic Review - Malaysia*, A/HRC/WG.6/4/L.16; A/HRC/11/30, 97 (Mar. 3 2009) *available at* http://www.refworld.org/docid/49f964f10.html.

[474]UN Human Rights Council, *Report of the Working Group on the Universal Periodic Review - United Arab Emirates*, A/HRC/10/75; A/HRC/WG.6/3/L.7, **9** (Jan 12 2009), *available at* http://www.refworld.org/docid/497476a40.html.

[475]Id at para. 41.

Finally, the Working Group Report on the Democratic Persons' Republic of Korea indicated that human rights violations have created severe problems of malnutrition and other generalized health issues for the population and older persons.[476] The Special Rapporteur on the Right to Health noted in his concluding remarks on the right to health for older persons that healthy aging is the beginning; the goal is truly "active and dignified aging, which should be planned and supported just like any other stage of the individual's life course."[477]

The 2009 Final Report of the Third Committee called upon Member States to "address the well-being and adequate health care of older persons by designing more effective prevention strategies, stronger laws and policies to address these problems and their underlying factors."[478] The report also called "upon governments to ensure conditions that enable families and communities to provide care and protection to persons as they age and evaluate improvement in the health status of older persons, including on a gender-specific basis."[479] The Third Committee has also declared that "states must ensure access to adequate health care services and address other underlying determinants of health, such as access to safe drinking water and food."[480] Similarly, the Report of the Secretary General on the follow-up to the Second World Assembly on Ageing highlighted a resolution of the fifty-eighth World Health Assembly "strengthening, active, and healthy aging"[481] by which it urged member states to strengthen national action in order to fulfill commitments to implementing the MIPAA and related regional plans of action relating to the health and well-being of older persons. The Assembly also requested the Director-General of WHO to strengthen the organization's capacity to incorporate work on aging throughout its activities and programs at all levels.[482]

---

[476]UN General Assembly, *Promotion and Protection of Human Rights: Human Rights Situations and Reports of Special Rapporteurs and Representatives: Report of the 3rd Committee: General Assembly, 63rd session*, A/63/430/Add.3, (Dec. 5 2008) *available at* http://www.refworld.org/docid/4c46bd212.html.

[477]UN Human Rights Council, *Thematic Study on the Realization of the Right to Health of Older Persons by the Special Rapporteur on the Right of Everyone to the Enjoyment of the Highest Attainable Standard of Physical and Mental Health, Anand Grover*, A/HRC/18/37, para. 71. (July 4 2011) *available at* http://www.refworld.org/docid/4e7322ac2.html.

[478]U.N. General Assembly, *Social Development: Report of the Third Committee*, A/64/432 **13** (Dec. 1, 2009) http://www.copac.coop/iyc/a-64-432-excerpt.pdf.

[479]U.N. General Assembly, *Follow-up to the Second World Assembly on Ageing*, A/C.3/64/L.6 8 (Oct. 7 2009) *available at* http://daccess-dds-ny.un.org/doc/UNDOC/LTD/N09/547/37/PDF/N09 54737.pdf?OpenElement.

[480]Conclusions and Recommendations, Sepulveda Report, 2010.

[481]*Fifty-eigth World Health Assembly* WHA58.16 (May 16–25 2005) http://apps.who.int/gb/ebwha/pdf_files/WHA58-REC1/english/A58_2005_REC1-en.pdf.

[482]U.N. Secretary General, *Second Review and Appraisal of the Madrid International Plan of Action on Ageing 2002*, E/CN.5/201/5 38, (Nov. 3 2011) *available at* http://www.un.org/ga/60/third/docslist.htm.

The follow-up to the World Summit for Social Development and the XXIV special session of the General Assembly review of the relevant UN plans and programs of action pertaining to the situation of social groups, established that

> access to preventive and curative care, the training of personnel and the adaptation of facilities to meet the special needs of older populations are key policy initiatives. Given the acknowledged importance of good health, efforts to expand and improve specific areas of health care for older persons have been established as a priority, although mainly in developed countries."[483] Furthermore, the follow-up report indicated that "the promotion of healthy living practices and equal access of older persons to health care and services that include disease prevention throughout the lifespan is the cornerstone of healthy aging.[484]

E. Rights of Older Persons as Members of Other Vulnerable Groups

As mentioned by the General Assembly regarding the realization of the right to health, General Comment No. 14 considers older persons to be a group especially vulnerable to human rights violations.[485] The condition of vulnerability requires special attention to ensure the necessary autonomy to guarantee individual choice of models of care, to the extent possible.[486]

Equality and non-discrimination are at the core of the recognition of the heightened need for state protection of vulnerable groups' rights. Historical discrimination, diminished representation in decision-making and barriers to political participation are all factors for singling out a group as vulnerable. The language of vulnerability is a common thread in human rights discourse.[487] The argument that

---

[483]U.N. Secretary General, *Second Review and Appraisal of the Madrid International Plan of Action on Ageing 2002*, E/CN.5/2012/5, 64,(Nov. 3 2011) *available at* http://daccess-dds-ny.un.org/doc/UNDOC/GEN/N11/579/60/PDF/N1157960.pdf?OpenElement.

[484]Id. at para. 65.

[485]UN Committee on Economic, Social and Cultural Rights (CESCR), *General Comment No. 14: The Right to the Highest Attainable Standard of Health (Art. 12 of the Covenant)*, E/C.12/2000/4, (Aug. 11 2000) **25** available at: http://www.refworld.org/docid/4538838d0.html.

[486]U.N. General Assembly, *Thematic Study on the Realization of the Right to Health of Older Persons by Special Rapporteur on the Right of Everyone to the Enjoyment of the Highest Attainable Standard of Physical and Mental Health, Anand Grover*, A/HRC/18/37 28 (July 4 2011) *available at* http://daccess-dds-ny.un.org/doc/UNDOC/GEN/G11/143/61/PDF/G1114361.pdf?OpenElement.

[487]*See*, e.g., UN Human Rights Council, *Compilation Prepared by the Office of the High Commissioner for Human Rights, in accordance with paragraph 15(b) of the Annex to Human Rights Council Resolution 5/1 - Bahrain*, A/HRC/WG.6/1/BHR/2, (Mar. 14 2008), *available at* http://www.refworld.org/docid/4856827f0.html.; UN Human Rights Council, *Report of the Working Group on the Universal Periodic Review - Russian Federation*, A/HRC/11/19, (Mar. 3 2009) *available at*: http://www.refworld.org/docid/49f964f3d.html (Algeria noting the need for further efforts to protect rights for vulnerable groups). South Africa's Working Group Report A/HRC/WG.6/1/ZAF/2, (Apr. 11 2008) *available at* http://www.hrw.org/reports/2008/southafrica0608/8.htm (CAT recommending that legal aid for vulnerable groups be strengthened).

older persons should be "recognized as a distinct group requiring special attention and care" is not a new one by any means.[488] However, the recognition has gained increasing attention in the past few years.[489] The General Assembly has restated that "a study undertaken by HelpAge International in 2010 found that only 93 out of 1,912 humanitarian assistance projects surveyed explicitly addressed older persons as a vulnerable group."[490]

It is worth noting that not all older persons are vulnerable in the same way, for example, not all children or persons with disabilities are considered to be in a situation of exclusion or marginalization by merely belonging to a 'vulnerable group'. But, in general, it is recognized that they may need special protection in many challenging circumstances in which their inherent vulnerability exposes them to more hardship than other groups. It is also important that the classification of vulnerability is identified, in a specific context, as an important if sometimes socially-constructed source of discrimination, isolation, poverty, and disempowerment. The conventions on the rights of women, children, and persons with disabilities, among others, were established on the recognition that these groups needed special assistance to ensure the protection and fulfilment of the rights enshrined in existing conventions for these persons.[491] The United Nations General Assembly emphasizes non-discrimination as a part of the UDHR, and the legal frameworks for the protection of vulnerable groups' rights to employment, food, social development, social integration, cultural participation and accessibility.

Multiple discriminations can be the source of compounded violations of older persons' rights, if we consider that older persons also make up a large portion of all vulnerable groups.[492] For example, the Special Rapporteur on the situation of human rights and fundamental freedoms of indigenous persons has noted that government reparations for wrongful actions against indigenous populations can and should be expedited for older persons, who are more likely to face circumstances of extreme need.[493] The HR Council recommends the continued raising of aware-

---

[488]Chung Working Paper, 2010, para. 45.

[489]*See, e.g.* Human Rights Council, *Thematic Study on the Realization of the Right to Health of Older Persons by the Special Rapporteur on the Right of Everyone to the Enjoyment of the Highest Attainable Standard of Physical and Mental Health, Anand Grover,* A/HRC/18/37 (July 4 2011) *available at* http://www2.ohchr.org/english/bodies/hrcouncil/docs/18session/A-HRC-18-37_en.pdf.

[490]U.N. Secretary General, *Second Review and Appraisal of the Madrid International Plan of Action on Ageing 2002,* E/CN.5/2012/5 63, (Nov. 3 2011) *available at* http://daccess-dds-ny.un.org/doc/UNDOC/GEN/N11/579/60/PDF/N1157960.pdf?OpenElement.

[491]Chung Working Paper, 2010, para. 5.

[492]General Comment 15 (Twenty-ninth session, 2002): The Right to Water (arts. 11 and 12 of the Covenant), ICESCR E/2003/22 (2002) 120 at 16(h), General Comment 5, and General Comment 6.

[493]UN Human Rights Council, *UN Human Rights Council: Addendum to the Report of the Special Rapporteur on the Situation of Human Rights and Fundamental Freedoms of Indigenous People, Study Regarding Best Practices Carried out to Implement the Recommendations Contained in the Annual Reports of the Special Rapporteur,* A/HRC/4/32/Add.4, (Feb. 26 2007), *available at* http://www.refworld.org/docid/4623896f2.html.

ness among traditional communities about the protection and promotion of the rights of older people, particularly the multiple discriminations faced by elderly women.[494] The Human Rights Committee has recommended human rights education in accordance with Articles 2(1) and 26 where vulnerable groups' rights are being exploited.[495]

The CESCR notes that economic and social disparities have particular effect on vulnerable groups, including, expressly, older persons,[496] who are particularly prone to unequal distribution of poverty in the first place.[497] It further urges efforts for greater economic, social and cultural rights for all vulnerable groups including, expressly, older persons.[498] The CESCR encourages participation,[499] and condemns legislation restricting the rights of vulnerable groups based on public opinion trends.[500]

A lack of state action to combat discrimination threatens vulnerable groups.[501] The HR Council emphasizes state facilitation of a policy of participation as a remedy for curtailed civil, cultural, economic, political, and social rights, and poverty in its Forum on Minority Issues.[502] The CAT emphasizes the elimination of police discrimination against all vulnerable groups, as well as their torture and ill treatment, and recommends steps to ensure the physical integrity of vulnerable

---

[494]UN Human Rights Council, *Report of the Working Group on the Universal Periodic Review - Burkina Faso*, A/HRC/10/80; A/HRC/WG.6/3/L.12, (Jan. 7 2009) *available at* http://www.refworld.org/docid/497476a80.html.

[495]*Report of the Human Rights Committee Volume I India*, ICCPR, A/52/40 vol. I (1997) 67 at 430 *available at* http://www.ccprcentre.org/wp-content/uploads/2012/04/A_52_40Vol-I_en.pdf. (last visited Oct. 23, 2014).

[496]Guatemala, ICESCR, E/1997/22 29 at 127 (1996) http://daccess-dds-ny.un.org/doc/UNDOC/GEN/G97/161/19/PDF/G9716119.pdf?OpenElement. (last visited Oct. 23, 2014).

[497]*Report of the UN Committee on Economic, Social and Cultural Rights, Twenty-eighth and Twenty-ninth Sessions (29 April - 17 May 2002, 11–29 November 2002)* CESCR, E/2003/22; E/C.12/2002/13, 39 at 221 (June 23 2003),*available at* http://www.refworld.org/docid/3f6b10ea4.html.

[498]Australia, ICESCR, E/1994/23 36 at 159 (1993) http://www.bayefsky.com/general/e_1994_23.pdf. (last visited Oct. 23, 2014).

[499]Poland, ICESCR, E/1993/22 32 at 132 (1992) http://www.bayefsky.com/general/e_1993_22.pdf. (last visited Oct. 23, 2014).

[500]United Kingdom of Great Britain and Northern Ireland (Hong Kong), ICESCR, E/1997/22 58 at 337 (1996) http://tbinternet.ohchr.org/_layouts/treatybodyexternal/Download.aspx?symbolno=E%2f1997%2f22%28SUPP%29&Lang=en. (last visited Oct. 23, 2014).

[501]*Human Rights Committee Annual Report to the U.N. General Assembly, Romania*, ICCPR, A/49/40 vol. I (1994) 28 at 140 http://www1.umn.edu/humanrts/hrcommittee/hrc-annual94.htm. (last visited Oct. 23, 2014).

[502]U.N. General Assembly, *Human Rights Council Forum on Minority Issues Second session Geneva, 12–13 November 2009 Note by the Independent Expert on Minority Issues, Gay McDougall, on Minorities and Effective Political Participation*
http://www.univie.ac.at/bimtor/dateien/greece_cat_2001_concob.pdf (last visited Oct. 23, 2014).

groups.[503] The United Nations Department of Economic and Social Affairs notes the need for equality and non-discrimination for vulnerable groups in its 2005 Report on the World Social Situation.[504] In Croatia's Working Group Report, the CERD emphasized the need for continuous monitoring of vulnerable or disadvantaged groups within minority communities to identify areas for improvement, and efforts targeted for the most vulnerable groups so that the government can measure and act accordingly.[505]

The Third Committee of the GA also asserted in the 2009 Final Report its ongoing commitment to vulnerable groups, reducing vulnerability of older persons, and developing cooperatives to enable the meeting of social service needs for vulnerable groups and, expressly, older persons.[506] Similarly, the UN Secretary General observed a number of actions pertinent to vulnerable groups. This included the need for targeted assistance, the need for improved law and social development programs that work toward social integration,[507] and they encouraged UN bodies to monitor the situations of vulnerable groups.[508]

Individual state comments during the UPRs often include praise for or recognition of shortcomings in connection with being a party to individual human rights treaties, individual states' domestic legal standards, mechanisms and institutions to monitor implementation and efforts to protect vulnerable groups. These state comments have included praise for preventing discrimination against vulnerable groups,[509] and note

---

[503]Greece, CAT, A/56/44 (2001) 38 at 88. http://www.unhchr.ch/tbs/doc.nsf/0/336a0d9ee8c62b8ec1256a4800558d6f?Opendocument. (last visited Oct. 23, 2014).

[504]Report on the World Social Situation, Dep't of Econ. and Soc. Affairs (2005) *available at* http://undesadspd.org/reportontheworldsocialsituation.aspx.

[505]Croatia, Comm. On the Elimination of Racial Discrimination, A/48/18 90 at 504 (1993) *available at* http://www.bayefsky.com/html/croatia_cerd_spec_dec.php. (last visited Oct. 23, 2014).

[506]Final Report of the Third Committee A/64/432 (2009) *available at* http://daccess-dds-ny.un.org/doc/UNDOC/GEN/N09/627/09/PDF/N0962709.pdf?OpenElement (last visited Oct. 23, 2014).

[507]*Report of the Secretary-General on the implementation of the outcome of the World Summit for Social Development and of the twenty-fourth special session of the General Assembly* A/60/117 (July 13 2005) *available at* http://daccess-dds-ny.un.org/doc/UNDOC/GEN/N05/418/73/PDF/N0541873.pdf?OpenElement.

[508]U.N. Secretary General, *Effective Promotion of the Declaration on the Rights of Persons Belonging to National or Ethnic, Religious, and Linguistic Minorities* A/60/333 (Sep. 6, 2005) http://daccess-dds-ny.un.org/doc/UNDOC/GEN/N05/481/36/PDF/N0548136.pdf?OpenElement.

[509]*See* UN Human Rights Council, *Report of the Working Group on the Universal Periodic Review - Bangladesh*, A/HRC/WG.6/4/L.4, A/HRC/11/18, (Mar. 3 2009), *available at* http://www.refworld.org/docid/49f964eb0.html; UN Human Rights Council, *Report of the Working Group on the Universal Periodic Review - Chile*, A/HRC/12/10; A/HRC/WG.6/5/L.9, (June 4, 2009) *available at* http://www.refworld.org/docid/4a9cdb920.html. (Czech Republic urging the approval of a draft refugee law according special attention to vulnerable groups to prevent discrimination).

the need of state action to protect human rights of vulnerable groups, including, expressly, older persons.[510]

During the UPR process, Macao Special Administrative Region (MSAR) of the Persons' Republic of China considered the rights of vulnerable groups, including, expressly, older persons.[511] The Philippines reports that it has constitutional provisions recognizing the rights of older persons.[512] Sri Lanka acknowledges the importance of development for social protection of vulnerable groups, including, expressly, older persons.[513] Some states have even requested reporting on specific efforts for vulnerable groups, including, expressly, older persons.[514] These comments provide an important window into the developing norms around the rights of older persons and how states view their international commitments when specifically addressing the rights of older persons.

1. The Rights of Older Refugees and Other Persons in Emergencies

CRPD in article 11 provides that persons with disabilities, as all people, have equal rights to humanitarian relief following a natural disaster or conflicts.[515] Older persons are also entitled to that right, on equal footing with others. Old age

[510]UN Human Rights Council, *Report of the Working Group on the Universal Periodic Review - Cameroon*, A/HRC/WG.6/4/L.7, A/HRC/11/21, (Mar. 3 2009), *available at* http://www.refworld.org/docid/49f964ebd.html.

[511]UN Human Rights Council, *Report of the Working Group on the Universal Periodic Review - China*, A/HRC/WG.6/4/L.11; A/HRC/11/25, (Mar. 3 2009), *available at* http://www.refworld.org/docid/49f964ed0.html.

[512]UN Human Rights Council, *Report of the Working Group on the Universal Periodic Review - The Philippines*, A/HRC/8/28; A/HRC/WG.6/1/PHL/4, (May 23 2008) *available at* http://www.refworld.org/docid/4857aa21d.html.

[513]UN Human Rights Council, *Report of the Working Group on the Universal Periodic Review - Sri Lanka*, A/HRC/8/46; A/HRC/WG.6/2/L.12, (June 5 2008) *available at* http://www.refworld.org/docid/485b96310.html.

[514]Mauritius' Working Group Report, A/HRC/11/28 (The Holy See requesting further information); A/HRC/12/3 Monaco's Working Group Report, (Philippines asking for specifics as to the practices and policies regarding vulnerable groups, including, expressly, older persons). UN Human Rights Council, *Report of the Working Group on the Universal Periodic Review - Mauritius*, A/HRC/WG.6/4/L.14; A/HRC/11/28, (Mar. 3 2009), available at: http://www.refworld.org/docid/49f964f1d.html (note from Korea in A/HRC/10/75 United Arab Emirates Report of Working Group (encouraging specific measures to ensure the rights of older persons and disabled). UN Human Rights Council, *Report of the Working Group on the Universal Periodic Review - United Arab Emirates*, A/HRC/10/75; A/HRC/WG.6/3/L.7, (Jan. 12 2009) *available at* http://www.refworld.org/docid/497476a40.html.

[515]CRPD, Article 11.

can result in decreased mobility, sight, hearing, and muscle strength, as well as in greater vulnerability to heat and cold.[516] Humanitarian actors have a duty to recognize the limited mobility of many groups. As the UNHCR guidelines for best practices regarding older persons in disaster and humanitarian crisis state: "The first key principle of good practice is to be aware of the presence of older persons in an emergency and take active steps to locate and identify them."[517] The guidelines also points out that "[l]imited mobility can create severe problems for older persons in crises."[518] According to research done by HelpAge International, four main problems were identified: incapacity to gain access to essential services; physical limitations; lack of transportation; and the loss of mobility aids, prostheses and glasses in the emergency.[519]

Article 4 of the ICCPR stipulates that in time of public emergency state parties may take measures derogating from certain of their obligations under the ICCPR. According to that provision, however, certain rights are non-derogable even in times of public emergency. When faced with situations of armed conflict, the HRC will necessarily examine whether these state parties complied with all of their ICCPR obligations and whether they have satisfied the conditions of Article 4 of the ICCPR.[520] Any derogation must be immediately brought to the attention of state parties through the intermediary of the Secretary-General and further notification is required upon the termination of the derogation.[521]

The Fourth Geneva Convention Relative to the Protection of Civilian Persons in Time of War defines humanitarian protections for civilians in a war zone, and prohibits the practice of total war.[522] Article 14 further states that in time of peace, and after the outbreak of hostilities, the states may establish in their own territory and, if the need arises in occupied areas, hospitals, and safety zones the means to

---

[516]*Humanitarian Action and Older Persons - An essential brief for humanitarian actors, Initiated by WHO and HelpAge International at the request of the Inter-Agency Standing Committee (IASC) Working Group*, Inter Agency Standing Comm., 3 (2008) *available at* http://www.google.com/url?sa=t&rct=j&q=&esrc=s&source=web&cd=1&ved=0CCcQFjAA&url=http%3A%2F%2Fwww.humanitarianinfo.org%2Fiasc%2Fdownloaddoc.aspx%3FdocID%3D4514%26type%3Dany&ei=0PPrUovlHO_IsAS6loHQDg&usg=AFQjCNFyiwgv37XBNA08RCVmiNUpb8xObg&sig2=mCQVCLHQYyNHSj5VMowGuA&bvm=bv.60444564,d.cWc.

[517]*See*, HelpAge International, Older People In Disasters and Humanitarian Crises, UNHCR, 3 *available at* http://www.refworld.org/pdfid/4124b9f44.pdf.

[518]Id. at 5.

[519]Id.

[520]U.N. General Assembly, *Report of the Human Rights Committee, Fifty-Sixth Session, Supplement No. 40*, A/56/40, *available at* http://www.ccprcentre.org/wp-content/uploads/2012/04/A_56_40Vol-I_en.pdf. (last visited Oct. 23, 2014). http://tb.ohchr.org/default.aspx searching "CCPR" and "Sessional/Annual Report of Committee" .

[521]Article 4, para. 3.

[522]*Geneva Convention relative to the Protection of Civilian Persons in Time of War*, Int'l Comm. Of the Red Cross (ICRC) 75 UNTS 287 (Aug. 12, 1949) *available at* http://www.icrc.org/ihl.nsf/385ec082b509e76c41256739003e636d/6756482d86146898c125641e004aa3c5.

protect from the effects of war, wounded, sick and as well as, inter alia, aged persons. Additionally, in accordance to Article 17, the parties to the conflict shall endeavor to conclude local agreements for the removal from besieged or encircled areas, of, inter alia, aged persons, and for the passage of ministers of all religions, medical personnel, and medical equipment on their way to such areas. A Report of the UNHCR emphasizes the obligation of all states and parties to armed conflicts to protect civilians in accordance with international humanitarian law, and invited states to promote a culture of protection, taking into account the particular needs of, among others, older persons.[523]

The rights of older refugees are specifically recognized in the Convention relating to the Status of Refugees, in particular the provisions related to the labor legislation and social security.[524] According to Article 24, the state parties shall guarantee to refugees lawfully staying in their territory the same treatment as is accorded to nationals in respect of, among others, social security including legal provisions in respect of old age, death, unemployment, family responsibilities and any other contingency, which, according to national laws or regulations, is covered by a social security.

UNHCR notes the particular need for ongoing protection of vulnerable groups within refugee groups. It further notes the particular need for ongoing protection of vulnerable groups within refugee groups, as well as the particular needs of vulnerable groups within refugee groups.[525] Older internally displaced persons (IDP's) in West Africa have been identified as requiring special attention.[526] The deprivation of older persons is one clear negative impact of food crises.[527] There, immediate food needs are closely related to the shortage of fertilizers and fuel.[528] Furthermore, the UNOCHA Guiding Principles on Internal Displacement, in Principle 4, indicate that

---

[523]U.N. General Assembly, *Report of the United Nations High Commissioner for Refugees, Questions Relating to Refugees, Returnees and Displaced Persons and Humanitarian Questions*, A/59/498 (Dec. 2 2004) *available at* (http://daccess-dds-ny.un.org/doc/UNDOC/GEN/N04/632/89/PDF/N0463289.pdf?OpenElement.

[524]U.N. High Commissioner for Refugees, *Convention and Protocol Relating to the Status of Refugees*, *available at* http://www.unhcr.org/protect/PROTECTION/3b66c2aa10.pdf. (last visited Oct. 23, 2014).

[525]U.N. General Assembly, *Report of the United Nations High Commissioner for Refugees, Questions Relating to Refugees, Returnees and Displaced Persons and Humanitarian Questions*, A/59/498 (Dec. 2 2004) *available at* (http://daccess-dds-ny.un.org/doc/UNDOC/GEN/N04/632/89/PDF/N0463289.pdf?OpenElement.

[526]UN Human Rights Council, *UN Human Rights Council: Addendum to the Report of the Representative of the Secretary-General on the Human Rights of Internally Displaced Persons, First Regional Conference on Internal Displacement in West Africa (Abuja - 26–28 April 2006)*, A/HRC/4/38/Add.4, (Feb. 20 2007), *available at* http://www.refworld.org/docid/461f8e2c2.html.

[527]UN Human Rights Council, *Report of the Special Rapporteur on the Situation of Human Rights in the Democratic People's Republic of Korea, Vitit Muntarbhorn*, A/HRC/10/18, at no. 49 (Feb. 24 2009) *available at* http://www.refworld.org/docid/49c8ef179843.html.

[528]Id.

> Certain internally displaced persons, such as children, especially unaccompanied minors, expectant mothers, mothers with young children, female heads of household, persons with disabilities and elderly persons, shall be entitled to protection and assistance required by their condition and to treatment which takes into account their special needs.[529]

Furthermore, older persons face situations in which they can suffer violations in an institutional, community, and family environment.[530] In an emergency, whether a natural disaster or armed conflict, the breakdown of social institutions can lead to abandonment, loss of family and community support, leaving them specially exposed to chaotic situations.[531] Their vulnerable condition can leave them marginalized and disenfranchised. The state must seek to adopt special measures where older persons face inaccessibility or limited of mobility to access essential services.[532] In situations of armed conflict, they are frequently targets of violence and reprisals. They are often among the least mobile, making them the least able to flee and the least able to get to places of safety and centers in camps to be registered for inclusion in food and health programs.[533]

Priority should be given to older persons in emergencies for food relief.[534] The right of older persons to have their basic needs addressed in disasters must be secured by the state concerned, with its ultimate responsibility to protect the rights of its people.[535]

In planning for displaced persons it is necessary to always keep in mind the needs and capabilities of older persons when designing programs that seek to overcome a crisis.[536] In emergency situations it is essential to ensure that older

[529]UNOCHA, Guiding Principles on Internal Displacement (2004) *available at* http://www.brookings.edu/~/media/Projects/idp/GPEnglish.pdf. (last visited Oct. 23, 2014).

[530]*A Global Response to Elder Abuse and Neglect: Building Primary Health Care Capacity to Deal with the Problem Worldwide: Main Report*, World Health Org., 2008, *available at* http://whqlibdoc.who.int/publications/2008/9789241563581_eng.pdf?ua=1 (last visited October 28, 2014).

[531]HelpAge International, *Older People in Disasters and Humanitarian Crises: Guidelines for Best Practice*, UNHCR, 9 (2000) *available at* http://www.refworld.org/docid/4124b9f44.html. (last visited Oct. 23, 2014).

[532]Id. at 5 and 7.

[533]Chung Working Paper, 2010, para. 20.

[534]UN Human Rights Council, *Report of the Special Rapporteur on the Situation of Human Rights in the Democratic People's Republic of Korea, Vitit Muntarbhorn*, A/HRC/10/18, at no. 11 (Feb. 24 2009) *available at* http://www.refworld.org/docid/49c8ef179843.html.

[535]UN Human Rights Council, *Report of the Special Rapporteur on the Situation of Human Rights in Myanmar, Tomás Ojea Quintana, on the implementation of Council resolutions S-5/1 and 6/33*, A/HRC/8/12, (Jun. 3 2008) *available at* http://www.refworld.org/docid/484d11142.html.

[536]*Humanitarian Action and Older Persons - An essential brief for humanitarian actors, Initiated by WHO and HelpAge International at the request of the Inter-Agency Standing Committee (IASC) Working Group*, Inter Agency Standing Comm., 4 (2008) *available at* http://www.google.com/url?sa=t&rct=j&q=&esrc=s&source=web&cd=1&ved=0CCcQFjAA&url=http%3A%2F%2Fwww.humanitarianinfo.org%2Fiasc%2Fdownloaddoc.aspx%3FdocID%3D4514%26type%3Dany&ei=0PPrUovlHO_IsAS6loHQDg&usg=AFQjCNFyiwgv37XBNA08RCVmiNUpb8xObg&sig2=mCQVCLHQYyNHSj5VMowGuA&bvm=bv.60444564,d.cWc. (last visited Oct. 23, 2014).

persons are able to participate in the decision-making process, design, and implementation of recovery programs. Income generation and repatriation programs must always take into account the specificities of older persons in caring for themselves and for other persons.[537] UNHCR has recognized the positive contribution of older persons in times of crisis[538] and has researched the amelioration of gaps in the protection of older refugees.[539]

The special needs of older IDP's should be identified and addressed in all phases of displacement.[540] Among the internally displaced, especially during wintertime, access to adequate housing is extremely important for older persons.[541] It is recommended that states should relax naturalization requirements, in particular language proficiency exams, for older persons.[542]

Older displaced persons have more serious health problems than the non-displaced due to problems of adaptation, poverty, and changing family care dynamics, limiting intergenerational support, among others.[543] UNHCR has called for a review of health services for older populations with non-communicable diseases in current UNHCR health policy.[544] In addition, UNHCR has sought to highlight issues of age-based violence, which is both a criminal and a health issue.[545] It has

---

537 Id. at 6.

538 Lindsey Judge, *The Rights of Older People: International Law, Human Rights Mechanisms and the Case for New Normative Standards*, 9 (Jan. 2009). http://www.globalaging.org/elderrights/world/2008/internationallaw.pdf.

539 UNHCR *Global Report 2007 Global Programme*, 2 http://www.unhcr.org/484923b52.pdf. (last visited Oct. 23, 2014).

540 UN Human Rights Council, *UN Human Rights Council: Addendum to the Report of the Representative of the Secretary-General on the Human Rights of Internally Displaced Persons, First Regional Conference on Internal Displacement in West Africa (Abuja - 26–28 April 2006)*, A/HRC/4/38/Add.4, (Feb. 20 2007) *available at* http://www.refworld.org/docid/461f8e2c2.html.

541 *Report of the Special Rapporteur on Adequate Housing, Government Communications* A/HRC/10/7/Add.1 at no. 72, (Feb. 17 2009) http://www2.ohchr.org/english/bodies/hrcouncil/docs/10session/A.HRC.10.7.Add.1.EFS.pdf.

542 UN Human Rights Council, *Report of the Special Rapporteur on Contemporary Forms of Racism, Racial Discrimination, Xenophobia and Related Intolerance, Doudou Diène: addendum: mission to Latvia*, A/HRC/7/19/Add.3, no. 88 (Mar. 5 2008), *available at* http://www.refworld.org/docid/47e11c0f2.html.

543 U.N. General Assembly, *Annual report of the Representative of the Secretary-General on the Situation of Human Rights on Internally Displaced Persons, Mission to Azerbaijan: at F*, A/HRC/24/13, at no. 70. (July 5, 2013) *available at* http://www.ohchr.org/EN/HRBodies/HRC/RegularSessions/Session24/Documents/A-HRC-24-13_en.pdf.

544 *Public Health Equity in Refugee and Other Displaced Persons Settings, Evaluation Report*, 14 (Apr. 2010) *available at* http://www.unhcr.org/4bdfe1699.pdf.

545 UNHCR Global Report 2008, Liberia 2, *available at* http://www.unhcr.org/4a2d29632.pdf. (last visited Oct. 23, 2014). UN High Commissioner for Refugees (UNHCR), *UNHCR Global Report 2008, Liberia*, (June 2009) available at http://www.refworld.org/docid/4bd800a40.html.

created self-reliance and income generation projects for West African refugees considering age,[546] and has created age-sensitive support for local integration.[547] It reported in 2006 that it was monitoring new refugee laws to see that they are in line with international protection standards and take into account gender and age-specific needs.[548] UNHCR also requires consideration for the needs of older persons who are returning refugees in order to ensure that they receive adequate protection, assistance and care throughout the repatriation and initial reintegration process.[549]

Furthermore, UNHCR has an Age, Gender and Diversity Mainstreaming Project, which is applied in every stage of relief to seek to ensure the particular protection needs of older refugees.[550] It was developed based on an evaluation of pilot project countries[551] to ensure that international standards of protection for older refugees are met.[552] Workshops[553] have trained staff members in new protocols[554] and best practices, which include emergency preparedness, age-disaggregated data collection, and border monitoring methods.[555] Technical experts on aging and operational data management have been fielded.[556] Standards and Indicators Reports continue tracking the mainstreaming methodology.[557] These measures have not changed the reality that older people face in displacement. All of the projects discussed above were implemented before a 2012 HelpAge

---

[546]UNHCR Global Report 2007, West Africa, 3 *available at* http://www.unhcr.org/4848ff602.pdf, (last visited Oct. 23, 2014).

[547]Id. Report No. 104 (LVI) – 2005 – Local Integration.

[548]UNHCR Global Report 2006 Sub-regional Report for Southern South America, 1 *available at* http://www.unhcr.org/4666d2570.pdf (last visited Oct. 23, 2014).

[549]Id. at 332. Report No. 101 (LV) – 2004.

[550]UNHCR, *Aggregated Executive Committee Reports includes Mentions of Older Persons of Concern*, 331 *available at* http://www.unhcr.org/3d4ab3ff2.pdf (last visited Oct. 23, 2014). (Includes all conclusions through October of 2008.) Report No. 100 (LV) – 2004.

[551]UNHCR, *Global Report 2005 Global Programme - Section on Older Refugees*, 86 *available at* http://www.unhcr.org/449267660.pdf (last visited Oct. 23, 2014).

[552]UNHCR Global Report 2007 Asia Summary, 3 *available at* http://www.unhcr.org/48480d782.pdf; UNHCR Global Report 2008, North America and the Caribbean, 5 http://www.unhcr.org/4a2d32ca2.pdf. (last visited Oct. 23, 2014).

[553]Id. at 5.

[554]UNHCR Global Report 2007 Asia, 3 *available at* http://www.unhcr.org/48480d782.pdf (last visited Oct. 23, 2014).

[555]*Border Monitoring Methodologies*, Peace Instit. Report (Nov. 2006) *available at* http://www.unhcr.org/4aa0d9839.pdf (last visited Oct. 23, 2014).

[556]UNHCR Global Report 2008, Internally Displaced Persons, 4 *available at* http://www.unhcr.org/4a2d199b2.pdf. (last visited Oct. 23, 2014).

[557]UNHCR Global Report 2007 Global Programme, 7 *available at* http://www.unhcr.org/484923b52.pdf (last visited Oct. 23, 2014).

International study finding that older persons received less than 1 % of humanitarian aid in 2010 and 2011.[558]

In its Policy on Older Refugees, 2000, UNHCR states that "older refugees will always have special needs,"[559] and affirmed that "the policy on older refugees has been formulated with the hope that future efforts on behalf of older persons will have become so well integrated into all aspects of protection and program planning and implementation."[560] The Policy also indicated that "the components of a policy aimed at older refugees shall be pre-requisites for ensuring their participation and well-being."[561] The Report further affirms the importance of age in ensuring the participation of refugees and other persons in the planning and implementation of programs and affirms the importance of giving a priority to addressing discrimination.[562]

The Report also called upon the governments concerned to focus attention and efforts on upholding protection to ensure proper consideration of the specific needs of older persons.[563] Additionally, the Report called for emphasis on the age and gender sensitive approaches and "support to boost the integration capacity of various groups with special needs",[564] including older refugees, and it also encouraged UNHCR "to develop and apply appropriate standards and indicators that account for age and gender considerations in local integration and self-reliance programs."[565]

The Report of the Secretary-General on human rights and mass exoduses notes that displaced persons are a category of most vulnerable persons partly because "the protection of uprooted persons is often problematic. As mass exoduses normally occur in situations of crisis and conflict, national authorities are either unable or unwilling to discharge their responsibility to protect their populations, even at the most basic level."[566] In addition to the mechanisms outlined above, and for the well-being of refugees, commitments to protection by governments have been monitored by the Special Rapporteur on the Human Rights of Internally Displaced

---

[558] *A Study of Humanitarian Financing for Older People and People with Disabilities*, HelpAge International, (2010–2011) *available at* http://www.helpage.org/what-we-do/emergencies/a-study-of-humanitarian-financing-for-older-people-and-people-with-disabilities/ (last visited Feb 19 2015).

[559] UN High Commissioner for Refugees (UNHCR), *UNHCR"s Policy on Older Refugees*, EC/50/SC/CRP.13, Annex II, (April 19 2000), *available at* http://www.refworld.org/docid/47036b502.html.

[560] Id.

[561] Id.

[562] Id. at para. 15.

[563] Id. at para. 43.

[564] Id. at para. 16.

[565] Id. at para. 16.

[566] UN General Assembly, *Human rights and mass exoduses, Report of the Secretary-General*, A/60/325, 17, (Sep. 1 2005) *available at* http://www.refworld.org/docid/43f30fb70.html.

Persons and the former Representative of the Secretary-General on the Human Rights of Internally Displaced Persons ("RSG IDP"). For example, the RSG IDP had been engaging in coordinated advocacy in favor of the protection and respect of the human rights of IDP's, and as it was reiterated in a report of the Secretary General, other areas of activity included dialogue with all sectors of society, strengthening the response to internal displacement, and mainstreaming the human rights of displaced populations in the UN system.[567]

In the Note by the Secretary-General transmitting the report of the RSG IDP, the Secretary-General highlighted the urgent needs of most the vulnerable groups among IDP's, including older persons. The representative indicated that national and local authorities, as well as international agencies and donors, must seek sustainable solutions for these persons, adopting policies regarding social housing, foster families or homes.[568]

The RSG IDP has addressed the vulnerability of older persons displaced by armed conflict on a number of occasions, including situations in Croatia, Bosnia and Herzegovina and Serbia and Montenegro, including Kosovo, in the 1990s.[569]

Similarly, in Serbia the RSG IDP noted that "thousands of IDPs continue to live in very difficult conditions in collective centers and irregular settlements," older persons being among them.[570] The RSG IDP urgently recommended "social housing, foster families or [in] homes respecting their right to human dignity."[571]

In Bosnia and Herzegovina, the RSG IDP expressed his concern about the deplorable living conditions of IDP's.[572] The report goes on to note that multiple discriminations along ethnic, gender, age, class and other lines result in an accumulation of rights deprivation. The RSG IDP recommended to the authorities that they mainstream the protection needs of vulnerable groups when formulating all

---

[567]U.N. Secretary General, *Representative of the Secretary-General on the Human Rights of Internally Displaced Persons*, A/60/338, (Sep. 7 2005) *available at* http://www2.ohchr.org/english/issues/idp/mandate.htm (The mandate of the Special Rapporteur on the Human Rights of Internally Displaced Persons, created by resolution of the Human Rights Council in 2010, has a similar mandate to that of its predecessor, the RSG IDP: *see* http://ap.ohchr.org/documents/dpage_e.aspx?si=A/HRC/RES/14/6).

[568]Id. at para. 10.

[569]U.N. Secretary General, *Specific Groups and Individuals: Mass Exodus and Displaced Persons, Report of the Representative of the Secretary-General on the Human Rights of Internally Displaced Persons, Walter Kalin,* U.N. Econ. And Soc. Council, E/CN/4/2006/71/Add.5 **61, 62,** 63 (Jan. 9 2006) *available at* http://daccess-dds-ny.un.org/doc/UNDOC/GEN/G06/101/33/PDF/G0610133.pdf?OpenElement.

[570]UN Commission on Human Rights, *Report on the Human Rights of Internally Displaced Persons, Addendum: Mission to Serbia And Montenegro*, E/CN.4/2006/71/Add.5, (Jan. 9 2006) *available at* http://www.refworld.org/docid/441181fcc1.html.

[571]RSG IDPs, 2006, at para. 70.

[572]UN Commission on Human Rights, *Report on the Human Rights of Internally Displaced Persons, Addendum, Mission to Bosnia and Herzegovina*, E/CN.4/2006/71/Add.4, (Dec. 29 2005) *available at* http://www.refworld.org/docid/441181f50.html.

policies and measures affecting them and devise specific measures to find durable solutions.[573]

In Croatia, the RSG IDP recommended that the government should ensure that IDP's living in collective centers are "provided realistic alternatives concerning their future status, with an identification of their particular needs."[574] The RSG IDP further recommended that the "government should ensure that public specialized facilities, such as social housing," be made available to older IDP's.[575]

In Georgia in 2006, the RSG IDP urged the government "to seek durable solutions targeting particularly vulnerable persons among IDP's, including the creation of adequate housing and appropriate institutional arrangements."[576] Those belonging to particularly vulnerable groups, including older persons without family support, were not being transitioned out of temporary shelters.[577] The RSG IDP also recommended to "design and adopt a comprehensive, rights-based policy which would support IDP's as they integrate into society and acquire adequate living conditions, while maintaining their option to return" and that "such a policy should seek to ensure the full enjoyment of all human rights of IDP's, as restated in the Guiding Principles on Internal Displacement."[578]

The RSG IDP has turned attention to other regions of the world such as Latin America, Africa, and Central Asia. In Colombia, the RSG IDP recommended the adoption of special measures for the health and assistance needs of older persons, in particular those who have to take care of children left with them by their parents.[579] In Sudan, the RSG IDP found that "returnees arriving at their destinations often remained without shelter, sufficient food, clean drinking water or access to medical services."[580] In this situation, it concluded that older persons remained particularly vulnerable.[581]

---

[573]Id. at para. 56.

[574]UN Commission on Human Rights, *Report on the human rights of internally displaced persons, Walter Kälin, Mission to Croatia*, E/CN.4/2006/71/Add.3, 48 (Dec. 29 2005) *available at* http://www.refworld.org/docid/441181f30.html.

[575]Id. at para. 48.

[576]Id. at para. 55.

[577]Id. at para. 54.

[578]Id. at para. 57.

[579]UN Human Rights Council, *UN Human Rights Council: Addendum to the Report of the Representative of the Secretary-General on the Human Rights of Internally Displaced Persons, Mission to Colombia*, A/HRC/4/38/Add.3, (RSG IDPs, 2007) 83 (Jan. 24 2007) *available at* http://www.refworld.org/docid/461e3f5a2.html.

[580]UN Commission on Human Rights, *Report of the Representative of the Secretary-General on the Human Rights of Internally Displaced Persons, Walter Kälin. Addendum MISSION TO THE SUDAN* ** (3–13 October 2005)*, E/CN.4/2006/71/Add.6, (RSG IDPs, 2006) 74 (Feb. 13 2006) *available at* http://www.refworld.org/docid/441182140.html. (last visited Oct. 23, 2014).

[581]Id at para. 74.

In Azerbaijan, the RSG IDP noted with concern that the special needs of internally displaced older persons were insufficiently addressed, in particular as they seemed to be at an economic disadvantage compared to their non-displaced peers, due to difficulties in adjusting and diminished family support due to the impoverishment of their children.[582] The RSG IDP concluded that specific surveys and needs assessments into the situation of older internally displaced persons and their access to counseling and appropriate medical care needed to be conducted, and noted that both general and specific government programs should pay special attention to particularly vulnerable groups among internally displaced persons,[583] recommending a differentiated approach for older persons.[584]

Much has been learned since the CESCR expressed its concern to Japan in 2001 that "many resettled earthquake victims, who were over 60 years of age lack community centers, access to health centers and outpatient nursing."[585] The 2009 Final Report of the Third Committee also called upon member states to adopt measures necessary to assist older persons in emergency situations.[586] Additionally, the related draft resolution further called upon member states to take concrete measures to further protect and assist older persons in emergencies, in accordance with MIPAA.[587]

2. *The Rights of Older Women*

Older women are often stigmatized because of their gender and age. The majority of older persons are women because of their longer life expectancy. In the more developed regions, the projected difference in life expectancy is even more

---

[582]UN Human Rights Council, *Report of the Representative of the Secretary-General on the Human Rights of Internally Displaced Persons, Walter Kälin: addendum: mission to Azerbaijan*, A/HRC/8/6/Add.2, (RSG IDPs, 2008) 72 (Apr. 15 2008) *available at* http://www.refworld.org/docid/48be9acb2.html (last visited Oct. 23, 2014).

[583]Id. at para. 71.

[584]UN Human Rights Council, *UN Human Rights Council: Addendum to the Report of the Representative of the Secretary-General on the Human Rights of Internally Displaced Persons, Mission to Colombia*, A/HRC/4/38/Add.3, (RSG IDPs, 2007) 78 (Jan. 24 2007) *available at* http://www.refworld.org/docid/461e3f5a2.html (last visited Oct. 23, 2014).

[585]*Report on the Twenty-Fifth, Twenty-Sixth, and Twenty-Seventh Sessions*, CESCR E/2002/22, E/C.12/2001/17 *available at* http://www.bayefsky.com/general/e_2002_22_2002.php (last visited Oct. 23, 2014).

[586]General Assembly, U.N. Doc A/64/432, 15, *available at* http://www.copac.coop/iyc/a-64-432-excerpt.pdf (last visited Oct. 23, 2014).

[587]U.N. General Assembly, *Follow-up to the Second World Assembly on Ageing*, A/C.3/64/L.6, (Oct. 7 2009) http://daccess-dds-ny.un.org/doc/UNDOC/LTD/N09/547/37/PDF/N0954737.pdf?OpenElement.

dramatic.[588] Older women face abuse and mistreatment, age-related discrimination, do not access inheritance and suffer economic exploitation.[589] Older women are disproportionately disadvantaged: despite making up an ever-greater proportion of the population, they are often poorer than older men. This is a result of gendered discrimination patterns throughout their lives.[590] On average, older women accumulate less wealth than men. It is more difficult for them to find work and they are paid less than older men. By the time women have reached old age, many have experienced multiple disadvantages resulting from biases to gender, widowhood, and old age, among others. Such situations lead many older women into poverty.[591] Further examination of multiple discriminations across the life course has been encouraged by the Report of the independent expert on the question of human rights and extreme poverty, in her March 2010 report on social security.[592] This Report, highlighted that

> [t]he majority of older women are excluded from formal social security and health insurance schemes as these are linked to paid, formal-sector employment. In developing countries, the great majority of women work all their lives in the informal sector or unpaid activities. In developed countries too, older women are more likely than men to be poor at old age.[593]

The ICCPR recognizes in its Article 3 that all civil and political rights set forth in the covenant are to be enjoyed equally by men and women. Additionally, the state parties to the ICCPR.[594] Also, the CRPD contains special provisions for women, in particular Article 6 which provides that "state parties recognize that women and girls with disabilities are subject to multiple discriminations, and in this regard shall take measures to ensure the full and equal enjoyment by them of all human rights and fundamental freedoms."

To ensure respect for the right to social security, the CESCR suggested that states should refrain from discriminating against women who choose to stay at home to raise a family. For example, states should acknowledge time spent at home as a homemaker as work experience when evaluating women for employment opportunities. Such decisions will assist in the amelioration of poverty among older women. To ensure this right, particularly in connection with Article 9

---

588 Chung Working Paper 2010, para. 21.

589 G.A. Res. 65/182, (Aug. 1–4 2011) *available at* http://social.un.org/ageing-working-group/documents/Chair_summary_2nd_session_OEWG_final.pdf.

590 *Follow-Up to the Second World Assembly on Ageing: Report of the Secretary-General, Submitted to the United Nations General Assembly*, A/64/127, 49 (July 2009) *available at* http://www.un.org/esa/socdev/documents/reports/Ageing.pdf.

591 Chung Working Paper 2010 para. 16.

592 UN Human Rights Council, *Report of the Independent Expert on the Question of Human Rights and Extreme Poverty*, A/HRC/14/31, (Mar. 31 2010) *available at* http://www.refworld.org/docid/4c061e242.html.

593 Sepulveda Report, 2010 at para. 20.

594 ICCPR at Article 3.

of the ICESCR, the CESCR considered that states should provide "non-contributory old-age benefits or other assistance for all persons."[595] This benefit would be given, regardless of sex, to those without adequate resources upon reaching a prescribed age specified in national legislation.[596] To fulfill this right, the CESCR indicated that states should promote the need for more men to stay at home to raise families.

The CESCR expressed its concern about women who, having stayed at home to raise a family, have made no contributions to a social security plan, and therefore do not have any source of income in their later years.[597] Unfortunately, many of these same women are also ineligible for widows' pensions.[598] Also, special measures of protection are expressly recognized in CEDAW Article 4.

CEDAW recognizes civil, political, social, economic, and cultural rights by requiring states "to take all appropriate measures, including legislation, to ensure the full development and advancement of women."[599] According to Article 3 women are entitled to the equal enjoyment and protection of all human rights and fundamental freedoms. These include rights embodied in other treaties such as the ICCPR like the right to equality (article 2), the right to life (article 6); the right not to be subjected to torture, or other cruel, inhuman or degrading treatment or punishment (article 7); the right to liberty and security of person (article 9), the right to equal protection under the law (article 26); and the right to be free from all forms of discrimination (article 26). Also, according to the ICESCR, women have the right to just and favorable conditions of work (article 7) and the right to the highest standard attainable of physical and mental health (article 12).

The CEDAW elaborates on the term "discrimination against women" as any exclusion or restriction made on the basis of sex which has the effect or purpose of impairing or nullifying the recognition, enjoyment or exercise by women irrespective of their marital status, on a basis of equality of men and women, of human rights and fundamental freedoms in the political, economic, social, cultural, civil or any other field.[600]

The Special Rapporteur on the right of everyone to the enjoyment of the highest attainable standard of physical and mental health, indicated in his study on the realization of the right to health of older persons that "differences between the

---

[595] *The Economic, Social, and Cultural Rights of Older Persons (Thirteenth Session, 1995)* Comm. on Econ. Soc. and Cultural Rights, U.N. Doc. E/1996/22, 21. (1996) http://www1.umn.edu/humanrts/gencomm/epcomm6e.htm (last visited Oct. 23, 2014).

[596] *See* Id. at para. 21 (stating that women should be the principal beneficiaries because they have a greater life expectancy and often do not have contributory pensions).

[597] *See* General Comment 6, para. 20 (stating that men and women should enjoy equal economic, social, and cultural rights).

[598] *See* Id. (emphasizing the critical situation of women who stay at home).

[599] CEDAW, Article 3.

[600] CEDAW, Article 1.

genders in respect of the aging process must also be acknowledged, given that global life expectancy at birth for women is currently 70 years, and is significantly higher than for men at 66 years," and that "older women are frequently excluded from social security and health insurance schemes that are linked to formal, paid employment."[601]

With regard to older women, Article 11 of the CEDAW guarantees "the right to social security, particularly in cases of retirement, unemployment, sickness, invalidity and old age and other incapacity to work, as well as the right to paid leave." As for other financial assistance, Article 13 of the CEDAW covers economic and social benefits, including the right family benefits, bank loans, mortgages, and other forms of financial credit. The CEDAW also acknowledges an especially vulnerable position of rural women asking state parties to take particular measures to ensure the application of the provisions, including their participation in the development planning, access to adequate health care facilities and equal access to economic opportunities through employment and enjoy adequate living conditions. As for marriage and family life, the CEDAW also emphasizes the obligation of the state parties to "eliminate discrimination against women in all matters relating to marriage and family relations."[602] Article 5 of the CEDAW requires states to take measures against pernicious cultural practices that discriminate against women.

In CEDAW General Recommendation No 24 on health, the CEDAW Committee stated that "there are societal factors, which are determinative of the health status of women and men, and for that reason, special attention should be given to the health needs and rights of women belonging to vulnerable and disadvantaged groups," such as, inter alia, "older women."[603]

Human rights experts addressing the CEDAW Committee regarding its General Recommendation No. 25, have indicated that, it "recognizes that discrimination against women can be compounded by other types of exclusion including age, and allows states to take temporary special measures to overcome the entrenched

[601] Human Rights Council, *Thematic Study on the Realization of the Right to Health of Older Persons by the Special Rapporteur on the Right of Everyone to the Enjoyment of the Highest Attainable Standard of Physical and Mental Health, Anand Grover,* A/HRC/18/3717 (July 4 2011), http://www2.ohchr.org/english/bodies/hrcouncil/docs/18session/A-HRC-18-37_en.pdf (as indicated by Chung, older women suffer disproportionately from depression, are more likely to become depressed, and are less likely to recover from depression (Chung Working Paper, 2010, para. 25).

[602] CEDAW at Article 16.

[603] *(Twentieth session, 1999): Article 2: Women and Health*, ICESCR, A/54/38/Rev.1 part I 3 at 6 (1999) *available at* http://www.bayefsky.com/themes/disability_general-comments.pdf (last visited Oct. 23, 2014); UN Committee on the Elimination of Discrimination Against Women (CEDAW), *CEDAW General Recommendation No. 24: Article 12 of the Convention (Women and Health)*, A/54/38/Rev.1, chap. I, (1999) *available at* http://www.refworld.org/docid/453882a73.html. (last visited Oct. 23, 2014).

nature of discrimination such women face."[604] It noted that certain groups of women might also suffer from multiple forms of discrimination based on additional grounds such as, inter alia, age. The recommendation stated that "such discrimination may affect these groups of women primarily, or to a different degree or in different ways than men. State parties may need to take specific temporary special measures to eliminate such multiple forms of discrimination against women and its compounded negative impact on them."[605]

In recent years, there has been growing pressure on the CEDAW Committee to pay closer attention to the rights of older women. Consequently, in 2010, the CEDAW Committee adopted the landmark General Recommendation No. 27 on the rights of older women and protection of their human rights,[606] The General Recommendation explores the relationship between the rights protected by CEDAW and aging.[607] It also acknowledges the multiple forms of discrimination that women face across their lives, outlines the scope of the state obligations under CEDAW with respect to the protection of older women and includes policy recommendations.[608] Ultimately, the General Recommendation provides guidance to states in the including of the situation of older women in their periodic reports to the Committee.[609] However, this guidance has not yet resulted in the routine inclusion of the situation of older women in reports.

Cultural practices of early marriage for women, in addition to their longer life expectancy, creates a situation in which women are more likely to be widowed than men, which in turn puts older women at higher risk of poverty. The CESCR expressed concern about the problem of the abandonment of older women and strongly recommended that the state party "undertake effective public education measures, including awareness-raising programs designed to eliminate gender-based prejudices and traditional practices that are harmful to women."[610] Land and inheritance rights in many countries deny women their right to own and keep

---

[604]Lindsey Judge, *The Rights of Older People: International Law, Human Rights Mechanisms and the Case for New Normative Standards* pg. 9 (Jan. 2009) *available at* http://www.globalaging.org//elderrights/world/2008/internationallaw.pdf.

[605]*Report of the Committee on the Elimination of Discrimination against Women: Thirtieth session*, UN Comm. on the Elimination of Discrimination Against Women (CEDAW), *(12–30 January 2004), Thirty-first session (6–23 July 2004)*, Thirtieth session, (2004): Article 4, Paragraph 1: Temporary Special Measures, A/59/38 part I 78 at 12. (2004) *available at* http://www.refworld.org/docid/4ef9ec682.html. (last visited Oct. 23, 2014).

[606]*General Recommendation No. 27 on older women and protection of their human rights*, CEDAW CEDAW/C/GC/27 (Dec. 16 2010). *available at* http://www.refworld.org/docid/4ed3528b2.html.

[607]Id. at para. 9.

[608]Id.

[609]Id. at para. 10.

[610]China, ICESCR, E/2006/22 25 at 148 and 178 (2005) *available at* http://daccess-dds-ny.un.org/doc/UNDOC/GEN/G06/422/64/PDF/G0642264.pdf?OpenElement (last visited Oct. 23, 2014).

property.[611] The CEDAW Committee has also denounced this problem and documented that widows in some countries are victims of "property grabbing," as a result of the application of domestic statutory or customary laws.[612] Witchcraft allegations and violence against older women are usually linked to disputes over property ownership and inheritance.[613] This can lead to impoverishment, abuse, violence and even murder.[614] The CEDAW Committee has expressed "serious concern about the number of older women who have been murdered for superstitious reasons by family members or by others in Zambia in recent years." The ongoing violence of the accusers occurs without recognition "that such violence constitutes a violation of the human rights of women under the Convention" in Zambia.[615]

Social pensions are also an important mechanism to confront the specific vulnerabilities of women but can also perpetuate inequalities affecting older women. In many countries, contributory social protection systems' coverage rates for older men are twice those for their female counterparts, therefore magnifying already existing gender inequalities.[616] Human rights experts consider that universal pensions can contribute to eliminate differences of treatment towards women. It can help older women with income, as well as women who provide unpaid services as caregivers. Evidently, there are structural changes that have to be addressed like lack of documentation of most women and attitudinal barriers in social services. [617] Likewise, in order to compensate women who have worked for many years without payment, non-contributory pensions can be implemented. These measures can be combined with other policies or programs such as social services, provision of land, and other resources. [618]

Women are often deprived of their status, property, and legal rights and excluded economically and socially. CEDAW explicitly condemns "discrimination" against women and calls on states to adopt measures that will eliminate such discrimination. It also considers the vulnerability of rural women due to their lack of access to health care and calls on states to implement measures that will allow

[611] *See HelpAge International, Ageing and Development*, Issue 26, (Sep. 2009) *available at* http://www.helpage.org/what-we-do/rights/ageing-and-development-26-older-peoples-rights/; HelpAge International, Final Evaluation Report of the Older Women Inheritance Rights Project, (May 2008) (In Tanzania, for example, 77 percent of the cases on disputes over inheritance and land rights were brought by older women).

[612] Committee, General Recommendation No. 27 on older women and protection of their human rights, *supra* note 558, para. 26.

[613] Chung Working Paper, 2010, para. 13.

[614] Chung Working Paper 2010 para. 11; Sepulveda Report, 2010 para 96.

[615] Zambia, CEDAW, A/57/38 part II 111 at 238 and 239 (2002) http://daccess-dds-ny.un.org/doc/UNDOC/GEN/N02/614/17/PDF/N0261417.pdf?OpenElement. (last visited Oct. 23, 2014).

[616] Sepulveda Report at para. 33.

[617] Sepulveda Report, 2010, at para. 95.

[618] Conclusions and Recommendations, Sepulveda Report 2010 at para. 115.

rural women to access health and social services and participate in the political and social process. General Comment No 25 recognizes discrimination compounded by age. General Comment No. 24 addresses the need to acknowledge the disproportionate amount of women that suffer from diseases associated with aging, due to their increased life expectancies.[619] General Recommendation 27 expands its analysis on the rights of older women and acknowledges the limited protection this group has received regarding the protection of their fundamental rights.[620]

The General Assembly Third Committee passed the resolution on implementation of the situation of women in rural areas in October, 2009, which included a provision to "ensure that the rights of older women in rural areas are taken into account with regard to their equal access to basic social services, appropriate social protection and social security measures, equal access to and control of economic resources, and empowerment of older women through access to financial and infrastructure services, with special focus on support to older women, including indigenous women, who often have access to few resources and are more vulnerable."[621] While the resolution is not legally binding, it is an important reflection of the norm of equal recognition and protection of older women, which could, arguably, be considered international customary law.

A Report of the Commission on the Status of Women asked the states "to develop multi-sector policies and programs and identify, strengthen and take all necessary measures to address the needs of women and girls, including older women and widows". The Commission on the Status of Women highlighted that "social and legal protection, increased access to financial and economic resources including microcredit and sustainable economic opportunities, education including opportunities to continue education, as well as access to health services, including affordable antiretroviral treatment, and nutritional support", as priority areas.[622] Further, the Report urged "governments and other relevant stakeholders to address the challenges faced by older women in accessing HIV prevention, treatment, care, and support."[623]

---

[619]Lindsey Judge, *The Rights of Older People: International Law, Human Rights Mechanisms and the Case for New Normative Standards* pg. 4 (Jan. 2009) *available at* http://www.globalaging.org//elderrights/world/2008/internationallaw.pdf.

[620]Committee, General Recommendation No. 27 on older women and protection of their human rights *supra* note 558.

[621]United Nations General Assembly, *Improvement of the Situation of Women in Rural Areas*, A/C.3/64/L.19 (Oct. 14 2009) *available at* http://daccess-dds-ny.un.org/doc/UNDOC/LTD/N09/559/46/PDF/N0955946.pdf?OpenElement.

[622]*Commission on the Status of Women Fifty-third session*, U.N. Econ. and Soc. Council E/2009/27 (Mar. 2–13 2009),*available at* http://daccess-dds-ny.un.org/doc/UNDOC/GEN/N09/283/98/PDF/N0928398.pdf?OpenElement; *Mainstreaming a Gender Perspective into All Policies and Programmes in the United Nations System*, U.N. Econ. and Soc. Council, E/2009/71 (May 12 2009) *available at* http://www.un.org/womenwatch/daw/documents/ecosoc2009.htm.

[623]Id. at para. 7.

The 2009 Final Report of the Third Committee called upon member states "to strengthen the gender perspective into all policy actions on aging," as well as to "eliminate and address discrimination based on age and gender." It further suggested that member states "engage with all sectors of society, including women's groups"... in changing negative stereotypes about ... older women in particular."[624] The Draft Resolution encouraged "governments to pay greater attention to building capacity to eradicate poverty among older women in particular, by mainstreaming aging issues into poverty eradication strategies and national development plans, and to include both aging-specific policies and age in mainstreaming efforts in their national strategies."[625] These recommendations, while not legally binding, are useful components of growing international recognition of the need to protect the human rights of older persons.

The Report of the Secretary General on the follow-up to the Second World Assembly on Ageing highlighted "special attention to the needs and concerns of older women in the review and appraisal of implementation of the Beijing Declaration and Platform for Action. It urged the consideration of the situation of older women in the context of the analysis of critical areas such as women and poverty, women and the economy, and women and health."[626]

The United Nations Declaration on the Elimination of Violence against Women[627] notes older women are one of the groups especially vulnerable to violence. The instrument defines the "violence against women" as "any act of gender-based violence that results in, or is likely to result in, physical, sexual, or psychological harm or suffering to women, including threats of such acts, coercion, or arbitrary deprivation of liberty, whether occurring in public or in private life"[628] In one instance, the CEDAW Committee commended a state party on the adoption of a law on domestic violence with protective measures for older persons but expressed the concern that the penalty is merely a fine.[629]

In the follow-up to and Progress in the Implementation of the Beijing Declaration and Platform for Action, the 23rd Special Session of the General Assembly addressed the situation of older women in society, indicating that the

[624]G.A. Res. A/64/432 12, *available at* http://www.copac.coop/iyc/a-64-432-excerpt.pdf. (last visited Oct. 23, 2014).

[625]U.N. General Assembly, *Follow-up to the Second World Assembly on Ageing*, A/C.3/64/L.6 (Oct. 7 2009) *available at* http://daccess-dds-ny.un.org/doc/UNDOC/LTD/N09/547/37/PDF/N0954737.pdf?OpenElement.

[626]U.N. General Assembly, *Report of the Secretary General on the Follow Up to the Second World Assembly on Ageing*, A/60/151 (July 22, 2005), para. 28, *available at* http://www.un.org/ga/60/third/docslist.htm.

[627]U.N. General Assembly, Declaration on the Elimination of Violence against Women, A/RES/48/104 (Dec. 20 1993) *available at* http://www.un.org/documents/ga/res/48/a48r104.htm.

[628]Id.

[629]Paraguay, Suppl. No. 38, A/60/38(Supp), 32nd and 33rd Sess., 279 *available at* (http://www.un.org/ga/60/third/docslist.htm) (last visited Oct. 23, 2014).

Political Declaration and MIPAA, "provided a broad spectrum of social, political and economic recommendations to improve the conditions of older women." It also mentioned "that the situation of older women must be a priority for policy action" and it also call upon states "to ensure that the needs, perspectives and experiences of older women are reflected in all development policies and programs."[630]

In the Report of the Secretary-General on the review of the implementation of the Beijing Declaration and Platform for Action, in 2010, it was noted that efforts have also been made to expand women's access to health information, as well as gender-sensitive health information to expand awareness about the availability of health services, with respect to, inter alia, healthy aging.[631] It mentioned the poverty of older women, and that the problem has been magnified by the fact that older women often assume responsibility for children orphaned by AIDS.[632] The Report also emphasized illiteracy which remains a key issue for women. It is particularly prevalent among women who are older, poor, disabled, from ethnic minorities, or from rural areas.[633] Finally, the Report identified as priority areas: research on barriers to women's access to decision-making, and the specific needs and priorities of vulnerable groups of women, including, inter alia, older women in relation to many of the critical areas of concern, especially in the area of health.[634]

In a unique connection, the Asia-Pacific High-level Intergovernmental Meeting to Review Regional Implementation of the Beijing Platform for Action at Fifteen Years concluded that "while women are largely responsible for natural resource management in the household, they have been excluded from developing, planning, and implementing environmental policies although they are, together with their children and older persons, among the most vulnerable to the consequences of climate change."[635] Likewise, in General Recommendation No. 27, the

---

[630]U.N. General Assembly, *Follow-up to and Progress in the Implementation of the Beijing Declaration and Platform for Action and the Outcome of ohe 23rd Special Session* A/58/166, 38, http://www.un.org/esa/documents/ac3.htm, (referring to resolution 57/177).

[631]U.N. Secretary General, *Review of the implementation of the Beijing Declarationand Platform for Action, the Outcomes of the Twenty-Third Special Session of the General Assembly and its Contribution to Shaping a Gender Perspective Towards the Full Realizationof the Millennium Development Goals*, E/CN.6/2010/2 94 (Feb. 8 2010) *available at* http://daccess-dds-ny.un.org/doc/UNDOC/GEN/N09/637/20/PDF/N0963720.pdf?OpenElement.

[632]Id. at para. 27.

[633]Id. at para. 67.

[634]Id. at paras 496 and 519.

[635]*Outcome Document of the Asia Pacific High Level Intergovernmental Meeting to Review Regional Implementation of the Beijing Platform for Action at Fifteen Years, and its Regional and Global Outcomes*, E/ESCAP/BPA/2009/CRP.1 (Nov. 18 2009) *available at* http://www.unescap.org/ESID/GAD/Events/HLM-2009/download/BPA09_CRP1E_edited_final.pdf (similarly in the Bangkok Declaration for Beijing, at para. 23).

CEDAW Committee concluded that older women have "limited access to resources and decision-making processes increase their vulnerability to climate change."[636] In its early days, much of the CEDAW Committee's attention to older women centered around older women's health and benefits programs.[637] Addressing decision-making processes of older women and their vulnerability to climate change highlights the CEDAW Committee's evolving understanding of the intersecting spheres of women's lives.

While still maintaining its observance of the situation of poverty many older women face,[638] it has more recently taken a broader approach to the rights of older women. It has started highlighting reports that lack attention to the rights of older

---

[636]Committee, General Recommendation No. 27 on older women and protection of their human rights, *supra* note 558, para. 25. *General recommendation No. 27 on older women and protection of their human rights*, CEDAW, CEDAW/C/GC/27, (Dec. 16 2010) *available at* http://www.refworld.org/docid/4ed3528b2.html.

[637]*See* Iceland, CEDAW, A/57/38 part I (2002) 27 at 249 and 250. (expressing concern about the change of the pension system, which negatively impacted on women more than men); Sri Lanka, CEDAW, A/57/38 part I (2002) 31 at 296 and 297 (highlighting the high percentage of households headed by women, many of whom are elderly and should be recognized as equal recipients and beneficiaries of development programs); Switzerland, CEDAW, A/58/38 part I (2003) 20 at 102 (positively noting the law reform on old age and survivors' insurance establishing a system of individual pensions independent of civil status, and requesting that Switzerland take into account the gender dimensions of the plan of action adopted by the Second World Assembly on Ageing in its next periodic report); Luxembourg, CEDAW, A/58/38 part I (2003) 47 at 295 (noting that pensions play a key role in preventing poverty among older women); Para. 15, A/56/38(SUPP) (CEDAW, 2001) http://daccess-dds-ny.un.org/doc/UNDOC/GEN/N01/534/56/PDF/N0153456.pdf?OpenElement (last visited Oct. 23, 2014) (expressing concern that elderly women may be marginalized within, as well as insufficiently covered by, the health insurance and pension systems); Paras. 330 and 333, Italy A/60/38(SUPP) (CEDAW, 2005) http://daccess-dds-ny.un.org/doc/UNDOC/GEN/N05/476/97/PDF/N0547697.pdf?OpenElement. (last visited Oct. 23, 2014). (requesting the provision of information on the health care of older women); France, CEDAW, A/58/38 part II (2003) 116 at 263 and 264 (recommending that France to research the needs of older women and develop measures that adequately address their health, economic and emotional situation in order to avoid poverty and isolation, as well as take women's unpaid and paid employment patterns and family responsibilities when altering legal and policy measures in order to avoid discrimination of effect); Luxembourg CEDAW, A/58/38(SUPP) (2003) 295. http://daccess-dds-ny.un.org/doc/UNDOC/GEN/N03/468/20/PDF/N0346820.pdf?OpenElement (commending Luxembourg for adopting new laws in support of the goal of gender equality, including the law on pensions as a key in preventing poverty among older people and particularly women).

[638]Ireland, CEDAW, A/60/38 part II (2005) 151 at 392 and 393.

women,[639] and requesting that state parties include more information on the implementation of aspects of the documents from the Second World Assembly on Ageing relating to relevant articles of the CEDAW in future periodic reports.[640]

According to the third working session of the UN Open-Ended Working Group on Ageing for the purpose of strengthening the protection of the human rights of older persons, "the reality of older persons has largely degraded, from elders being respected and serving in their communities as mediators, to experiencing increasing amount of violence and abuse today, especially older women." It also referred to "several examples of abuse and violence, including gang rape, …are increasingly reported in some regions of Africa."[641]

Older rural women remain among the most impoverished and vulnerable in any population.[642] The CEDAW Committee has stated "its concern about the absence of detailed information on the situation of older women in rural areas" and urged

---

[639]*See* U.N. General Assembly, *Report of the Committee on the Elimination of Discrimination Against Women*, A/55/38(SUPP), CEDAW, (2000) http://daccess-dds-ny.un.org/doc/UNDOC/GEN/N00/611/52/PDF/N0061152.pdf?OpenElement. (last visited Oct. 23, 2014); Nicaragua; CEDAW Committee, A/HRC/WG.6/1/NLD/3, 11, 314, and 315 (Mar. 13 2008); UN Human Rights Council, *Summary prepared by the Office of the High Commissioner for Human Rights, in accordance with paragraph 15(c) of the Annex to Human Rights Council Resolution 5/1 - The Netherlands*, A/HRC/WG.6/1/NLD/3, (Mar. 13 2008), *available at* http://www.refworld.org/docid/4857a6f30.html; *Concluding Comments of the Committee on the Elimination of Discrimination against Women: Indonesia*,UN Committee on the Elimination of Discrimination Against Women (CEDAW), CEDAW/C/IDN/CO/5, (Aug. 10 2007) *available at* http://www.refworld.org/docid/46d281000.html [accessed 31 January 2014]; Kenya, para. 45, CEDAW/C/KEN/CO/6 (CEDAW, 2007) *Concluding Comments of the Committee on the Elimination of Discrimination against Women: Kenya*, UN Committee on the Elimination of Discrimination Against Women (CEDAW), CEDAW/C/KEN/CO/6, (Aug. 10 2007) available at http://www.refworld.org/docid/46d280ff6.html. *Concluding observations of the Committee on the Elimination of Discrimination against Women: Madagascar*, CEDAW, CEDAW/C/MDG/CO/5, (Nov. 7 2008) *available at* http://www.refworld.org/docid/494ba8cfd.html.
UN Committee on the Elimination of Discrimination against Women: Concluding Comments, Kazakhstan, CEDAW/C/KAZ/CO/2, (Feb. 2 200) *available at* http://www.refworld.org/docid/45f6cddb2.html.

[640]U.N. General Assembly, *Report of the Committee on the Elimination of Discrimination Against Women*, A/60/38(SUPP), CEDAW (2005), http://daccess-dds-ny.un.org/doc/UNDOC/GEN/N05/476/97/PDF/N0547697.pdf?OpenElement. (last visited Oct. 23, 2014) (citing Pars Samoa, 68; Algeria, 165; Croatia, 207; Gabon, 253; Paraguay, 295; Turkey, 385; Lao People's Democratic Republic, 116).

[641]G.A. Res. 65/182 *Open-ended Working Group on Ageing* (Aug. 21–24 2012) *available at* http://social.un.org/ageing-working-group/documents/Chairssummary3rdsessionOEWGfinal.pdf. (last visited Oct. 23, 2014).

[642]*Concluding Observations of the Committee on the Elimination of Discrimination Against Women: Portugal*, CEDAW, CEDAW/C/PRT/CO/7, (Apr. 1 2009) *available at* http://www.refworld.org/docid/494ba8d12.html. (last visited Oct. 23, 2014).

the states to "ensure that they benefit from policies and programs in all areas, in particular access to health, education, social services and decision-making."[643]

Further, the CEDAW Committee has highlighted and recommended that states take measures to increase the literacy levels of older women.[644] It has also noted with concern disproportionate numbers of illiterate older women, highlighted a lack of information in the reports on older women, and requested more information on them.[645] The CEDAW Committee, for example, also expressed its concern about the extremely high rate of illiteracy among older women in Malawi and recommended that the state party make every effort to improve the literacy level.[646]

Also, the CEDAW Committee refers to systemic indirect discrimination against women in employment resulting in older women with lower incomes than older men.[647] The CEDAW Committee expressed particular concern about the unemployment situation of women older than 40 years,[648] encouraging labor legislation and use of temporary special measures.[649]

In 2012, the UNHCHR noted that "older women face obstacles to inheriting housing, land and property. In other regions, long-term older residents may be harassed by landowners or real estate agents to force them to evacuate their housing so that it can be redeveloped or sold at a higher price."[650] In the follow-up to the Second World Assembly on Ageing Report of the Secretary-General it was pointed out that "the higher incidence of poverty among older seniors is explained by the fact that their pensionable income was lower than that of younger groups of workers"… "another important factor is the significantly higher proportion of women among older seniors"… "in 27 out of 30 OECD countries, older women are at greater risk of poverty than older men."[651]

---

[643]Suriname, CEDAW, A/57/38 part II (2002) 82 at 37, 65 and 66. http://daccess-dds-ny.un.org/doc/UNDOC/GEN/N02/614/17/PDF/N0261417.pdf?OpenElement. (last visited Oct. 23, 2014).

[644]Id. at para. 314, Romania, CEDAW Committee.

[645]Portugal, CEDAW, A/57/38(SUPP), 339–347,(2002) http://daccess-dds-ny.un.org/doc/UNDOC/GEN/N02/614/17/PDF/N0261417.pdf?OpenElement. (last visited Oct. 23, 2014).

[646]*UN Committee on the Elimination of Discrimination against Women: Concluding Comments, Malawi*, CEDAW, CEDAW/C/MWI/CO/5, (Feb. 3 2006) *available at* http://www.refworld.org/docid/453778440.html.

[647]*UN Committee on the Elimination of Discrimination against Women: Concluding Comments, Serbia*, CEDAW, CEDAW/C/SCG/CO/1, **31** (June 11 2007) *available at* http://www.refworld.org/docid/468b55902.html.

[648]Croatia, CEDAW, A/60/38(SUPP) **194** (2005) ***available at*** http://daccess-dds-ny.un.org/doc/UNDOC/GEN/N05/476/97/PDF/N0547697.pdf?OpenElement.

[649]Croatia, CEDAW, A/60/38 part I 30 at 194 and 195 (2005) (encouraging temporary special measures in accordance with Article 4, paragraph 1, of the CEDAW).

[650]*Report of the United Nations High Commissioner for Human Rights*, E/20/251, 45, (Apr. 20 2012) http://www.un.org/ga/search/view_doc.asp?symbol=E/2012/51.

[651]U.N. Secretary General, *Follow-up to the Second World Assembly on Ageing*, A/67/188, 26, (July 26, 2012) *available at* http://social.un.org/ageing-working-group/documents/A_67_188_E.pdf.

In a report on Mozambique, the CEDAW Committee expressed concern over cultural beliefs about older women, and accusations of witchcraft against them, which are often correlated with land-grabs, and requested that the government challenge those views.[652] A separate Report of the Secretary-General on the improvement of the situation of women in rural areas concluded that "in many countries, customary laws and practices allow women to access land and enable them to engage in and benefit from agricultural, livestock and forest based production. However, when land becomes a marketable asset, women's access rights to land may be violated or ignored by family and community members, particularly in the case of widowed and divorced women."[653]

In its report on France, the CEDAW Committee expressed that it remains "concerned that specifically targeted measures to combat the marginalization of older immigrant and migrant women are missing."[654] Furthermore, the Committee encourages the state party to develop a consistent overall policy in order to improve the situation of older women and to develop comprehensive programs and policies targeted specifically at the elimination of marginalization of older women."[655]

3. The Rights of Older Persons with Disabilities

In 2012, the Report of the United Nations High Commissioner for Human Rights stated that "while aging cannot be equated to a disability in and of itself, old age can lead to disabilities." Likewise, it mentioned that "the World Report on Disability 2011 confirms the relationship between old age and disability across all regions. Old age and disability are factors that, separated or combined, can make a person vulnerable to a range of human rights violations."[656]

Persons with disabilities form a specific group that is equally varied and complex as older persons. Where the two groups overlap, there are special protections

---

[652]*See UN Committee on the Elimination of Discrimination against Women: Concluding Comments, Mozambique*, CEDAW, CEDAW/C/MOZ/CO/2, **42** (June 11 2007) *available at* http://www.refworld.org/docid/4688b4fc2.html.

[653]*A Gender Perspective on Land Rights: Equal Footing*", Food and Agriculture Organization of the United Nations, A/60/165, **16** (2005) (http://www.un.org/ga/60/third/docslist.htm) (last visited Oct. 23, 2014).

[654]*Concluding comments of the Committee on the Elimination of Discrimination against Women: France*, CEDAW, CEDAW/C/FRA/CO/6, (Apr. 8 2008) *available at* http://www.refworld.org/docid/4885cf98d.html.

[655]Id. at para. 39.

[656]*Report of the United Nations High Commissioner for Human Rights*, U.N. Econ. and Soc. Council, E/2012/51, 58 (Apr. 20 2012) http://www.un.org/ga/search/view_doc.asp?symbol=E/2012/51.

in place for older persons with disabilities. The major treaty in this regard is the Covenant on the Rights of Persons with Disabilities (CRPD).[657] In particular, the principles of equality and non-discrimination are reaffirmed as the state parties pledge to prohibit all discrimination based on disability, and guarantee that persons with disabilities enjoy equal and effective legal protection against discrimination on all grounds. Additionally, article 5 provides that the state parties are to "take all appropriate steps to ensure that reasonable accommodation is provided" to the group.[658]

The CRPD establishes standards pertaining to accessibility to enable persons with disabilities of all ages to participate in society and live independently. States are required to take appropriate measures to ensure that persons with disabilities have access, on an equal basis with others, to the physical environment, transportation, information and communications technologies and systems, as well as other facilities and services open or provided to the public.[659] The paradigm shift towards supported decision-making provided for by article 12 of the CRPD places autonomy and independence of the individual at the core of this notion.[660]

Likewise, article 12 outlines the states' duty to provide "appropriate and effective safeguards" that include "measures relating to the exercise of legal capacity with respect to the rights, will and preferences of the person,… free of conflict of interest and undue influence, … proportional and tailored to the person's circumstances, and… subject to regular review by a competent, independent and impartial authority or judicial body."[661] This is a great advantage for older persons with disabilities; it respects the important right to liberty, as discussed in previous sections of this chapter. The protection of legal capacity for a disabled older person to make decisions for him or herself is foundational to the exercise of all other rights.

Older persons with disabilities deserve specific attention as they are subject to hospitalization and institutionalization on the account of their age, their disability, or both factors.[662] Once admitted, the vast majority of older persons with disabilities spend the remaining part of their lives in institutions. Older persons with disabilities can often find themselves in a situation of powerlessness and vulnerability, in which they may be under the total control of another person.[663]

---

[657] *Convention on the Rights of Persons with Disabilities*, http://www.un.org/disabilities/default.asp?id=259. (last visited Oct. 23, 2014).

[658] Id. at Article 5.

[659] Id. at Article 9.

[660] *Report of the United Nations High Commissioner for Human Rights*, U.N. Econ. and Soc. Council, E/2012/51, 21, (Apr. 20, 2012) http://www.un.org/ga/search/view_doc.asp?symbol=E/2012/51.

[661] Art 12, CRPD.

[662] Id. at para. 28.

[663] Id. at para. 59.

The Madrid Plan of Action called the attention of the international community to the growing challenge of responding to the mental health needs of rapidly aging populations. Mental health problems are a leading cause of disability and reduced quality of life in older age. The timely and accurate diagnosis of such problems and their appropriate treatment can, inter alia, avoid unnecessary institutionalization. Strategies to cope with mental health diseases include medication, psychological support, cognitive training programs, training for caregiver family members, and care provider staff and specific models for in-patient care.[664] Also, these safeguards must be proportional to the effects that such measures have on the person's rights and interests.

As for the rights to property, the state parties have a duty to ensure persons with disabilities are not arbitrarily deprived of their property.[665] Older persons who have accumulated some form of relative wealth who experience disability of any kind are at a magnified risk for abuse in this regard.

The Draft Resolution I "Realizing the Millennium Development Goals for persons with disabilities through the implementation of the World Programme of Action concerning Disabled Persons and the Convention on the Rights of Persons with Disabilities" called for further action and "providing assistance to developing countries, in particular to least developed countries, with special attention to persons with disabilities in vulnerable circumstances."[666]

The UN Secretary General's Report on the review and appraisal of the World Programme of Action Concerning Disabled Persons has stated that institutions, services, and development activities must be focused on where persons live.[667] This is problematic for rural areas, where services for all are often scarce. Older persons often have disabilities, and since many live in rural areas due to the urbanization of middle-aged people, the successful promotion of sustainable livelihoods for persons with disabilities must involve their full and effective participation and increased access to social and economic opportunities.[668]

According to the follow-up to the World Summit for Social Development and the XXIV special session of the General Assembly, "policies and programmes targeting older persons with disabilities are reported in a number of developed

[664]U.N. Secretary General, *Second Review and Appraisal of the Madrid International Plan of Action on Ageing*, E/CN.5/2012/5 78, (Nov. 3 2011) *available at* http://daccess-ddsny.un.org/doc/UNDOC/GEN/N11/579/60/PDF/N1157960.pdf?OpenElement.

[665]Id. at Article 12.

[666]U.N. General Assembly, *Resolution Adopted by the General Assembly on 18 December 2008*, A/63/424 p. 23 (Feb. 11 2009) *available at* http://www.un.org/en/ga/search/view_doc.asp?symbol=A/RES/63/151&Lang=E.

[667]U.N. Secretary- General, *Social Development, Including Questions Relating to the World Situation and to Youth, Ageing, Disable Persons, and the Family*, E/2003/5 P. 5, (2003) *available at* http://www.un.org/disabilities/default.asp?id=40.

[668]Id.

countries. Some of these initiatives come in the wake of the ratification of the Convention on the Rights of Persons with Disabilities. The Australian National Disability Agreement of 2009 includes provisions targeting older persons with disabilities, as did its 2007 revision of the Home and Community Care Act of 1985. Serbia also established a new project in 2009 to improve services at the local level, which includes provisions for older persons with disabilities."[669]

4. The Rights of Older Migrant Workers

The key human rights instrument for this group is the Convention on the Protection of the Rights of All Migrant Workers and Members of Their Families ("Convention on Migrant Workers"), which sets out the scope of application of the instrument in Article 1 that shall cover all migrant workers and members of their families without distinction of any kind such, inter alia, age or "other status."[670] As for the time limit of application, the Convention on Migrant Workers states in Article 1 that it "shall apply during the entire migration process," including also the entire period of stay and remunerated activity in the State of employment as well as return to the State of origin or the State of habitual residence.[671] The 2010 Report on human rights and extreme poverty emphasized that "Migrant workers can be at risk of losing income" and social security benefits in their old age," because "persons who have migrated to work in their adulthood may have contributed to pension systems where they worked, when they retire and return to their countries of origin they may be unable to reap the benefits of their prior contributions."[672]

In the Convention on Migrant Workers state parties undertake the obligation to ensure all the rights provided for in this instrument without distinction of any kind, including also age and 'other status.'[673] Age is expressly prohibited as grounds for discrimination against migrant workers, which is a notable characteristic of this treaty if we consider that almost all of the key human rights instruments fail to identify age as a prohibited ground for discrimination.[674]

---

[669]U.N. Secretary General, *Second Review and Appraisal of the Madrid International Plan of Action on Ageing* 83, (Nov. 3 2011) *available at* http://daccess-dds-ny.un.org/doc/UNDOC/GEN/N11/579/60/PDF/N1157960.pdf?OpenElement.

[670]G.A. Res. 45/158, A/45/49 *International Convention on the Protection of Migrant Workers and their Families*, Article 7, (2009) *available at* http://www1.umn.edu/humanrts/instree/n8icprmw.htm (last visited Oct. 23, 2014).

[671]Id.

[672]Sepulveda Report, 2010. at para. 23.

[673]Id.

[674]UN SG Report on Aging, *supra* note 342, para. 17.

In the Human Rights Council's UPR process, the rights of older migrants in Malta were highlighted. There are concerns that particularly vulnerable persons, including older persons, poor treatment in detention and detention periods for those groups in some cases exceeded twelve months.[675] The report recommended that Malta explore alternative policies concerning the treatment of asylum seekers. Further, the Report positively noted that the reviewed state's key priorities include social welfare and the care for older persons.[676] Saudi Arabia's Working Group Report noted the state's efforts in eliminating discrimination against migrant workers and recommended to extend human rights equally to all migrant workers, regardless of their age and gender.[677] Additionally, it is worth noting that the HR Council's Special Rapporteur of human rights of migrants condemned placement of older persons in migration detention.[678]

F. UN Specialized Agencies

There are a number of specialized international organizations within and affiliated to the United Nations. The International Labour Organization (ILO), the World Health Organization (WHO) and its regional organization, the Pan-American Health Organization (PAHO), and the World Bank all have produced relevant work and guidance relating to older persons' rights.

1. International Labour Organization

The ILO is the international organization responsible for drawing up and overseeing international labor standards. It brings together representatives of governments, employers, and workers to jointly shape policies and programs promoting decent work for all.[679] The ILO conventions most pertinent to the lives of older persons focus on anti-discrimination, training, full employment, and social

[675]UN Human Rights Council, *Report of the Working Group on the Universal Periodic Review - Malta*, A/HRC/12/7; A/HRC/WG.6/5/L.6, 44 (Jun. 4 2009) *available at* http://www.refworld.org/docid/4a9cdb940.html.

[676]Id. at para. 56.

[677]UN Human Rights Council, *Report of the Working Group on the Universal Periodic Review - Saudi Arabia*, A/HRC/WG.6/4/L.9; A/HRC/11/23, 67 (Mar. 4 2009) available at: http://www.refworld.org/docid/49f964f31a.html.

[678]UN Human Rights Council, *Report of the Special Rapporteur on the human rights of migrants, Jorge Bustamante: addendum: mission to the United States of America*, A/HRC/7/12/Add.2, no. 31 (Mar. 5 2008), *available at* http://www.refworld.org/docid/47d647462.html.

[679]*About the International Labor Organization*, Int'l Labor Org., http://www.ilo.org/global/about-the-ilo/lang--en/index.htm. (last visited Oct. 23, 2014).

security benefits. More recently, the ILO has broadened its scope to include issues of the prevention of gender discrimination,[680] and the creation of a "Social Protection Floor" to provide social security for all.[681]

The *Older Workers Recommendation, 1980, (No. 162)*,[682] aims to protect the rights of older workers with equality of treatment and gives practical measures that should be implemented for employment, social security, and retirement. The ILO's understanding of age and older persons is much more far-reaching than this. It notes that "age" is a relatively new ground for discrimination.[683]

States have expanded the grounds on which discrimination is prohibited beyond those expressly mentioned in the *Discrimination (Employment and Occupation) Convention, 1958 (No. 111)*,[684] which allows states to adopt special measures to protect against age discrimination; including the *Workers with Family Responsibilities Convention, 1981 (No. 156)*,[685] that protects workers who take care of older persons and protects older workers who started work late in their lives because they were caring for families; the *Employment Policy Convention, 1964, (No. 122)*[686] that sets standards for employment policy regarding vulnerable groups, including older people; and the *Human Resources Development Convention, 1975, (No. 142)*,[687] which indicates that career guidance and vocational training should not discriminate on the basis of age.

Several ILO conventions and declarations recognize the right to social security, including the Declaration of Philadelphia on the aims and purposes of the ILO to pursue "the extension of social security measures to provide a basic income to all

---

[680] *See*, *Equality at Work: Tackling the Challenges. Global Report Under the Follow-up to the ILO Declaration on Fundamental Principles and Rights at Work* Report of the Director-General, 147–52, (2007) *available at* http://www.ilo.org/wcmsp5/groups/public/---dgreports/---dcomm/---webdev/documents/publication/wcms_082607.pdf. (last visited Oct. 23, 2014).

[681] *Building Social Protection Floors and Comprehensive Social Security Systems*, Int'l Labor Org., *available at* http://www.ilo.org/gimi/gess/ShowTheme.do?tid=1321.

[682] *Older Workers Recommendation*, Section II, paragraph 5(g) (1980) *available at* http://www.ilo.org/ilolex/cgi-lex/convde.pl?R162 (last visited Oct. 23, 2014).

[683] *Report VI, Gender Equality, Report on Equality at work*, Int'l Labor Org. (2009) *available at* http://www.ilo.org/wcmsp5/groups/public/---ed_norm/---relconf/documents/meetingdocument/wcms_105119.pdf. (last visited Oct. 23, 2014).

[684] *Discrimination (Employment and Occupation), 1958 Convention (No. 111*, Int'l Labor Org., Article 5, p.9. *available at* http://www.ilo.org/dyn/normlex/en/f?p=1000:12100:0::NO::P12100_ILO_CODE:C111 (foreseeing the possibility of special measures to meet the particular needs of workers based on age).

[685] *Workers with Family Responsibilities Convention*, Int'l Labor Org., (1981) *available at* http://www.ilo.org/ilolex/cgi-lex/convde.pl?C156 (last visited Oct. 23, 2014).

[686] *Employment Policy Convention*, Int'l Labor Org. (1964) *available at* http://www.ilo.org/ilolex/cgi-lex/convde.pl?C122 (last visited Feb. 19, 2015).

[687] *Human Resources Development Convention*, Int'l Labor Org. (1975) http://www.ilo.org/ilolex/cgi-lex/convde.pl?C142 (last visited Oct. 23, 2014).

in need of such protection and comprehensive medical care."[688] The main standards are established by the *Convention on Minimum Standards of Social Security, 1952, (No. 102)*,[689] which establishes worldwide minimum standards for social security equality and anti-discrimination, and the *Convention on Invalidity, Old-Age and Survivors' Benefits, 1967, (No. 128)*.[690]

The VI report of the International Labour Conference, 98th Session, 2009 referred to other instruments regarding older persons such as:

> the *Employment Policy Convention, 1964 (No. 122)*, which "promotes full, productive and freely chosen employment' for all who are available for and seeking work', 'with a view to stimulating economic growth and development, raising standards of living, and overcoming unemployment and underemployment'. Although it does not specifically mention either women or older workers, the *Human Resources Development Convention, 1975 (No. 142)*, can be used to support vocational training and lifelong learning for all adults. The recently adopted *Human Resources Development Recommendation, 2004 (No. 195)*, and the *Employment Relationship Recommendation, 2006 (No. 198)*, each contain specific references. Used together, they make a strong case for the inclusion of older women in formal work.[691]

ILO's report on *Gender equality at the heart of decent work*[692] highlights discrimination and its detrimental effects on women, in particular. In its conventions and reports, equality and non-discrimination come through as major themes.

2. *World Health Organization*

The World Health Organization's policies for older persons follow the goals and standards set forth in MIPAA[693] for preventive health care,[694] active participation

---

[688]Normlex, Information on Labor Standards, Int'l Labor Org. http://www.ilo.org/ilolex/english/constq.htmojo. (last visited Oct. 23, 2014) (The Declaration of Philadelphia was adopted at the 26th Conference of the International Labour Organisation in 1944 and was added as annex to the ILO's constitution. The Declaration focused on a series of aims and purposes of the ILO).

[689]*Equal Remuneration Convention, (No. 100)*, (1951) *available at* http://www.ilo.org/dyn/normlex/en/f?p=NORMLEXPUB:12100:0::NO:12100:P12100_ILO_CODE:C100; *Discrimination (Employment and Occupation) Convention, 1958 (No. 111)* http://www.ilo.org/dyn/normlex/en/f?p=1000:12100:0::NO::P12100_ILO_CODE:C111 (last visited Oct. 23, 2014) (addressing discriminatory practices and principles of equal pay for work of equal value).

[690]*Invalidity, Old-Age, and Survivors' Benefits Convention (No. 128)*, Int'l Labor Org., (1967) *available at* http://www.ilo.org/ilolex/cgi-lex/convde.pl?C128 (last visited Oct. 23, 2014).

[691]*Gender Equality at the Heart of Decent Work*, Int'l Labor Conference, 386 (2009) *available at* http://www.ilo.org/wcmsp5/groups/public/---ed_norm/---relconf/documents/meetingdocument/wcms_105119.pdf (last visited Feb. 19 2015).

[692]*International Labour Conference, 98th Session, 2009, Gender Equality at the Heart of Decent Work*, Int'l Labor Org., (2009) *available at* http://www.ilo.org/wcmsp5/groups/public/---ed_norm/---relconf/documents/meetingdocument/wcms_105119.pdf. (last visited Oct. 23, 2014).

[693]*A Global Response to Elder Abuse and Neglect*, World Health Organization, (2008) *available at* http://www.who.int/ageing/publications/ELDER_DocAugust08.pdf.

[694]*Strengthening Active and Healthy Ageing*, WHA58.16, (2005) http://apps.who.int/gb/ebwha/pdf_files/WHA58/WHA58_16-en.pdf. (last visited Oct. 23, 2014).

in society,[695] and social security.[696] The WHO's areas of focus in its policy documents and reports, are on primary health care, HIV/AIDS, and elder abuse.[697]

The WHO works toward the elimination and prevention of elder abuse by advocating for strengthened laws explicitly protecting older persons, ending customs of abuse, such as the abandonment of widows and abuse predicated on allegations of witchcraft, and prevention strategies for caregivers.[698] It concluded that the problem of elder abuse cannot be properly solved if the essential needs of older persons–access to health care, food and shelter, and security, – are not met.[699] It emphasizes state-specific health, participation and security as the pillars of healthy aging, and encourages consensus building that addresses the concerns of multiple sectors.[700]

Its approved resolution in the World Health Assembly, *Strengthening Active and Healthy Ageing (2005),*[701] stressed the need for public health policies geared toward aging. Policy documents from the WHO regarding older persons have included reports on violence against older persons,[702] recommending stronger national laws preventing elder abuse, access to essential materials and information, and participation in society. These recommendations tie in closely with the right to be free from violence, the right to information, and the right to participation explored above. Another aspect of the crime of elder abuse is the limitation of primary health care capacity to address it in the medical setting. The WHO has sought to raise awareness and training for primary health care providers to detect and intervene in elder abuse or neglect.[703] This intervention represents a path for which medical and legal professionals could work together to support the rights of older persons to live lives with the highest attainable standard of health, free from violence.

---

[695] *Active Ageing-A Policy Framework*, World Health Org. (2002) *available at* http://whqlibdoc.who.int/hq/2002/WHO_NMH_NPH_02.8.pdf (last visited Feb. 19, 2015).

[696] Id. at 44–46.

[697] International Plan of Action on Ageing: WHO Report on Implementation, World Health Org., (Nov. 2004) *available at* http://apps.who.int/gb/archive/pdf_files/EB115/B115_29-en.pdf.

[698] *World Health Report on Violence and Health - Abuse of the Elderly*, World Health Org., *available at* http://www.who.int/violence_injury_prevention/violence/global_campaign/en/chap5.pdf (last visited Feb. 19, 2015).

[699] Id.

[700] Id at 55.

[701] WHA 58.16: Resolution on Strengthening Active and Healthy Ageing, *available at*: http://www.monitoringris.org/documents/norm_glob/WHA58_16-en.pdf.

[702] *World Health Report on Violence and Health - Abuse of the Elderly*, World Health Org., *available at* http://www.who.int/violence_injury_prevention/violence/global_campaign/en/chap5.pdf (last visited Feb. 19, 2015).

[703] *A Global Response to Elder Abuse and Neglect: Building Primary Health Care (PHC) Capacity to Deal with the Problem Worldwide: Main Report*, World Health Organizaiton, (2008) *available at* http://www.who.int/ageing/publications/ELDER_DocAugust08.pdf (last visited Oct. 23, 2014) (Annex 4: Summaries of Country Reports; again, a bunch of materials of elder abuse in the references).

Health, participation, and security are all highlighted by the WHO as problems for older persons in emergencies.[704] The WHO has sought to build institutional capacity and commitment of governments, the capacity of health care systems, develop expertise, and promote active aging as a strategy to reduce vulnerability and develop resiliency.[705]

Overall, the WHO's approach can be characterized by strengthening of health care delivery systems and increased attention to often overlooked health issues of older persons such as their vulnerability to HIV/AIDS, and vulnerability to abuse.

3. PAHO

PAHO policies consider United Nations Principles for Older Persons (independence, participation, care, self-fulfillment, and dignity) of 1991 when creating public policies on aging and health for the region of Latin America and the Caribbean. In PAHO's Sanitary Conference and the Directing Council, the human rights of older persons are expressly incorporated in policies and decisions. PAHO has focused on the issues of specialized health delivery systems and preparedness for older persons due to their uniquely vulnerable status.

It encourages the adoption of national policies, strategies, plans, and programs that increase older persons' access to health programs and services, including health promotion and disease prevention programs, in order to achieve active aging. It has successfully promoted dialogue to increase support for initiatives targeted at promoting older persons' health such as supporting capacity building for training the human resources for the health needs of older persons.[706] Care and support are important for this growing population, and primary care is one key to their health.[707]

PAHO advocates for the promotion and protection of older persons' basic human rights by adopting legal frameworks and implementation mechanisms in collaboration with the Permanent Council of the Organization of American States ("OAS"). Along with other organizations there, it has examined the possibility of

---

[704] *Active Ageing-A Policy Framework Policy Response*, World Health Organization, (2008) *available at* http://whqlibdoc.who.int/hq/2002/WHO_NMH_NPH_02.8.pdf (last visited Oct. 23, 2014).

[705] David Hutton, *Older People in Emergencies: Considerations for Action and Policy Development*, by, World Health Org., (2008) *available at* http://www.who.int/ageing/publications/Hutton_report_small.pdf (last visited Feb. 19, 2015).

[706] CD49.R15 *Plan of Action on the Health of Older Persons, Including Active and Healthy Aging*.

[707] *Health and Ageing26[th] Pan American Sanitary Conference* CSP26/13 (Eng.) (July 10 2002) *available at* http://www.paho.org/english/gov/csp/csp26-13-e.pdf.

preparing an inter-American convention on the rights of older persons.[708] In its policy documents, health and access to health care are legally protected rights, taking into account social and economic determinants of health.[709] It emphasizes gender equality, and explicitly includes a human rights perspective.[710] It has increasingly advocated for a human rights framework with binding applicability.[711]

PAHO's overarching body is the Sanitary Conference, which decides on the plans and policies of the Directing Council. In reaction to the creation of MIPAA, it urges national plans of action on aging.[712] It further discusses a plan of action on aging, including a collaboration of the PAHO Directing Council with the OAS Permanent Council on the meeting of experts and the potential convention on the rights of older persons.[713] It has called for implementation of public policies and legal frameworks to improve the determinant factors of health and reduce disparities among vulnerable groups, including older people.[714]

PAHO has also looked at how older persons have some special health considerations that are often overlooked. It has focused on a regional response to demographic changes and the increased need to prepare for population aging.[715] The response should include specific, primary care for different life stages.[716] They have a right to research and benefit from research,[717] specialized mental health

---

[708] *142nd Session of the Executive Committee Health of Older Persons, Including Active and Healthy Aging: Regional Strategy*, World Health Org., (June 9 2008) *available at* http://new.paho.org/hq/index.php?option=com_docman&task=cat_view&Itemid=139&gid=1303&orderby=dmdate_published&ascdesc=DESC (Annex Table 1 and 2, practices of nations in Americas).

[709] *Family and Community Health*, World Health Org., CD49/20 (Aug 17 2009) *available at* http://www2.paho.org/hq/dmdocuments/2009/CD49-20-e.pdf.

[710] *Plan of Action on the Health of Older Persons, Including Active and Healthy Aging*, 49th Directing Council, World Health Org., (Sep. 28-Oct. 2 2009) *available at* http://new.paho.org/hq/index.php?option=com_content&task=view&id=1640&Itemid=1425&lang=en.

[711] *Health of Older Persons, including Active and Healthy Aging: Regional Strategy 142nd Session of the Executive Committee*, (Document CE142/14) at para. 106., (June 23–27 2008) *available at* http://www.paho.org/english/gov/ce/ce142index-e.htm.

[712] *Pan American Sanitary Conference Resolution on Health and Ageing*, World Health Org., (Sep. 23–27 2002) *available at* http://www.paho.org/english/gov/csp/csp26.r20-e.pdf.

[713] Id.

[714] *CD43.R11, 43rd Directing Council, Resolution on Health Promotion in the Americas.*

[715] *Plan of Action on the Health of Older Persons, Including Active and Healthy Aging*, 144th Sess. of the Exec. Comm., Document CE144/9, ., (June 22–26 2009) *available at* http://new.paho.org/hq/index.php?option=com_content&task=view&id=1281&Itemid=1258&lang=en.

[716] *Family and Community Health* 144th Sess. of the Exec. Comm., Document CE144/21 (June 22–26 2009) 200 *available at* http://new.paho.org/hq/index.php?option=com_content&task=view&id=1281&Itemid=1258&lang=en.

[717] Policy on Research for Health (Document CD49/10) 46; *Resolution on Regional Program on Bioethics* 42nd Directing Council, CD42.R6 (Sep. 25–29 2000) *available at* http://www.paho.org/english/gov/cd/cd42index.htm.

services,[718] and vaccinations.[719] It has noted that some diseases, like tuberculosis, are more likely to affect older people.[720] It has also noted the need to enhance prevention efforts directed toward particularly affected and vulnerable groups.[721]

PAHO has taken a wide-ranging approach to health. It is one of the only regional health organizations to discuss disability and aging, and how disability can arise from aging.[722] It emphasizes that disability prevention in aging is an important component of health.[723] It has addressed such wide-ranging topics such as older persons being particularly vulnerable to climate change,[724] to chronic disease,[725] to gender equality in sexual and reproductive health policies.[726]

4. World Bank

The Bank's mandate does not include any reference to human rights, but it does inform a broad and comprehensive interpretation of legal empowerment of the

---

[718]Strategy and Plan of Action on Mental Health (Document CD49/11); *Resolution on Mental Health*, World Health Org., CD43/15 http://www.paho.org/english/gov/cd/cd43index-e.htm (last visited Oct. 23, 2014).; CD49.R15, Plan of Action on the Health of Older Persons, Including Active and Healthy Aging.

[719]*Influenza Pandemic: Preparation in the Hemisphere* 44th Directing Council, CD44.R8, (Sep. 22–26 2003) *available at* http://www.paho.org/english/gov/cd/cd44index-e.htm.

[720]*Regional Strategy for the Control of Tuberculosis for 2005–2015* (Documents CE136/17 and CE136/17, Corrig.), 136th Sess. of the Exec. Comm., at 214. (June 20–24 2005) *available at* http://www.paho.org/english/gov/ce/ce136index-e.htm.

[721]*43rd Directing Council, Resolution on Acquired Immunodeficiency Syndrome (AIDS) in the Americas* CD43.R16, World Health Org., *available at* http://iris.paho.org/xmlui/bitstream/handle/123456789/1431/cd43.r16-e.pdf?sequence=1. (last visited Oct. 23, 2014).

[722]*Disability: Prevention and Rehabilitation in the Context of the Right to the Enjoyment of the Highest Attainable Standard of Physical and Mental Health and Other Related Rights* 47th Directing Council, CD47.R1 (Sep. 25–29 2006) *available at* http://www.paho.org/english/gov/cd/cd47index-e.htm.

[723]*Disability: Prevention and Rehabilitation in the Context of the Enjoyment of the Highest Attainable Standard of Health and other Related Rights* (Document CE138/15) 138th Sess. of the Exec. Comm., para. 144. (June 19–23 2006) *available at* http://www.paho.org/english/gov/ce/ce138index-e.htm.

[724]*Roundtable on Climate Change and its Impact on Public Health: A Regional Perspective* (Documents CD48/16, CD48/16, Add. I, Rev. 2, and CD48/16, Add. II) 48th Directing Council, 10829 (Sep-Oct. 3 2008) *available at* http://new.paho.org/hq/index.php?option=com_content&task=view&id=770&Itemid=733&lang=en.

[725]CE138.R1 *Chronic Diseases, including Diet and Physical Activity.*

[726]*Subcommittee on Women, Health, and Development* (Documents CE136/7 and CE136/8) 136th Session of the Executive Committee, 32. (June 20–24 2005) *available at* http://www.paho.org/english/gov/ce/ce136index-e.htm.

poor that encapsulates both poverty reduction and governance initiatives[727] within the legal limits imposed by its Articles of Agreement.[728] Rights-based approaches (RBA) to development are based on the fundamental principles of equality and non-discrimination, and play an increasing role in the Bank's development work, as shareholders pay increasing attention to human rights, and as it becomes a clearer path to improving development.[729]

The World Bank has issued a number of policy papers addressing human rights in development, poverty reduction, and principles of equality and non-discrimination,[730] and analyzing where development activities and human rights are interrelated. The international human rights framework can help inform a broad interpretation of legal empowerment of the poor that encapsulates both poverty reduction and governance initiatives.[731]

In considering the CRPD, the World Bank has published a guide, noting that development is based on the fundamental principles of equality and non-discrimination, and its mandate "does not preclude the integration of disability issues as reflected in the CRPD in support of its operations."[732] These approaches, while not directly implicating older persons, are principled on equality and non-discrimination for the advancement of development, which would include the older people.

---

[727] Id. at p.37.

[728] Ana Palacio, *The Way Forward - Human Rights and the World Bank, by Special Report, Development Outreach*, World Bank Institute at P. 36. ("*Political human rights in particular have traditionally been considered to lie beyond the permitted range of considerations under the Articles of Agreement, which bar decisions based on political considerations or political systems, as well as interference in domestic political affairs of its members. The World Bank's role is a facilitative one, in helping our members realize their human rights obligations. In this sense, human rights would not be the basis for an increase in Bank conditionalities, nor should they be seen as an agenda that could present an obstacle for disbursement or increase the cost of doing business.*").

[729] *See generally*, Katherine Guernsey, Marco Nicoli and Alberto Ninio, *Convention on the Rights of Persons with Disabilities: Its Implementation and Relevance for the World Bank*, The World Bank, (June 2007) http://siteresources.worldbank.org/SOCIALPROTECTION/Resources/SP-Discussion-papers/Disability-DP/0712.pdf.

[730] Klaus Decker, Siobhán McInerney-Lankford and Caroline Sage, *WORKING PAPER: Human Rights and Equitable Development: "Ideals", Issues and Implications* (2006) *available at* http://siteresources.worldbank.org/INTWDR2006/Resources/477383-1118673432908/Human_Rights_and_Equitable_Development_Ideals_Issues_and_Implications.pdf (last visited Oct. 23, 2014).

[731] Ana Palacio, *The Way Forward - Human Rights and the World Bank*, World Bank Institute Special Report *available at* http://siteresources.worldbank.org/EXTSITETOOLS/Resources/PalacioDevtOutreach.pdf (last visited Oct. 23, 2014).

[732] *See* Katherine Guernsey, Marco Nicoli and Alberto Ninio *Convention on the Rights of Persons with Disabilities: Its Implementation and Relevance for the World Bank*, The World Bank (June 2007) *available at* http://siteresources.worldbank.org/SOCIALPROTECTION/Resources/SP-Discussion-papers/Disability-DP/0712.pdf.

a. International Development Association

IDA projects have the primary goal of domestic economic growth to increase the living standards of the population: "The content of country I-PRSP's [Interim Poverty Reduction Strategy Papers] and PRSP's [Poverty Reduction Strategy Papers] has stressed the central importance of achieving strong and sustained economic growth as a *sine qua non* of effective poverty reduction."[733] IDA views improvement in the well-being of disadvantaged groups as ensuring an overall increase in living standards of a country.[734] It pays extra attention to the disadvantaged human and social capital-building,[735] to reduce economic and social vulnerability.[736] This directly affects the well-being of older persons as a group, and as members of other vulnerable groups facing multiple discriminations. The IDA's focus is often on disadvantaged groups to increase the average well-being of their communities.[737]

The IDA has a number of policy publications, which take into account the situations of vulnerable groups, or situations which affect older persons directly. It prioritizes grants for older persons' households[738]; it has also set aside funds targeted vulnerable groups and older persons, specifically[739]; sought to entice elderly persons to use public transit[740]; program fosters local participation in development processes, for those groups typically left out.[741]

According to Progress Report of IDA12 implementation some PRSP's focus "more generally on reducing regional disparities and improving the socio-economic status of women and other disadvantaged groups (such as the elderly or

---

[733]*Progress Report on IDA Implementation*, Int'l Dev. Ass'n, (Feb. 2001) *available at* http://siteresources.worldbank.org/IDA/Resources/Seminar%20PDFs/progress%20report.pdf.

[734]*Nepal: Building Peaceful, Prosperous and Just New Nepal*, Int'l Dev. Ass'n, *available at* http://siteresources.worldbank.org/IDA/Resources/IDA-Nepal.pdf?cid=ISG_E_WBWeeklyUpdate_NL (last visited Oct. 23, 2014).

[735]Id.

[736]Eastern Caribbean States: Reducing Vulnerability and Supporting Growth and Competitiveness, Int'l Dev. Ass'n, *available at* http://siteresources.worldbank.org/IDA/Resources/IDA-OECS.pdf (last visited Oct. 23, 2014).

[737]*Tamil Nadu Empowerment and Poverty Reduction Project*, Int'l Dev. Ass'n, *available at* http://web.worldbank.org/WBSITE/EXTERNAL/EXTABOUTUS/IDA/0,contentMDK:22301573~menuPK:4754051~pagePK:51236175~piPK:437394~theSitePK:73154,00.html (last visited Oct. 23, 2014).

[738]Public Works and Grants Create a Safety Net, Int'l Dev. Ass'n, (Aug. 2009) *available at* http://web.worldbank.org/WBSITE/EXTERNAL/EXTABOUTUS/IDA/0,contentMDK:21395349~menuPK:3266877~pagePK:51236175~piPK:437394~theSitePK:73154,00.html.

[739]*Getting People and Traffic Moving Again in Lagos*, Int'l Dev. Ass'n, s. (last visited Oct. 23, 2014). http://web.worldbank.org/WBSITE/EXTERNAL/EXTABOUTUS/IDA/0,contentMDK:22229765~menuPK:4754051~pagePK:51236175~piPK:437394~theSitePK:73154,00.html.

[740]*Community-Driven Development*, Int'l Dev. Ass'n, *available at* http://siteresources.worldbank.org/IDA/Resources/IDA-CDD.pdf. (last visited Oct. 23, 2014).

[741]International Development Association Report, *available at* http://siteresources.worldbank.org/IDA/Resources/Seminar%20PDFs/progress%20report.pdf (last visited Oct. 23, 2014).

indigenous peoples)."[742] For example, the *Tamil Nadu Engenders Empowerment and Voice*[743] (Tamil Nadu Empowerment and Poverty Reduction Project) include government efforts to focus on enhancing the lives of disadvantaged groups; the *Honduras 2.5 Million Benefit From Social Investment Fund*[744] was designed to develop social assistance programs for disadvantaged groups; the *Access to Technical Education in India*[745] included vulnerable groups in technician education though formal and informal education and training and the *Nepal: Building a Peaceful, Prosperous and Just New Nepal*[746] concentrates on expanding and honing social programs to increase opportunities and wellbeing, especially for the poor and excluded.

While it is difficult to see a unified approach to aging in the policies of the IDA, it is clear that issues affecting the elderly and other vulnerable groups are to be included in its focus.

G. The Human Rights of Older Persons and Millennium Development Goals

The Millennium Development Goals (MDG's) are eight benchmarks to combat important challenges in development.[747] The goals are codified in the United Nations Millennium Declaration.[748] They include such crucial international development issues as: curbing poverty and hunger; providing primary education to all; empowering women; reducing child mortality; improving maternal health; stopping the spread of HIV, AIDS, malaria, and other diseases; ensuring environmental sustainability; and focusing on international cooperation as a means of eradicating poverty

---

[742]*Progress Report on IDA 12 Implementation*, 26 (Feb. 2001) *available at* http://siteresources.worldbank.org/IDA/Resources/Seminar%20PDFs/progress%20report.pdf.

[743] *Tamil Nadu Empowerment and Poverty Reduction Project*, Int'l Dev. Ass'n, *available at* http://web.worldbank.org/WBSITE/EXTERNAL/EXTABOUTUS/IDA/0,,contentMDK:22301573~menuPK:4754051~pagePK:51236175~piPK:437394~theSitePK:73154,00.html (last visited Oct. 23, 2014).

[744]Honduras 2.5 Million Benefit From Social Investment Fund, Int'l Dev. Ass'n, *available at* http://web.worldbank.org/WBSITE/EXTERNAL/EXTABOUTUS/IDA/0,,contentMDK:21940442~pagePK:51236175~piPK:437394~theSitePK:73154,00.html (last visited Oct. 23, 2014).

[745]*Access to Technical Education in India*, Int'l Dev. Ass'n, *available at* http://web.worldbank.org/WBSITE/EXTERNAL/EXTABOUTUS/IDA/0,,contentMDK:21387625~menuPK:3266877~pagePK:51236175~piPK:437394~theSitePK:73154,00.html. (last visited Oct. 23, 2014).

[746]*Nepal: Building a Peaceful, Prosperous and Just New Nepal*,Int'l Dev. Ass'n, *available at* http://siteresources.worldbank.org/IDA/Resources/IDA-Nepal.pdf?cid=ISG_E_WBWeeklyUpdate_NL (last visited Oct. 23, 2014).

[747]*See* The Millennium Assembly of the United Nations, *Millennium Summit*, A/Res/55/2 (Sep. 18 2000) *available at* http://www.un.org/en/ga/search/view_doc.asp?symbol=A/RES/55/2.

[748]Id.

worldwide.[749] Every country in the world, along with national and international development organizations, has agreed to seek to achieve these goals by 2015.[750]

While the focus of the MDG's is laudable, the goals are regrettably silent in their treatment of older persons. The goals have also largely sidestepped the human rights narrative. Below we will discuss the inextricable linkages the MDG's, human rights and older persons, ending with some visions of what MDG's inclusive of older persons might look like.

After a decade of high-level United Nations conferences and summits, world leaders met at the United Nations Headquarters in New York in September 2000 to adopt the United Nations Millennium Declaration.[751] The leaders made a commitment to create a global partnership that would seek to achieve the goals of the Millennium Declaration, which deal chiefly with reducing extreme poverty by 2015.[752] The goals articulated in that meeting became known as the Millennium Development Goals.[753] Two years later, the United Nations Secretary General commissioned the Millennium Project, which aimed to create a plan that nations could follow to achieve the MDG's.[754] Soon thereafter, the United Nations Millennium Campaign was launched. The Campaign sought to motivate all members of the international community, including governments, civil society organizations, and the private sector, to act in achieving the MDG's.[755] In the 2010 MDGs summit,[756] the highlight of the meeting was the adoption of the global action plan titled, "Keeping the Promise: United to Achieve the Millennium Development Goals."[757] Furthermore, focusing on the accelerated improvement of the health of women and children, a number of the summit attendees pledged $40 billion to achieve the goals in these areas by 2015.

The MDG's and human rights overlap in several ways. The essence of the Millennium Declaration is the notion that "men and women have the right to live their lives and raise their children in dignity, free from hunger and from the fear of violence, oppression or injustice."[758] Human rights, which are universal, indivisible,

---

[749]*See* United Nations Development Program, *Human Rights and the Millennium Development Goals: Making the Link, Page 6, available at* http://www.undp.org/content/dam/aplaws/publication/en/publications/environment-energy/www-ee-library/water-governance/human-rights-and-the-millennium-development-goals-making-the-link/Primer-HR-MDGs.pdf. (last visited Oct. 23, 2014).

[750]*See* United Nations, *Background: United Nations Millennium Development Goals,* (2013) http://www.un.org/millenniumgoals/bkgd.shtml. (last visited Oct. 23, 2014).

[751]Id.

[752]Id.

[753]Id.

[754]Id.

[755]Id.

[756]Id.

[757]Id.

[758]*See* United Nations, *United Nations Millennium Declaration*, A/res/55/2, 6. *available at* http://www.un.org/millennium/declaration/ares552e.htm. (last visited Oct. 23, 2014).

and inalienable, refer to the rights possessed by all persons by virtue of their common humanity, to live in freedom and dignity.[759] Though the core of the MDG's is making concrete improvements in human development, both have in common the purposes of promoting human development and welfare, and respecting the dignity of all people.[760] They also share the key elements of participation, empowerment, and national ownership.[761] Further, both the MDG's and human rights are tools with which governments can be held accountable.[762]

Finally, the MDG's and human rights are linked in that they are interdependent and mutually-reinforcing frameworks.[763] In other words, through concrete actions, the MDG's strive to increase the wellbeing of all people, especially in terms of economic and social rights. In many respects, bettering the wellbeing of all persons leads to respect for their human rights.[764] In turn, human rights serve as legal and moral support for efforts to achieve the MDG's.[765]

In addition, a human rights approach to attaining the MDG's is useful in terms of creating informed policy responses, monitoring, and evaluation.[766] Following this approach, practitioners should identify the rights that have been violated in the development process, understand the reasons for discrimination that disallow certain individuals or groups to access certain processes, and identify how to remedy the situation.[767]

Though the MDG's focus on the increased wellbeing of all people, they specifically set forth targets for poverty reduction and enhanced welfare for women and children. In contrast, older persons are not mentioned.[768] As a vulnerable group, due to their susceptibility to discrimination and abuse, as well as their role as caregivers for large populations of women and children, older persons should be included in any future MDG's or similar development target structures.

Further, the lack of mention of older persons is staggering in view of the rapid population aging taking place in the world. Statistics highlight the importance of prioritizing older persons in development strategies; by 2050, almost 2 billion

---

[759]United Nations Development Program, *Human Rights and the Millennium Development Goals: Making the Link*, Page 6, http://www.undp.org/content/dam/aplaws/publication/en/publications/environment-energy/www-ee-library/water-governance/human-rights-and-the-millennium-development-goals-making-the-link/Primer-HR-MDGs.pdf. (last visited Oct. 23, 2014).

[760]Id. at 7.

[761]Id.

[762]Id.

[763]Id.

[764]Id.

[765]Id. at 16.

[766]Id. at 11.

[767]Id.

[768]*See United Nations Millennium Declaration*, United Nations, A/Res/55/2, *available at* http://www.un.org/millennium/declaration/ares552e.htm (last visited Oct. 23, 2014).

persons will be 60 years old or above, and eighty percent of these persons will live in developing countries.[769] Population aging presents a unique set of challenges for countries worldwide, including, among others, a rise in ageism, which denies older persons the rights and opportunities available to other adults, strains on pension and social security systems, increasing demand for healthcare.[770] All of these challenges intersect with the MDG's' focus on lifting persons around the world out of extreme poverty and providing the opportunity for wellbeing. It follows that the wellbeing of older persons will have a strong influence on the attainment of the MDG's. Indeed, the President of the General Assembly, speaking on the United Nations International Day of Older Persons in 2003, echoed this sentiment, stating that "it is difficult to imagine how countries may progress if such a significant proportion of the population is not taken into account in the development process."[771]

The Madrid Plan of Action (MIPAA) clarifies the urgency and importance for increased concern of the elderly in international development instruments. MIPAA resolves to ensure that older persons live in security and dignity. It also recognizes that the specific concerns of older persons must be addressed within the major objectives of the international development agenda.[772] This has not yet been accomplished.

The appraisal of a Post-2015 agenda opens the door for advocacy efforts in support of the inclusion of the rights of older people. Several justifications grounded in human rights law exist for the insertion of a goal or goals dealing specifically with older people. First, older persons face stereotyping and discrimination based on age, which leads to their marginalization. As such, they are denied the dignity and rights established in Article 1 of the Universal Declaration of Human Rights.[773] In addition, older women in particular suffer discrimination based on their age.[774] They are most vulnerable to verbal, sexual, and psychological aggression and to financial abuse, based on compounded societal disadvantages.[775] In spite of all of this, international human rights conventions do not adequately

---

[769] *World Economic and Social Survey 2007: Development in an Ageing World*, United Nations Department of Economic and Social Affairs, *3*, *available at* http://www.un.org/en/development/desa/policy/wess/wess_archive/2007wess.pdf (last visited Oct. 23, 2014).

[770] *Ageing and Life-Course*, World Health Org., *available at* http://www.who.int/ageing/en/ (last visited Oct. 23, 2014).

[771] *See* UN General Assembly, *President of the 58th Session of the Statements*, (Oct. 1 2003) *available at* http://www.un.org/ga/president/58/speeches/031001-2.htm.

[772] *World Economic and Social Survey 2007: Development in an Ageing World*, United Nations Department of Economic and Social Affairs, E/2007/50/Rev. 1, pg 32, *available at* http://www.un.org/en/development/desa/policy/wess/wess_archive/2007wess.pdf (last visited Oct. 23, 2014).

[773] *Preamble, The Universal Declaration of Human Rights*, http://www.un.org/en/documents/udhr/. (last visited Oct. 23, 2014).

[774] Open-Ended Working Group on Ageing, United Nations (Apr. 18–21 2011) *available at* http://social.un.org/ageing-working-group/documents/Al-%20Suwaidi%20OEWG%20on%20Ageing.pdf.

[775] Id.

protect older people, and they remain on the margins of development work around the world.

Without an international convention, states do not have a comprehensive guide to articulate enforceable obligation in order to better protect this often-vulnerable group. This not only creates a gap in protection, but also a lack of accountability and, very importantly, clear focus of the international community on rights of older persons. Although the Convention on the Rights of Persons with Disabilities references the elderly, it does so solely in the context of persons with disabilities.[776] An international convention to protect and enforce human rights for the elderly would create a binding comprehensive framework that defines the rights of older persons, identifies responsibilities of member states, and creates solutions through which the rights of older persons can be adequately protected.[777]

In concert with an international convention that protects the rights of older persons, the universal goals could fill in the gaps in terms of practical, tangible methods of ensuring the elderly of their rights and raising their level of wellbeing. The Post-2015 agenda must take account of all people, especially vulnerable groups, in terms of creation and delivery of the goals.

H. Conclusion

Older persons' rights have been expressly recognized in a few economic, social and cultural rights,[778] but the world's understanding of economic, social and cultural rights has changed dramatically. There is increasing recognition of the interrelatedness of civil and political rights with economic, social and cultural rights. For an example of this relationship, consider the (civil and political) rights to a fair trial and to non-discrimination. These may protect older persons from certain types of forced retirement and slow judicial proceedings regarding (economic and social rights to) health or social benefits. Rights to these benefits can be immediately necessary to protect other rights.[779] Violence and abuse, inhumane treatment and violations of the right to property in social pensions cases are all violations of civil and political rights, though they are no often discussed in the context of what they mean for the lives of older persons in their social and economic environment.

Population change, climate change, and migration to urban areas will all change the social landscape in the coming decades. The negative effects of these global trends on the lives of older persons can be combated by promoting the full enjoyment of their human rights and fundamental freedoms. Aging should be viewed as part of the life course, and not as a condition apart from the rest of

---

[776] Id.

[777] Id.

[778] Diego Rodriguez-Pinzon and Claudia Martin, *supra* note 254.

[779] *See* Id. at 921.

society.[780] Older persons' full participation will result in enhanced quality of life for individuals, stronger societies and economies, and it lays the groundwork for improving the economic and social conditions prevalent in society and the eradication of poverty.[781]

In the public domain, older persons may be denied equity in opportunities available to them and in resources allocated to them. Older persons are often not active participants in society. Barriers precluding the participation of older persons include poverty; poor health; low educational levels; lack of transportation and access to services; negative stereotypes about aging; and overt or subtle age discrimination.[782] Independence, participation, care, self-fulfillment, and dignity are rarely addressed within the legal framework of the UN.[783] Policy structures across different countries have focused on health, nutrition, long-term care, and social security. A more proactive, inclusive and rights-based approach for older persons[784] will improve the outlook for both individuals and society.

[780]Kelly Report, *supra* note 30, at 11.

[781]UN SG Report on Aging, *supra* note 342, paras. 20, 21.

[782]Kelly Report, *supra* note 30, at 6.

[783]*The Rights of Older Persons in Asia*, International Federation on Ageing and HelpAge International (2009), *available at* http://www.globalaging.org/elderrights/world/2009/humanrightsasia.pdf. (last visited Oct. 23, 2014).

[784]Id., at 12.

The Human Rights of Older Persons in the European Institutions: Law and Policy

A. Background

The proportion of older people in Europe is the highest in the world and it is expected to remain as such until 2050.[1] As of 2011, 17, 5 per cent of the population was on average 65 or older in the then 27 members of the European Union.[2] According to United Nations projections, 35 percent of Europe's population is projected to be 60 years or over in 2050, while 27 percent is expected to be 65 years or over at the same time.[3] Moreover, the segment of those 80 years or above will grow at a faster pace than any other group of the older population, almost tripling between 2011 and 2060.[4] At the same time, projections forecast that the working-age population will shrink, whereas the older population will continue to increase, thereby raising the old-age dependency ratio from 28 percent in 2010 to 58 percent

---

[1]United Nations Department of Economic and Social Affairs, Population Division, World Population Ageing (2009), at 13.

[2]European Commission, Eurostat, Population structure and ageing, From Statistics Explained, at 5, available at http://epp.eurostat.ec.europa.eu/statistics_explained/index.php/Population_structure_and_ageing, (last visited Dec. 7, 2014). Croatia joined the European Union on July 1, 2013 increasing the number of state members of the European Union to 28.

[3]United Nations Department of Economics and Social Affairs, Population Division, see *supra* note 1.

[4]European Commission, Eurostat, Population structure and ageing, From Statistics Explained, see *supra note* 2. In 2011, this segment of the older population constituted 4.8 percent, and it is expected to rise to 12 percent in 2060.

C. Martin et al., *Human Rights of Older People*, Ius Gentium: Comparative Perspectives on Law and Justice 45, DOI 10.1007/978-94-017-7185-6_3

in 2060.[5] The labor force participation rate among older Europeans is the lowest in the world.[6]

This data has urgent implications for the way European countries structure policies regarding physical accessibility, access to justice, daily care, health systems and markets for older people.

Europe, broadly understood,[7] is not a homogenous region of countries; its Western states enjoy far greater economic and social protections than those in the East. Illiteracy remains a problem for older people in some Southern European and Eastern European countries, but overall it is quite low.[8] Other measures show distinctions between countries as well. Life expectancy in some Eastern European countries is shorter than in other parts of Europe, where life expectancy at birth is above 70 years old.[9] In Belarus and the Republic of Moldova, for example, life expectancies range between just 66 and 69 years.[10]

The enormous and growing population of older people in Europe has a variety of protections. Europe has one of the most advanced international court systems in the world, and it has a long tradition of human rights. The first international treaty for human rights was adopted by the Council of Europe in 1950.[11] It also has a long history of recognizing that older persons have special needs, as is shown by the adoption of a myriad of social policies, particularly over the past two decades. What the region is lacking are specific legal safeguards tailored to the situation of older persons, protecting their rights. Older people have received comparatively little attention within its human rights system. Programs, plans of action, and economic measures have been explored in greater depth than legal frameworks and

---

[5]The Economist, European Demography, Working-age shift, available at http://www.economist.com/news/finance-and-economics/21570752-growth-will-suffer-workers-dwindle-working-age-shift (last visited Dec. 29, 2014). The old-age-dependency ratio has been defined as the ratio of the number of older people aged 65 and over, compared to the number of people of working age between 15–64 years old. See European Commission, Eurostat, Population structure and ageing, From Statistics Explained, Glossary, available at http://ec.europa.eu/eurostat/statistics-explained/index.php/Glossary:Old-age-dependency_ratio (last visited Dec. 29, 2014).

[6]United Nations Department of Economics and Social Affairs, Population Division, see *supra* note 1, at 39.

[7]For this chapter the notion of Europe comprises the state members of the Council of Europe. Some of these states are also members of the European Union.

[8]*Id.*, at 36.

[9]*Id.*, at 6. Average life expectancy at birth for 2008–10 in the 27 member states of the European Union reached 75.3 years for men and 81.7 years for women. See Health at a Glance: Europe 2012, OECD iLibrary, available at http://www.oecd-ilibrary.org/sites/9789264183896-en/01/01/index.html;jsessionid=mkm2g2m1v5q2.x-oecd-live-02?contentType=&itemId=/content/chapter/9789264183896-4-en&containerItemId=/content/serial/23056088&accessItemIds=/content/book/9789264183896-en&mimeType=text/html (last visited Dec. 7, 2014).

[10]*Id.*

[11]European Convention for the Protection of Human Rights and Fundamental Freedoms, opened for signature Nov. 4, 1950, 213 U.N.T.S. 222, E.T.S. No. 005, commonly referred to as the "*European Convention on Human Rights*," and referred to hereinafter as the "*European Convention*."

rights. In August of 2011, at the United Nations' Working Group on the Human Rights of Older Persons, the European Union (EU) spoke against a human rights treaty for older people, citing existing mechanisms, policies, and programs.[12]

This chapter will provide an overview of the regional human rights framework in Europe. It will then examine some of the regional policies intended to guide national policies. In the main section on specific rights, policy of the EU and Council of Europe (CoE) will be interwoven with legal analysis. For this region in particular, policies speak volumes about the intention of protection of older persons' rights. This further highlights the duality of the situation many older people in Europe find themselves in: excellent policies, with little in the way of recognition of their rights.

B. Regional and Human Rights Framework

Two major political structures exist in the European landscape with the power to shape the protection of human rights: the Council of Europe (CoE) and the European Union (EU). Even if the CoE has been the most relevant regional political organization with a mandate to promote and protect human rights, in recent years the EU has adopted new instruments and policies regarding the protection of fundamental rights which have contributed to further guaranteeing individual rights in Europe. Each has its own governing body, governing human rights documents, and court. Both have an important role to play in the future of ageing in Europe and the protection of older persons' rights.

1. Council of Europe

The CoE is made up of eight core institutions, including: the Committee of Ministers, the Secretary General, the Parliamentary Assembly, the Secretariat, the Commissioner for Human Rights, the Congress of Local and Regional Authorities for Europe, the European Court of Human Rights, and the Conference of International Non-Governmental Organizations. Among these institutions, three of them have decision-making powers: the Committee of Ministers, comprised primarily of the ministers for foreign affairs of the member states; the Parliamentary Assembly, made up of delegates from the national parliaments of each member state; and the Congress of Local and Regional Authorities of Europe, comprised of designated local or regional authorities from the member states.[13]

---

[12]See *Opening Statement of the European Union*, August 1–4, 2011, available at http://social.un.org/ageing-working-group/documents/SDOC3776.pdf (last visited Dec. 29, 2014).

[13]Anne Theodore Briggs, *Waking Sleeping Beauty: The Revised European Social Charter*, 7 Human Rights Brief 2, available at: http://www.wcl.american.edu/hrbrief/07/2walking.cfm (last visited Dec. 7, 2014).

The rights of older persons have been present on the agenda of the Committee of Ministers for some time. In particular, the Committee of Ministers has issued a number of recommendations containing, *inter alia*, guiding principles to be considered when adopting measures regarding older persons, as well as standards regarding older persons with disabilities and elderly migrants.[14] In 2014, the Committee of Ministers adopted a specific *Recommendation to member states on the promotion of the human rights of older persons (Recommendation on the human rights of older persons).*[15] This Recommendation incorporates specific principles on the protection of the human rights of older persons and recommends states to comply with them in national legislation and practice.[16] Even if the principles are non-binding, the *Recommendation on the human rights of older persons* encourages states to examine, within five years, the implementation of the principles in cooperation with the CoE.[17] The CoE Parliamentary Assembly has also adopted its own recommendations and passed resolutions concerning older persons within the context of social policy in Europe.[18] Furthermore, the Commissioner for Human Rights has contributed to the debate on the rights of older persons through background papers and other documents.[19]

---

[14]See, *inter alia, Recommendation R(94)9 of the Committee of Ministers to member states concerning elderly people,* adopted on October 10, 1994*; Recommendation CM/Rec (2009)6* of the Committee of Ministers to member states *on ageing and disability in the 21st century: sustainable frameworks to enable quality of life in an inclusive society, adopted on July 8, 2009, and Recommendation CM/Rec (2011)5 of the Committee of Ministers to member states on reducing the risk of vulnerability of elderly migrants and improving their welfare, adopted on May 25, 2011.*

[15]*Recommendation CM/Rec(2014)2 of the Committee of Ministers to member states on the promotion of the human rights of older persons,* adopted on February 19, 2014, available at http://www.coe.int/t/dghl/standardsetting/hrpolicy/Publications/cmrec_2014_2en.pdf (last visited Dec. 29, 2014).

[16]*Id.*, Preamble of the Recommendation.

[17]*Id.*

[18]See, in this regard *Recommendation 1254 (1994) on the medical and welfare rights of the elderly: ethics and policies*, adopted on November 10, 1994; *Recommendation 1428 (1999) on the future of senior citizens: protection, participation and promotion*, adopted on September 23, 1999; *Recommendation 1619 (2003), on the rights of elderly migrants*, adopted on September 8, 2003; *Recommendation 1796 (2007), on the situation of the elderly persons in Europe* (2007), adopted on May 24, 2007. Also, see *Resolution 1008 (1993), on social policies for elderly persons and their self-reliance*, adopted on September 3, 1993; *Resolution 1502 (2006), on demographic challenges for social cohesion*, adopted on May 29, 2006; and *Resolution 1793 (2011), on promoting active ageing-capitalising on older people's working potential*, adopted on January 28, 2011.

[19]See Viewpoint of the Commissioner for Human Rights of April 8, 2008: "Aged people are too often ignored and denied their full human rights"; and Background paper and conclusions: Seminar on protection of human rights and the special situation of elderly people in retirement homes or institutions, October 21–23, 2001. These documents can be found at Selection of relevant Council of Europe texts concerning the human rights of the elderly, CDDH-AGE (2012)02, February 2, 2012.

The *European Convention on Human Rights* (*European Convention*) has been the primary human rights instrument of the CoE since it was signed in 1950 and entered into force in 1953.[20] Seventeen Additional Protocols have been adopted since then to incorporate new rights or to amend the functioning of the petition mechanism.[21] The *European Convention* is of particular importance within the context of international human rights for several reasons. It was the first comprehensive treaty in the world in this field. It established the first international complaints procedure as well as the first international court for the determination of human rights matters. Further, it remains the most developed judicial system of all the human rights systems. Even if the *European Convention* does not protect older persons specifically, the rights enshrined in that treaty are essential for the protection of fundamental guarantees of all individuals, including those who have reached older age. The *European Convention* protects mostly civil and political rights.

The CoE's European Court of Human Rights (ECtHR), created under the *European Convention*, has existed in its current iteration since 1998 when Protocol 11 came into force providing individuals in the 47 (as of 2014) ratifying states the opportunity to directly apply to the ECtHR for interpretation of human rights standards protected by that treaty.[22] Since its creation in 1959, the ECtHR has issued 17,000 judgments.[23] The ECtHR now plays roles as a constitutional court, and a court above national constitutional courts across Europe. This increasingly broad interpretation of its role could benefit older people, who face discrimination and who are the focus of increasing anxiety over the economic impact of their rights to benefits.

Another relevant instrument within the CoE is the 1996 *Revised Social Charter* (*Revised Social Charter*), which replaced the 1961 *European Social Charter*.[24] In

[20] *European Convention for the Protection of Human Rights and Fundamental Freedoms*, see *supra* note 11.

[21] The last additional Protocol adopted was Protocol 16. It was adopted on October 2, 2013 and has not come yet into force. For more information on the Protocols see http://conventions.coe.int/Treaty/Commun/ListeTraites.asp?MA=3&CM=7&CL=ENG (last visited Dec. 7, 2014).

[22] Protocol 11 eliminated the European Commission on Human Rights and concentrated the protection role in the existing ECtHR, see the text of Protocol 11 at http://conventions.coe.int/Treaty/en/Treaties/Html/155.htm (last visited Dec. 29, 2014).

[23] See Overview 1959–2013, ECHR, Council of Europe (2014), at http://www.echr.coe.int/Documents/Overview_19592013_ENG.pdf (last visited Dec. 29, 2014).

[24] The European Social Charter (CETS No 035) was adopted on October 10, 1961 and came into force on February 2, 1965. It was amended by three subsequent Protocols in 1988, 1991 and 1995. The *Revised Social Charter* (CETS No 163) was adopted on 5 March, 1996 and entered into force on July 1, 1999. The Revised Charter embodies all the rights protected by the 1961 Charter, the 1988 Protocol and the amendments in the monitoring procedure established by the 1995 Protocol. Also, the new instrument incorporated additional rights. Once state parties to the 1961 Charter ratify the 1996 Revised Charter, the new instrument will replace the obligations arising out of the prior treaty. See in general Explanatory Report to the *Revised Social Charter*, available at http://conventions.coe.int/Treaty/Commun/QueVoulezVous.asp?NT=163&CM=8&CL=ENG (last visited Dec. 7, 2014).

that instrument, states agree to adopt positive measures to ensure that elderly persons enjoy human rights as full members of society. In particular, states must provide adequate resources to enable them to live decent lives and play an active part in public, social, and cultural life.[25]

The CoE European Committee of Social Rights (ECSR) is a body tasked with monitoring whether state parties are in conformity in law and in practice with the provisions of the *Revised Social Charter*.[26] It is comprised of independent experts elected by the Committee of Ministers. Upon ratification of this instrument, states must select six rights to be bound by from a list of nine provisions provided in Part II of the treaty.[27] Moreover, they must select an additional number of articles or numbered paragraphs of the same section of the *Revised Social Charter* to be bound by, for a total of no less than sixteen articles or sixty-three numbered paragraphs.[28] With respect to the remaining rights from the Charter that are not selected, states still must pursue implementation even if they are not legally obliged to do so.[29] The ECSR supervises compliance through two mechanisms: (1) a reporting procedure according to which it reviews national reports submitted by state parties and adopts conclusions; and (2) a collective complaint procedure on

[25]Article 23 of the *Revised Social Charter* states: With a view to ensuring the effective exercise of the right of elderly persons to social protection, the Parties undertake to adopt or encourage, either directly or in co-operation with public or private organisations, appropriate measures designed in particular:

- to enable elderly persons to remain full members of society for as long as possible, by means of:
  a. adequate resources enabling them to lead a decent life and play an active part in public, social and cultural life;
  b. provision of information about services and facilities available for elderly persons and their opportunities to make use of them;
- to enable elderly persons to choose their life-style freely and to lead independent lives in their familiar surroundings for as long as they wish and are able, by means of:
  a. provision of housing suited to their needs and their state of health or of adequate support for adapting their housing;
  b. the health care and the services necessitated by their state;
- to guarantee elderly persons living in institutions appropriate support, while respecting their privacy, and participation in decisions concerning living conditions in the institution.

[26]The Committee still holds jurisdiction to monitor compliance with the two prior instruments to the *Revised Social Charter*, the 1961 Social Charter and the 1988 Additional Protocol. See Article 2 of the Rules of the European Committee of Social Rights, available at http://www.coe.int/t/dghl/monitoring/socialcharter/ESCRrules/Rules_rev_en.pdf (last visited Dec. 29, 2014).

[27]See Article A, paragraph b, Part III of the *Revised Social Charter, supra* note 24. The nine provisions enshrined the following: the right to work, the right to organize, to right to bargain collectively, the right of children and young persons to protection, the right to social security, the right to social and medical assistance, the right of the family to social, legal and economic protection, the right of migrant workers and their families to protection and assistance, and the right to equal opportunities and equal treatment in matters of employment and occupation without discrimination on the grounds of sex.

[28]See Article A, paragraph c, Part III of the *Revised Social Charter, supra* note 24.

[29]See Article A, paragraph 1 a, Part III of the *Revised Social Charter*, *supra* note 24.

the basis of which the Committee reviews complaints alleging violations to the applicable treaties and adopts decisions on the matter.[30] The right of elderly persons to social protection is not listed among the rights from which states must select the initial six provisions to be bound by, but it is a right that may still be selected among the final sixteen provisions to which state parties must abide. Out of 33 ratifications as of 2013, 18 states have decided to be bound by Article 23 of the *Revised Social Charter*,[31] which states that "Every elderly person has the right to social protection."[32] So far, the ECSR has had the opportunity to assess compliance with that provision and further define the scope of the rights protecting older persons through the reporting procedure and two collective complaints submitted against Finland and one against Ireland.[33]

2. The European Union

The EU has a long history of promulgating social policy for older persons. The EU's principal executive body, the European Commission, has issued a number of communications relevant to economic and social well-being of older persons.[34] It has highlighted the importance of the "exchange of information and experience and the transfer of knowledge on the contribution of older people."[35] As the EU sought to move to a single-market system, it analyzed the social and economic implications both of an aging population and the consequences for older persons in the region.[36] It has communicated on issues as varied as "early retirement for farmers, free movement, and right of residence …and simplification of the liquidation of pensions."[37]

---

[30]See information on this issue at http://www.coe.int/t/dghl/monitoring/socialcharter/ecsr/ecsrdefault_EN.asp (last visited Dec. 7, 2014).

[31]These states include: Andorra, Bosnia and Herzegovina, Finland, France, Ireland, Italy, Malta, Montenegro, The Netherlands, Norway, Portugal, Serbia, Slovak Republic, Slovenia, Spain, Sweden, Turkey and Ukraine.

[32]See Article 23 of the *Revised Social Charter*, *supra* note 25.

[33]For the collective complaints see ECSR, *The Central Association of Carers in Finland v. Finland*, Complaint No. 70/2011, December 4, 2012, *The Central Association of Carers in Finland v. Finland*, Complaint No. 71/2011, 4 December 2012 and *International Federation of Human Rights Leagues (IFHR) v. Ireland*, Complaint No. 42/2007, June 3, 2008.

[34]Commission of the European Communities, *Proposal for a Council Decision on Community Actions for the Elderly, 1990*, available at http://aei.pitt.edu/3713/1/3713.pdf (last visited Dec. 29, 2014).

[35]See http://www.europarl.europa.eu/comparl/libe/elsj/charter/art25/default_en.htm (last visited Dec. 7, 2014).

[36]Commission of the European Communities, *Proposal for a Council Decision on Community Actions for the Elderly*, 1990, *supra* note 34.

[37]See http://www.europarl.europa.eu/comparl/libe/elsj/charter/art25/default_en.htm (last visited Dec. 7, 2014).

The primary human rights instrument of the European Union is the *Charter of Fundamental Rights of the European Union* (*Charter of Fundamental Rights*). The Charter was adopted as a non-binding instrument[38] and became binding for European Union institutions and Member states within the scope of European Union law in 2009, after the Treaty of Lisbon came into force.[39] It applies exclusively to the states that are members of the EU, and includes civil, political, economic, and social rights.[40] In relation to the protection of older persons, Article 25 of the *Charter of Fundamental Rights* protects the rights of older persons "to lead a life of dignity and independence and to participate in social and cultural life."[41] The explanatory report to the *Charter of Fundamental Rights* indicates that this right must be interpreted in light of Article 23 of the *Revised Social Charter* and the *Community Charter of the Fundamental Social Rights of Workers*,[42] and that participation in social life includes the right to participate in political life.[43]

---

[38]*The Charter of Fundamental Rights of the European Union* was adopted on December 7, 2000. For a full text of the Charter, see http://eur-lex.europa.eu/LexUriServ/LexUriServ.do?uri=OJ:C:2010:083:0389:0403:en:PDF (last visited Dec. 7, 2014).

[39]The Treaty of Lisbon amended the Treaty on European Union and the Treaty establishing the European Community. Article 6 of the Treaty of Lisbon states:
1. The Union recognises the rights, freedoms and principles set out in the *Charter of Fundamental Rights* of the European Union of 7 December 2000, as adapted at Strasbourg, on 12 December 2007, which shall have the same legal value as the Treaties.
The provisions of the Charter shall not extend in any way the competences of the Union as defined in the Treaties.
The rights, freedoms and principles in the Charter shall be interpreted in accordance with the general provisions in Title VII of the Charter governing its interpretation and application and with due regard to the explanations referred to in the Charter, that set out the sources of those provisions.
2. The Union shall accede to the European Convention for the Protection of Human Rights and Fundamental Freedoms. Such accession shall not affect the Union's competences as defined in the Treaties.
3. Fundamental rights, as guaranteed by the European Convention for the Protection of Human Rights and Fundamental Freedoms and as they result from the constitutional traditions common to the Member States, shall constitute general principles of the Union's law. The text of the treaty is available at http://bookshop.europa.eu/is-bin/INTERSHOP.enfinity/WFS/EU-Bookshop-Site/en_GB/-/EUR/ViewPublication-Start?PublicationKey=FXAC07306 (last visited Dec. 7, 2014). For further discussion on this issue see in general Sionaidh Douglas-Scott, *The European Union and Human Rights after the Treaty of Lisbon*, HRLR 11 (2011).

[40]For a full text of the *Charter of Fundamental Rights of the European Union*, see *supra* note 38.

[41]Article 25 of the *Charter of Fundamental Rights of the European Union* states: The Union recognizes and respects the rights of the elderly to lead a life of dignity and independence and to participate in social and cultural life.

[42]Text of the explanations relating to the complete text of the Charter as set out in CHARTE 4487/00 CONVENT 50, at 24, available at http://www.europarl.europa.eu/charter/pdf/04473_en.pdf (last visited Dec. 29, 2014).

[43]*Id.*

The *explanatory report* further indicates that the *Charter of Fundamental Rights* is intended to be precedent for the *Court of Justice of the European Union* (CJEU).[44]

Another provision that is relevant for the protection of the rights of older persons in Europe is Article 19(1) of the *Treaty on the Functioning of the European Union* ("TFEU"), which authorizes the European Council to take action to combat discrimination on the basis of age, among other grounds.[45] This provision was the basis for the European Council to adopt Directive 2000/78/EC[46] which created a framework to address equal treatment in employment and occupation.[47] The CJEU, with power to interpret and apply EU law, has ruled on several cases regarding alleged discrimination on the basis of age based on Directive 2000/78.[48]

As mentioned above, the human rights mechanisms in the region of Europe are quite developed. The system encompasses all human rights, which are interrelated and interdependent. For an organized approach, however, this chapter treats some of the most salient rights of older persons individually and in light of the most traditional classification of civil and political rights and economic, social and cultural rights. Also, given the lack of an international treaty in Europe that embodies the protection of the human rights of older persons, the categories of civil and political rights as well as economic, social, and cultural rights allow for identification of the diverse sources of existing policies, rights enshrined in international treaties and other instruments, as well as international case-law applicable to that group.

C. Civil and Political Rights of Older Persons in Europe

Within the scope of civil and political rights, the following rights protected by the *European Convention* may have particular relevance for the protection of the rights of older persons: Article 2 (right to life); Article 3: (prohibition of torture, inhuman

---

[44] *Id.* Articles 51 and 52.

[45] Article 19(1) of the TFEU specifically states: Without prejudice to the other provisions of the Treaties and within the limits of the powers conferred by them upon the Union, the Council, acting unanimously in accordance with a special legislative procedure and after obtaining the consent of the European Parliament, may take appropriate action to combat discrimination based on sex, racial or ethnic origin, religion or belief, disability, age or sexual orientation. Available at http://eur-lex.europa.eu/LexUriServ/LexUriServ.do?uri=OJ:C:2012:326:FULL:EN:PDF (last visited Dec. 7, 2014).

[46] In actuality, the Directive was adopted on the basis of Article 6 of the Treaty of the European Union, which became Article 19 of the TFEU after the adoption of the Treaty of Lisbon.

[47] Council Directive 2000/78/EC of 27 November 2000 establishing a general framework for equal treatment in employment and occupation, Official Journal L 303, 02/12/2000 P. 0016 – 0022, available at http://eur-lex.europa.eu/LexUriServ/LexUriServ.do?uri=CELEX:32000L0078:en:HTML (last visited Dec. 29, 2014).

[48] See, *inter alia*, Case C-144/04, *Mangold v. Helm*, November 22, 2005; Case C-411, *Félix Palacios de la Villa v. Cortefield Servicios SA.*, February 15, 2007; C-141/11, *Torsten Hörnfeldt v Posten Meddelande AB*, July 5, 2012.

or degrading treatment or punishment); Article 5 (right to liberty and security of person); Article 6 (right to a fair and public hearing in criminal and civil matters); Article 8 (right to respect for private and family life, home and correspondence); Article 10 (freedom of expression); Article 13 (right to an effective remedy); Article 14 (prohibition of discrimination), Article 1 of Protocol 1 (right to peaceful enjoyment of possessions), Article 2 of Protocol 4 (freedom of movement), and Article 1 of Protocol 12 (general prohibition of discrimination).

The *Charter of Fundamental Rights'* first Article enshrines the principle of human dignity stating that this principle is inviolable and must be respected and protected.[49] The right to life (Article 2), the right to integrity of the person (Article 3), the prohibition of torture and inhuman or degrading treatment or punishment (article 4), are interrelated with the right to dignity. The *Charter of Fundamental Rights* also affords protection to fundamental freedoms such as the right to liberty and security (Article 6), respect for private and family life (Article 7), protection of personal data (Article 8), freedom of expression and information (Article 11) and right to property (Article 17). Chapter III embodies the principles of equality before the law (Article 20), non-discrimination (Article 21), equality between men and women (Article 23), and, as stated before, specifically protects the rights of the elderly (Article 25).

1. The Right to Recognition Before the Law

The right to recognition before the law is defined as the right to be a person or subject before the law, and as such, to be able to exercise all the rights and obligations existing within the framework of the legal system.[50] In practice, this right involves enjoying the legal capacity to make binding decisions and have those decisions acknowledged and respected by the law.[51] Even if in principle this right is recognized in full weight to adults, older persons may have this right unduly curtailed based on the perception that due to age, they are unable to take care of their welfare or manage their assets.

The right to recognition before the law is not expressly protected by the *European Convention*.[52] Notwithstanding, the ECtHR has found that the deprivation

[49] *Charter of Fundamental Rights of the European Union,* supra note 38, Article 1.

[50] European Commissioner for Human Rights, *Who Gets to Decide? Right to legal capacity for persons with intellectual and psychosocial disabilities*, CommDH/Issue Paper (2012)2, February 20, 2012.

[51] *Id.*

[52] See Manfred Nowak, *UN Covenant on Civil and Political Rights, CCPR Commentary*, Article 16, Recognition of Legal Personality, Engels Publishers (2005), at 369. According to Nowak, this right was not included in the European Convention because at the time of drafting the Committee of Experts of the Council of Europe considered that this right could be deduced from other rights protected by the treaty.

of legal capacity involves an interference with the right to private life protected by Article 8 of the *European Convention*.[53] For such interference to be in compliance with that treaty, it must be adopted in accordance with existing domestic law, pursue a legitimate aim, and be proportionate to the aim that the measure seeks to achieve.[54] The Court has concluded that the deprivation of legal capacity in proceedings that failed to ensure basic fair trial rights constituted disproportionate measures in violation of Article 8 of the *European Convention*.[55] The cases involved the application of guardianship rules that stripped legal capacity without allowing the affected party to participate in the proceedings or appeal the decision for review by a superior court.[56] Moreover, the guardianship rules failed to provide for a system of periodic review of the measures stripping out legal capacity.[57] Even if these cases related to persons with mental disabilities, the holding of the cases and the reasoning employed by the Court could be analogously applied to older persons facing similar situations.

The right to recognition before the law is ensured in other international human rights treaties such as the International Covenant on Civil and Political Rights (ICCPR) (Article 16), the American Convention on Human Rights (ACHR) (Article 3), and the Convention on the Rights of People with Disability (CRPD) (Article 12). Article 12 of the CRPD has shifted the paradigm regarding the recognition of the right to legal capacity of persons with disabilities from a model of guardianship to a system of supported decision making which emphasizes the autonomy and independence of the individual.[58] Under this new paradigm, the CRPD recognizes that persons with disabilities "enjoy legal capacity on an equal basis with others in all aspects of life."[59] As a result, states must refrain from stripping individuals with disabilities of their legal capacity, but rather must provide them with access to the necessary support required to allow these persons to make decisions that have legal value.[60] In providing that support, states must ensure,

[53]ECtHR, *Case of Shtukaturov v. Russia*, 44009/05, Judgment (Merits and Just Satisfaction), March 27, 2008, 90; *Case of Salontaji-Drobnjak v. Serbia*, 36500/05, January 13, 2009, 144.

[54]*Id*, 85 and 140, respectively.

[55]*Id*., 96 and 145, respectively.

[56]*Id*., 91–5 and 144, respectively.

[57]ECtHR, *Case of Salontaji-Drobnjak v. Serbia*, *supra* note 53, 134.

[58]Office of the High Commissioner for Human Rights, *Normative standards in International human rights law in relation to older persons, Analytical Outcome Paper*, August 2012, at p. 22. Also see generally General Comment No. 1, Article 12: Equal recognition before the law, CRPD/C/GC/1, May 19, 2014.

[59]Article 12(2) of the CRPD.

[60]Article 12(3) of the CRPD and General Comment No. 1, *supra* note 58, at 4.

through adequate safeguards, that persons with disabilities enjoy their legal capacity without being subject of abuse from those providing the support.[61]

While the practice and case-law of the international bodies supervising compliance of the ICCPR and the ACHR may be used as a guidance, the approach followed by Article 12 of the CRPD and the interpretations of the Committee on the Rights of Persons with Disability will be of extreme relevance to defining the scope of the right of older persons to recognition before the law in the European context. Most European countries have ratified this treaty, as well as the EU.[62]

In fact, Article 12 has already served as the model to draft the principles on legal capacity[63] in the *Recommendation on the promotion of the human rights of older persons* adopted by the Committee of Ministers of the Council Europe in 2014.[64] Article III on Autonomy and Participation ensures the rights of older persons to live independently, have self-determination, and have autonomy to handle issues that concern them, particularly those related to property, income, finances, place of residence, health, medical treatment or care and funeral arrangements.[65] These rights would only be subject to limitations in light of the specific situation and taking into account the proportionality of the measures compared to the aims that they intend to achieve.[66] The general principles of respect for human dignity, individual autonomy, and self-determination had already been recognized prior by *Recommendation No. R(94)9* of the Committee of Ministers.[67]

In the same vein, Article III of the *Recommendation on the promotion of the human rights of older persons* also protects the right of older persons to legal

---

[61]Article 12(4) of the CRPD. In particular, this provision states: 4. States Parties shall ensure that all measures that relate to the exercise of legal capacity provide for appropriate and effective safeguards to prevent abuse in accordance with international human rights law. Such safeguards shall ensure that measures relating to the exercise of legal capacity respect the rights, will and preferences of the person, are free of conflict of interest and undue influence, are proportional and tailored to the person's circumstances, apply for the shortest time possible and are subject to regular review by a competent, independent and impartial authority or judicial body. The safeguards shall be proportional to the degree to which such measures affect the person's rights and interests.

[62]Available at http://treaties.un.org/Pages/ViewDetails.aspx?src=TREATY&mtdsg_no=IV-15&chapter=4&lang=en (last visited Dec. 7, 2014).

[63]See *Explanatory Memorandum to the Recommendation on the promotion of the human rights of older persons*, 26, available at http://www.coe.int/t/dghl/standardsetting/hrpolicy/Publications/cmrec_2014_2en.pdf (last visited Dec. 29, 2014).

[64]*Recommendation CM/Rec(2014)2 of the Committee of Ministers to member states on the promotion of the human rights of older persons, supra* note 15.

[65]*Id.*, Article III.9.

[66]*Id.*

[67]*Recommendation R(94)9 of the Committee of Ministers to member states concerning elderly people, supra* note 14, section on rights and duties.

capacity on an equal basis with others.[68] Furthermore, the principles envision a system of supported decision making to exercise legal capacity and require states to adopt appropriate legislation to allow older persons to regulate their personal affairs in case they are unable to express their wishes later in their life due to old age or disability.[69] The issue of supported decision making legal frameworks has also been on the radar of the ESCR. The ESCR has consistently required states to provide information on this issue in their national reports.[70] In general, states had failed to address the matter in their reports.[71] However, some European states have recently taken measures to implement the system of supported decision making in the exercise of legal capacity by reforming their domestic laws.[72] In 2009, the Committee of Ministers issued *Recommendation CM/Rec (2009)*11 adopting principles concerning continuing powers of attorney and advance directives for incapacity and recommended states to introduce domestic legislation implementing these principles.[73]

Even if Article III of the *Recommendation on the human rights of older persons* ensures older persons the right to legal capacity as a principle, this provision also allows for exceptional restrictions of this right "when it may be required for protection purposes."[74] The restrictions must be established by law and be applied on a case-by-case basis.[75] Additionally, states must set forth appropriate and effective safeguards to prevent abuse of older persons in the exercise of the legal capacity, particularly with respect to the appointment of third parties to support decision making.[76] States may draw from Article 12(4) of the CRPD to legislate on the safeguards to prevent abuse. In that light, the *Explanatory Memorandum to the Recommendation on the promotion of the human rights of older persons (Explanatory Memorandum to the Recommendation on the human rights of older persons)* provides that states "should ensure that restriction measures respect the rights, will and preferences of the older person concerned, are free of conflict of

[68]*Recommendation CM/Rec(2014)2 of the Committee of Ministers to member states on the promotion of the human rights of older persons, supra* note 15, Article III.12.

[69]*Id*, Article IIII.13 and 14.

[70]See, *inter alia*, ECSR, Conclusions 2013, France, at 43; Conclusions 2013, Andorra, at 37.

[71]*Id.*

[72]Belgium and Czech Republic are among those states. See in this regard, Article III, Autonomy and participation, section Good practices in the *Recommendation CM/Rec(2014)2 of the Committee of Ministers to member states on the promotion of the human rights of older persons, supra* note 15.

[73]*Recommendation CM/Rec(2009)11* of the Committee of Ministers to member states on principles concerning continuing powers of attorney and advanced directives for incapacity.

[74]*Recommendation on the promotion of the human rights of older persons, supra* note 15, Article III.15.

[75]*Explanatory Memorandum to the Recommendation on the promotion of the human rights of older persons, supra* note 63, 26.

[76]*Recommendation on the promotion of the human rights of older persons, supra* note 15, Article III.15.

interest and undue influence, are proportional and tailored to the person's circumstances, apply for the shortest time possible and are subject to regular review by a competent, independent and impartial authority or judicial body."[77]

Another important issue regarding the exercise of legal capacity relates to the right of older persons to provide free and informed consent with respect to medical care and their placement in residential, institutional, or psychiatric care. Article VI, Care, of the *Recommendation on the promotion of the human rights of older persons* enshrines the principle entitling older persons to receive medical care only with their free and informed consent.[78] This provision also allows for older persons to freely withdraw such consent at any time.[79] This principle is mostly based on the provisions of the *Convention for the Protection of Human Rights and Dignity of the Human Being with regard to the Application of Biology and Medicine*, adopted by the CoE in 1997.[80] Furthermore, the *Charter of Fundamental Rights* also ensures the right to free and informed consent regarding the provision of medical care, as part of the broader right to the integrity of the person.[81]

In the same vein, Article VI, Care states that in principle older persons should only be placed in residential, institutional or psychiatric care upon their free and informed consent.[82] In both cases, however, the *Recommendation on the promotion of the human rights of older persons* envisions the existence of circumstances that may justify exceptions to the principle of free and informed consent.[83] In those cases, effective safeguards must be in place to prevent abuse and guarantee that older persons would not be stripped of their right to consent disproportionally.[84]

Even if the right to recognition before the law is not protected by the *European Convention*, the ECtHR has been willing in some cases to incorporate the protection of this right under the scope of Article 8 of that treaty, ensuring the right to private life. The cases have been decided with respect to persons with mental disabilities, but the reasoning and holding of those judgments could be applied analogously to cases on older persons that might be litigated before that tribunal. Moreover, the principles incorporated in the *Recommendation on the promotion of*

[77] *Explanatory Memorandum to the Recommendation on the promotion of the human rights of older persons, supra* note 63, 26.

[78] *Recommendation CM/Rec(2014)2 of the Committee of Ministers to member states on the promotion of the human rights of older persons, supra* note 15, Article VI.36.

[79] *Id.*

[80] Council of Europe, *Convention for the Protection of Human Rights and Dignity of the Human Being with regard to the Application of Biology and Medicine: Convention on Human Rights and Biomedicine* (ETS no. 164), April 4, 1997.

[81] See Article 3 of the *Charter of Fundamental Rights of the European Union, supra* note 38.

[82] *Recommendation CM/Rec(2014)2 of the Committee of Ministers to member states on the promotion of the human rights of older persons, supra* note 15, Article VI.43.

[83] *Id.*, Article VI.37–39 and 43.

[84] *Id.*

the human rights of older persons, though non-binding, have become a powerful instrument to advocate for changes in national legislation and practice. Ultimately, these principles may also be useful to interpret provisions of the *European Convention* or the *Revised Social Charter* when the ECTHR or the ECSR is faced with complaints regarding the rights of older persons.

2. The Right to Equality and Non-discrimination

The right to non-discrimination under the *European Convention*'s Article 14 guarantees equality in the "[t]he enjoyment of the rights and freedoms set forth in this Convention."[85] Thus, Article 14 is not a free-standing equality clause; as an anti-discrimination provision, it applies only in relation to other substantive rights protected by the *European Convention* and its Protocols. The ECtHR has provided a broad interpretation of the scope of substantive rights for purposes of the application of Article 14.[86] In particular, it has consistently stated that as long as the facts at issue fall within the ambit of one or more of those provisions, and an unjustified difference in treatment exist, there is no need for a violation of the substantive right to be demonstrated for Article 14 to apply.[87] For example, the ECtHR has interpreted the right to respect for private and family life to encompass the protection of issues not enumerated by this provision, inter alia: adoption,[88] names, gender identification, sexual orientation, and sexual life.[89] Moreover, it has followed a similar approach with respect to the right to property under Article 1 of Protocol 1 and has found that issues related to social security and pensions fall under the ambit of that right.[90] Those provisions, in connection to Article 14, were argued in the existing case law developed by the ECtHR with respect to older persons. Moreover, this broad interpretation will be relevant to argue in the future discrimination cases regarding older persons, where other rights protected by the *European*

[85]Article 14 of the *European Convention* states: The enjoyment of the rights and freedoms set forth in this Convention shall be secured without discrimination on any ground such as sex, race, colour, language, religion, political or other opinion, national or social origin, association with a national minority, property, birth or other status.

[86]*Handbook on European non-discrimination law*. European Union Agency for Fundamental Rights and European Court of Human Rights-Council of Europe (2011), at 61–62.

[87]ECtHR, *Case of X and Others v. Austria*, 19010/07, Judgment (Merits and Just Satisfaction), February 19, 2013.

[88]ECtHR, *Gas and Dubois v. France*, 25951/07, Judgment (Merits and Just Satisfaction), March 15, 2012.

[89]ECtHR, *E.B. v. France*, 43546/02, Judgment (Merits and Just Satisfaction), January 22, 2008, 41.

[90]See in this regard, ECtHR, *Gaygusuz v Austria*, 17371/90, Judgment (Merits and Just Satisfaction), September 16, 1996, 41; *Andrejeva v. Latvia*, 55707/00, Judgment (Merits and Just Satisfaction), February 18, 2009, 77–80.

Convention are involved. Issues concerning personal liberty, freedom of expression, and access to justice are good examples of situations where older persons may face unequal treatment.

Article 14 prohibits discrimination "on any ground such as sex, race, colour, language, religion, political or other opinion, national or social origin, association with a national minority, property, birth or other status."[91] Since the list is not exhaustive, the ECtHR has used the notion of "other status" to consider a number of other grounds when analyzing whether a particular difference in treatment constituted discrimination.[92] Age, for example, has been considered a discrimination ground under the notion of "other status" in the case of *Schwizgebel v. Switzerland*, where the ECtHR concluded that Swiss domestic laws imposing a maximum age difference between adopter and adoptee were not unreasonable or disproportionate given the lack of common ground on this issue in the other European countries.[93] Under this precedent, older persons, discriminated against on the basis of age, can claim a violation of Article 14 under the notion of "other status."

Protocol 12 to the *European Convention* creates a free-standing anti-discrimination provision that does not require linkage to another right protected by that Convention to be asserted.[94] Article 1 of the Protocol ensures the enjoyment of any right secured by law[95] without discrimination on account of the same grounds listed in Article 14 of the *European Convention*.[96] In the Explanatory Report to Protocol 12, the CoE explained the reasons for not including age as an explicitly protected ground:

[91]See text of Article 14 of the European Convention, *supra* note 85.

[92]For an extensive discussion on this matter, see Janneke Gerards, *The Discrimination Grounds of Article 14 of the European Convention on Human Rights*, Human Rights Law Review (2013).

[93]ECtHR, *Schwizgebel v. Switzerland*, 25762/07, Judgment (Merits), June 10, 2010.

[94]Protocol 12 to the European Convention for the Protection of Human Rights and Fundamental Freedoms, ETS No. 177, November 4, 2000.

[95]According to the Explanatory Report to the Protocol, the scope of Article 1 goes beyond the protection of rights set forth by law and includes cases where discrimination is claimed: "(i) in the enjoyment of any right specifically granted to an individual under national law; (ii) in the enjoyment of a right which may be inferred from a clear obligation of a public authority under national law, that is, where a public authority is under an obligation under national law to behave in a particular manner; (iii) by a public authority in the exercise of discretionary power (for example, granting certain subsidies); (iv) by any other act or omission by a public authority (for example, the behavior of law enforcement officers when controlling a riot). See *Explanatory Report on Protocol 12 to the Convention for the Protection of Human Rights and Fundamental Freedoms*, Council of Europe, 22, available at http://conventions.coe.int/Treaty/en/Reports/Html/177.htm (last visited Dec. 29, 2014). The ECtHR followed this interpretation in the *Case of Savez Crkava "Rijec Zivota" and Others v. Croatia*, 7798/08, Judgment (Merits and Just Satisfaction), March 9, 2011, 103–105.

[96]Article 1 provides: The enjoyment of any right set forth by law shall be secured without discrimination on any ground such as sex, race, colour, language, religion, political or other opinion, national or social origin, association with a national minority, property, birth or other status.

> The list of non-discrimination grounds in Article 1 is identical to that in Article 14 of the Convention. This solution was considered preferable over others, such as expressly including certain additional non-discrimination grounds (for example, physical or mental disability, sexual orientation or age), not because of a lack of awareness that such grounds have become particularly important in today's societies as compared with the time of drafting of Article 14 of the Convention, but because such an inclusion was considered unnecessary from a legal point of view since the list of non-discrimination grounds is not exhaustive...[97]

The first ECtHR judgment finding a violation of Protocol No. 12 was delivered in 2009.[98] The Court concluded that, notwithstanding the difference in scope of application between Article 14 of the *European Convention* and Article 1 of Protocol 12, the notion of discrimination is identical under both provisions.[99] Therefore, the case law developed by this tribunal under Article 14 will guide the interpretation of Article 1 of Protocol 12.

The ECtHR has defined discrimination as treating persons in similar or analogous situations differently, without an objective and reasonable justification.[100] In other words, the difference in treatment would not be justified "if it does not pursue a legitimate aim or if there is not a reasonable relationship of proportionality between the means employed and the aim sought to be realised."[101] States enjoy an initial margin of appreciation to assess whether the difference in treatment falls within the standards required by the *European Convention*. However, "[t]he scope of the margin of appreciation will vary according to the circumstances, the subject matter and the background."[102] With respect to distinctions based on so called "suspect grounds", including sex, sexual orientation, race and ethnic origin and nationality, the ECtHR has found that to justify a difference in treatment on account of those bases, states must provide "very weighty" reasons.[103] In those cases, the state's margin of appreciation will be narrower and the defense claimed by states will be strictly scrutinized.[104] Only with respect to race or ethnicity, the ECtHR does not accept any justification because "no difference in treatment which is based exclusively or to a decisive extent on a person's ethnic origin is capable of

---

[97] *Explanatory Report on Protocol 12 to the Convention for the Protection of Human Rights and Fundamental Freedoms*, *supra* note 95, 20.

[98] ECtHR, *Sejdić and Finci v. Bosnia and Herzegovina*, 27996/06 and 34836/06, Judgment (Merits and Just Satisfaction), December 22, 2009. The Court also found a violation of Article 1 of Protocol 12 in the *Case of Savez Crkava "Rijec Zivota" and Others v. Croatia*, *supra* note 95.

[99] ECtHR, *Sejdić and Finci v. Bosnia and Herzegovina*, *supra*, 55–6.

[100] ECtHR, *Case of Carson and Others v. United Kingdom*, 42184/05, Judgment (Merits), March 16, 2010, 61.

[101] *Id.*

[102] ECtHr, *Case of Andrle v. The Czech Republic*, 6268/08, Judgment (Merits), June 6, 2011, 49.

[103] European Commission Justice, *The Prohibition of Discrimination under European Human Rights Law: Relevance for the EU non-discrimination directives—an update*, Part I, Sect. 1.1.2., at 15–7 (2011).

[104] ECtHR, *Sejdić and Finci v. Bosnia and Herzegovina*, *supra* note 98, 44.

being objectively justified in a contemporary democratic society built on the principles of pluralism and respect for different cultures."[105]

Age is not considered so far a suspect ground, though it is perceived that due to the influence of the case law of the CJEU, this ground together with disability "are rapidly rising in the hierarchy of prohibited grounds."[106] This assertion, however, may be subject to challenges in cases in which the age distinction occurs in the context of social policy, for example with respect to mandatory retirement, pension rights, or other social security benefits. The ECtHR has consistently stated that a wide margin of appreciation is provided to states in relation to measures that come within the notion of economic or social strategy.[107]

Beyond direct discrimination, the ECtHR has also contemplated cases on indirect discrimination in its case-law. Even if this case-law is incipient and less developed than the standards set by the CJEU on this issue, it is expected that indirect discrimination will take a more central role in the jurisprudence of the ECTHR in the future, given the increasingly close interaction between these two European tribunals.[108] Specifically, in the *Thlimmenos* case, the ECtHR expanded its definition of discrimination under Article 14 to include situations in which "States without and objective and reasonable justification fail to treat differently persons whose situations are significantly different."[109] Moreover, the ECtHR has also characterized as discrimination a general policy or measure that disproportionately affects a particular group, placing that group at a particular disadvantage.[110] In both cases, the policy or measure is an apparently neutral rule or criterion and consequently the discrimination is not the result of intentional action or inaction. Although the discrimination of older persons has not been addressed by existing case law, this group may be disproportionally affected by otherwise neutral measures or policies due to their vulnerability in certain contexts. Thus, the prohibition of indirect discrimination remains essential to ensure the right to equality and non-discrimination of older persons under Article 14 of *European Convention* and Article 1 of Protocol 12.

In addition to the *European Convention*, Article E of the *Revised Social Charter* enshrines a non-discrimination clause that ensures equality in the enjoyment of the

---

[105] *Id.*

[106] European Commission Justice, *The Prohibition of Discrimination under European Human Rights Law. Relevance for the EU non-discrimination directives – an update*, *supra* note 103, at 20.

[107] See, *inter alia*, ECtHR, *Case of Carson and Others v. United Kingdom*, *supra* note 100, 61; *Case of Andrle v. The Czech Republic*, *supra* note 102, 50.

[108] *The Prohibition of Discrimination under European Human Rights Law*, European Commission, Directorate-General for Justice, May 2011, part 1.1.3.

[109] ECtHR, *Case of Thlimmenos v Greece*, 34369/97, Judgment (Merits and Just Satisfaction), April 6, 2000, 44.

[110] ECtHR, *Case of Hugh Jordan v. The United Kingdom*, 24746/94, Judgment (Merits and Just Satisfaction), May 4, 2001, 154; *Case of D.H. and Other v. The Czech Republic*, 57325/00, Judgment (Merits and Just Satisfaction), November 13, 2007, 175.

rights protected by that treaty.[111] Like Article 14 of the *European Convention*, Article E is not an autonomous right, but complements the protection afforded by the substantive provisions of the *Revised Charter*.[112] Article E does not require a violation of the substantive provisions of that treaty, but it can only be applied if the claims at issue fall within the ambit of one or more of those rights.[113] In contrast to the *European Convention*, the *Revised Social Charter* protects economic, social, and cultural rights. Therefore, Article E ensures non-discrimination in the access to social services, health, and housing, among others.

Article E ensures non-discrimination on the basis of the same prohibited grounds listed in Article 14 of the *European Convention* and Article 1 of Protocol 12. Like those provisions, the language of Article E states that discrimination would be prohibited on the basis of any "such grounds as" and, after listing specific grounds, concludes with the phrase "or other status." Consequently, this language implies that the enumeration of grounds on which discrimination is prohibited is not exhaustive.[114] In the same vein, the ECSR has found in several cases that Article E prohibits discrimination on the basis of disability because this ground is covered under the reference to "other status."[115] Even if the ECSR has not yet ruled on a case singling out age as a prohibited ground, it is expected that it will reach the same conclusion as the ECtHR, given the fact that such Committee uses the case law of the ECtHR to guide its interpretations in matters of discrimination. Moreover, the *Revised Social Charter* provides specific protections to older persons under Article 23. The existence of this provision gives strong support to the claim that Article E prohibits discrimination on the basis of age.

In keeping with prevailing human rights case law, the *Appendix to the Revised European Social Charter* provides that only "differential treatment based on an objective and reasonable justification" does not constitute discrimination.[116] The ECSR has followed the two-prong standard developed by the ECtHR to assess whether a differential treatment may be justified. Thus, for a measure to be objective and reasonable it must pursue a legitimate aim and show a relation of proportionality between the means employed and the aims that the measure seeks to

[111] *Revised Social Charter*, *supra* note 24, Article E stating that "the enjoyment of the rights set forth in this Charter shall be secured without discrimination on any ground such as race, colour, sex, language, religion, political or other opinion, national extraction or social origin, health, association with a national minority, birth or other status".

[112] ECSR, *Autisme-Europe v. France*, Complaint 13/2002, November 4, 2003.

[113] ECSR, *Confédération Française Démocratique du Travail (CFDT) v. France*, Complaint 50/2008, September 9, 2009, 3.

[114] This is confirmed by the Explanatory Report to the *Revised Social Charter*, *supra* note 24, 136.

[115] See, *inter alia*, ECSR, *Mental Disability Advocacy Center (MDAC) v. Bulgaria*, Complaint No. 41/2007, June 3, 2008, 49.

[116] *Appendix to the Revised European Social Charter, Part. V, Article E*, *supra* note 24.

achieve.[117] States enjoy a margin of appreciation to make the initial assessment, but it is up to the ECSR to decide whether the distinction is justified.[118] Even if the ECSR does not seem to have embraced the notion of "suspect grounds," it has held, in light of the case law of the ECtHR that distinctions based exclusively or to a large extent on a person's ethnic origin can never be justified in a contemporary democratic society.[119]

Additionally, the ECSR has concluded that Article E of the *Revised Charter* prohibits indirect discrimination. Specifically, it has repeatedly stated that "such discrimination may arise by failing to take due and positive account of all relevant differences or by failing to take adequate steps to ensure that the rights and collective advantages that are open to all are genuinely accessible by and to all."[120] In applying that standard, the ECSR has found that failure to provide sufficient funds in the state's budget to support the education of children and adolescents with autism violated their right to non-discrimination because that measure indirectly disadvantaged that group.[121] Likewise, it concluded that failure to take into account the situation of adults with disabilities to ensure access to public services on an equal footing with others constituted indirect discrimination in violation of Article E of the *Revised Social Charter*.[122]

As stated before, Article 23 of the *Revised Social Charter* recognizes the right of older persons to social protection. Even if the language of the provision does not expressly contemplate the principle of non-discrimination, the ECSR has consistently required states to adopt a legal framework to protect older persons from discrimination on account of age, when monitoring compliance through its reporting procedure.[123] In particular, the ECSR has provided that under Article 23 of the *Revised Social Charter*, states have a duty to combat age discrimination in a range of areas, including in access to goods, facilities, and services.[124] This provision does not cover age discrimination in the area of employment since this protection is afforded under Articles 1, paragraph 2 and 24 of the *Revised Social Charter*.[125]

---

[117]ECSR, *European Roma Rights Centre (ERRC) v. France*, Complaint No. 51/2008, October 19, 2009, 82.

[118]*Id.*

[119]ECSR, *Centre on Housing Rights and Evictions (COHRE) v. Italy, Complaint No.* 58/2009, June 15, 2010, 37–38.

[120]ECSR, *Autism-Europe v. France, supra* note 112, 52; *European Roma Rights Centre (ERRC) v. France, supra* note 117, 83.

[121]ECSR, *European Action of the Disabled (AEH) v. France*, Complaint No. 81/2012, September 11, 2013, 143–145.

[122]ECSR, *International Federation for Human Rights (FIDH) v. Belgium, Complaint No* 75/2011, March 18, 2013, 208–209.

[123]See, ECSR, *Conclusions 2013, Andorra,* January 2014, at 37; *Conclusions 2013, The Netherlands,* January 2014, at 36.

[124]ECSR, *Conclusions 2009, Italy,* January 2010, at 33; *Conclusions 2013, Ireland,* January 2014, at 23.

[125]ECSR, *Conclusions 2013, France,* January 2014, at 43.

Article 19(1) of the *Treaty on the Functioning of the European Union* ("TFEU") (former Article 13 of the *Treaty Establishing the European Community*) authorizes the European Council to take action to combat discrimination on the basis of age, among other grounds.[126] This power was used as the legal basis for the adoption of Directives 2000/43/EC and 2000/78/EC, which expanded the framework to address equal treatment under EU law. While Directive 2000/43/EC implements the principle of equal treatment irrespective of racial or ethnic origin,[127] Directive 2000/78/EC establishes a general framework to ensure equal treatment in employment and occupation.[128] The most important distinction between these two Directives is that Directive 2000/43/EC prohibits discrimination on account of race or ethnic origin not only in employment and occupation, but also with respect to social protection, social advantages, education, and access to supply of goods and services. Even if Directive 2000/78/EC applies only in the context of employment and occupation, its contribution has been to expand protection against discrimination on the basis of age, sexual orientation, disability, religion or belief. In 2008, the European Council proposed extending the scope of this Directive to ensure the principle of equality on the basis of age, among other grounds, to sectors outside the field of employment and occupation, including the supply of goods and services, to no avail.[129] In addition, the *Charter of Fundamental Rights* includes a number of provisions related to equal protection, in particular Article 21 which prohibits discrimination on account of, *inter alia*, age.[130]

With respect to the relationship of these provisions and their scope of application, the CJEU has consistently stated that Directive 2000/78 gives expression to the EU general principle of non-discrimination on grounds of age[131] and the prohibition of non-discrimination enshrined in Article 21 of the *Charter of Fundamental*

---

[126]For a text of the provision, see *supra* note 45.

[127]Council Directive 2000/43/EC of 29 June 2000 implementing the principle of equal treatment between persons irrespective of racial or ethnic origin, Official Journal L 180, 19/07/2000, available at http://eur-lex.europa.eu/LexUriServ/LexUriServ.do?uri=CELEX:32000L0043:en:HTML (last visited Dec. 7, 2014).

[128]Council Directive 2000/78/EC of 27 November 2000 establishing a general framework for equal treatment in employment and occupation, Official Journal L 303, *supra* note 47.

[129]For a background discussion on this issue and the feasibility of adopting such a Directive, see European Parliament, *Implementing the principle of equal treatment between persons, Complementary Impact Assessment of the proposed horizontal Directive on Equal Treatment*, January 2014, available at http://www.europarl.europa.eu/RegData/etudes/etudes/join/2014/514088/IPOL-JOIN_ET%282014%29514088_EN.pdf (last visited Dec. 29, 2014).

[130]Article 21 of the *Charter of Fundamental Rights* states: "Any discrimination based on any ground such as sex, race, colour, ethnic o social origin, genetic features, language, religion, or belief, political or any other opinion, membership of a national minority, property, birth, disability, age or sexual orientation shall be prohibited."

[131]See, *inter alia*, Case C-555/07, *Kücükdeveci v. Swedex GmbH & Co. KG*, January 19, 2010, 21; and Case C-447/09, *Prigge v. Deutsche Lufthansa AG*, September 13, 2011, 38.

Rights.[132] Ultimately, for the general principle of EU law, Directive 2000/78 and Article 21 of the *Charter of Fundamental Rights* to apply, the cases must fall within the scope of European Union law.[133]

Article 2 of Directive 2000/78 provides that the principle of equal treatment means that direct[134] and indirect discrimination[135] are prohibited on the bases of the grounds listed in Article 1, including age. However, Article 6 allows for differences in treatment on account of age when the distinction at issue is objectively and reasonably justified by a legitimate aim, and the means of achieving such aim are appropriate and necessary. This provision specifically lists as legitimate aims some social policy objectives, including those related to employment policy, the labor market, and vocational training.[136] Additionally, Article 2(5) of the Directive authorizes additional restrictions to the principle on equal treatment when there is a need to ensure public order, security and health, the prevention of criminal offences, and the protection of the rights and freedoms of others, in a democratic society.[137] Finally, Article 4(1) of the Directive provides that discrimination on the

[132]Case C-476/11, HK Danmark v. Experian A/S, September 26, 2013, 31–32.

[133]*Kücükdeveci v. Swedex GmbH & Co. KG*, *supra* note 131, 23.

[134]According to Article 2(2) direct discrimination shall exist "where one person is treated less favourably than another is, has been or would be treated in a comparable situation, on any of the grounds referred to in Article 1." Council Directive 2000/78/EC of 27 November 2000 establishing a general framework for equal treatment in employment and occupation, Official Journal L 303, *supra* note 47.

[135]The same provision defines indirect discrimination as a situation "where an apparently neutral provision, criterion or practice would put persons having a particular religion or belief, a particular disability, a particular age, or a particular sexual orientation at a particular disadvantage compared with other persons…" Council Directive 2000/78/EC of 27 November 2000 establishing a general framework for equal treatment in employment and occupation, Official Journal L 303, *supra* note 47.

[136]Article 6(1) also gives some examples where distinctions on the basis of age would be justified, namely: the setting of special conditions on access to employment and vocational training, employment and occupation, including dismissal and remuneration conditions, for young people, older workers and persons with caring responsibilities in order to promote their vocational integration or ensure their protection; the fixing of minimum conditions of age, professional experience or seniority in service for access to employment or to certain advantages linked to employment; and the fixing of a maximum age for recruitment which is based on the training requirements of the post in question or the needs for a reasonable period of employment before retirement. Council Directive 2000/78/EC of 27 November 2000 establishing a general framework for equal treatment in employment and occupation, Official Journal L 303, *supra* note 47.

[137]Article 2(5) provides: 5. This Directive shall be without prejudice to measures laid down by national law which, in a democratic society, are necessary for public security, for the maintenance of public order and the prevention of criminal offences, for the protection of health and for the protection of the rights and freedoms of others. Council Directive 2000/78/EC of 27 November 2000 establishing a general framework for equal treatment in employment and occupation, Official Journal L 303, *supra* note 47.

basis of age may be justified on account of "a genuine and determining occupational requirement."[138]

EU Member states have transposed into national law the Directive's provisions on age-based discrimination, as requested by Article 18 of such Directive. This has generated an increased litigation on the matter not only at the domestic level, but also at the CJEU level, through the preliminary reference procedure provided by Article 267 of the TFEU.[139] The CJEU has approached cases alleging breaches of Directive 2000/78 by establishing first whether the alleged discrimination involves a distinction on the basis of age that falls within the scope of Article 2 of such Directive.[140] Next, the CJEU determines whether the distinction may be justified by assessing whether the aim adduced by the state is legitimate in light of the language of Article 6(1). Lack of precision in domestic legislation regarding the aim pursued by a measure does not invalidate the possibility that it can serve as a justification for an age distinction, as long as the general context of the law enables a court to make an appropriate review of the legitimacy of the aim and establish whether the means used to achieve it are appropriate and necessary. Moreover, the aims listed in the provision are not exhaustive and states have some flexibility to justify restrictive measures on the basis of other valid goals.[141] However, those

---

[138]Article 4(1) states: 1. Notwithstanding Article 2(1) and (2), Member States may provide that a difference of treatment which is based on characteristic related to any of the grounds referred to in Article 1 shall not constitute discrimination where, by reason of the nature of the particular occupational activities concerned or of the context in which they are carried out, such a characteristic constitutes a genuine and determining occupational requirement, provided that the objective is legitimate and the requirement is proportionate. Council Directive 2000/78/EC of 27 November 2000 establishing a general framework for equal treatment in employment and occupation, Official Journal L 303, *supra* note 47.

[139]Article 267 of the TFEU provides: The Court of Justice of the European Union shall have jurisdiction to give preliminary rulings concerning:
(a) the interpretation of the Treaties; (b) the validity and interpretation of acts of the institutions, bodies, offices or agencies of the Union;
Where such a question is raised before any court or tribunal of a Member State, that court or tribunal may, if it considers that a decision on the question is necessary to enable it to give judgment, request the Court to give a ruling thereon.
Where any such question is raised in a case pending before a court or tribunal of a Member State against whose decisions there is no judicial remedy under national law, that court or tribunal shall bring the matter before the Court.
If such a question is raised in a case pending before a court or tribunal of a Member State with regard to a person in custody, the Court of Justice of the European Union shall act with the minimum of delay.
Available at http://eur-lex.europa.eu/LexUriServ/LexUriServ.do?uri=CELEX:12008E267:EN:HTML (last visited Jan. 20, 2014).

[140]See, *inter alia*, *Félix Palacios de la Villa v. Cortefield Servicios SA*, *supra* note 48, 50–51; Case C-45/09, *Gisela Rosenbladt v. Oellerking Gebäudereinigungsges.mbH.*, October 12, 2010, 37; HK Danmark v. Experian A/S, *supra* note 132, 35–36.

[141]Case C-388/07, *Age Concern England v. Secretary of State for Business, Enterprise and Regulatory Reform*, March 5, 2009, 43; Case C-45/09, *Gisela Rosenbladt v. Oellerking Gebäudereinigungsges.mbH.*, *supra* note 140, 40.

goals must have a public interest nature and cannot be based on purely individual reasons.[142] Moreover, they must fall within the category of social policy objectives.[143]

Once the CJEU decides that the measure at issue pursues a legitimate aim, it must assess whether the means used to achieve or implement the aim are appropriate and necessary. The CJEU has acknowledged that states have a broad discretion in choosing the means to implement the goals pursued by the restriction,[144] but they must not defeat the implementation of the principle of non-discrimination on account of age.[145] In *Fuchs and Köhler*, for example, the CJEU concluded that domestic legislation establishing compulsory retirement of prosecutors at the age of 65 was appropriate and necessary because it allowed the recruitment of young lawyers and improve personnel management.[146] In contrast, the CJEU found that the termination of pilot contracts at the age of 60 years on the basis of a collective agreement was not appropriate and necessary because national and international legislation allowed pilots to fly until the age of 65.[147]

The CJEU follows a similar analysis when establishing whether a distinction on the basis of age can be justified in light of the restrictions authorized by Articles 2(5) and 4(1) of Directive 2000/78. In the first case, after ascertaining that the goal of the distinction corresponds to one of the aims listed in Article 2(5), the CJEU must assess whether the means used to achieve the stated goal are necessary in a democratic society. In the *Case of Petersen*, for example, the Court accepted that states may set an age limit to dental practice in order to protect the health of the patients.[148] However, it found that setting the limit at age 68 for dentists practicing under the state panel system was not necessary because dentists in private practice were allowed to practice beyond that age.[149] Consequently, patients could be treated without concern to their health by dentists older than 68 in private practice, but not by dentists providing services under the state panel system, which made the measure inconsistent.[150]

---

[142]Case C-388/07, *Age Concern England v. Secretary of State for Business, Enterprise and Regulatory Reform*, *supra*, 46.

[143]Case C-447/09, *Prigge v. Deutsche Lufthansa AG*, September 13, 2011, *supra* note 131, 81. In this case the ECJ found that air traffic safety did not fall within the legitimate aims envisioned by Article 6(1).

[144]Case C-555/07, *Kücükdeveci v. Swedex GmbH & Co. KG*, *supra* note 131, 21; Joined Cases C-250/09 and C-268/09, *Vasil Ivanov Georgiev v Tehnicheski universited-Sofia, filial Plovdiv*, 50.

[145]Case C-499/08, *Ingeniørfoeningen i Danmark v. Region Syddanmark*, October 12, 2010, 33.

[146]Case C-159/10, *Gerhard Fuchs and Peter Köhler v. Land Hessen*, July 21, 2011, 68.

[147]Case C-447/09, *Prigge v. Deutsche Lufthansa AG*, supra note 131, 64.

[148]Case C-341/08, *Domnica Petersen v. Berufungsausschuss für Zahnärtzte für den Bezirk westfalen-Lippe*, January 12, 2010, 52.

[149]*Id*, 58–62.

[150]*Id.*, 62.

With respect to the application of Article 4(1), states must demonstrate the existence of a characteristic related to age that constitutes a genuine and determining occupational requirement to justify an otherwise prohibited distinction.[151] Next, states must show that the objective pursued by the domestic legislation introducing the distinction is legitimate and that the occupational requirement is proportionate to the aim that it intends to achieve.[152] In the *Case of Wolf*, the CJEU assessed whether the age limit of 30 years for the recruitment of intermediate career post in the fire service constituted discrimination on the basis of age. The CJEU accepted that given the high physical demands required to perform an intermediate post in the fire service, the maximum recruitment age constituted a genuine and determining occupational requirement.[153] Next, it concluded that the goal pursued by the state to ensure the operational capacity and proper functioning of the professional fire service was a legitimate one.[154] Finally, the CJEU found that performing high demanding physical activities involving fire-fighting and rescue duties, which are at the core of the intermediate career in the fire service, related to age because they could only be performed by younger officers. It was estimated that these activities could only be performed until 45 years of age. Therefore, recruiting officials of an older age would not permit them to be assigned to perform physically demanding tasks and would require that they carry out other duties. Consequently, since the management of the professional fire service requires, for the intermediate career, a correlation between the physically demanding posts not suitable for older officials and the less physically demanding posts suitable for those officials, the limit set at 30 years to be recruited for such positions was found proportionate.[155]

Directive 2000/98 and Article 21 of the *Charter of Fundamental Rights* have brought prominence to discrimination on account of age because age appears expressly as one of the prohibited grounds in both documents. Much has been argued on whether such acknowledgment implies that age can be considered as a "suspect ground" under EU law. The fact that age discrimination can be objectively justified under Article 6(1) while discrimination on account of other grounds cannot has been considered to weight against such status.[156] Likewise, a wide margin of discretion provided to states to determine what constitutes a legitimate objective of public policy seems to suggest that the scrutiny the CJEU applies in cases of age

[151]Case C-447/09, *Prigge v. Deutsche Lufthansa AG*, supra note 131, 66.

[152]Case C-229/08, *Colin Wolf v. Stadt Frankfurt am Main*, January 12, 2010, 36.

[153]*Id.*, 40.

[154]*Id*, 39.

[155]*Id.*, 41–44.

[156]Colm O'Cinneide, *Age Discrimination and the European Court of Justice: EU Equality Law Comes to Age*, (2009–10) 2 Revue des Affaires Européennes 253, at 258; *European Commission, The Evolution and Impact of the Case-Law of the Court of Justice of the European Union on Directives 2000/43/EC and 2000/78/EC*, European Network of Legal Experts in the non-discrimination field, November 2012, Part III.

discrimination would be less stringent.[157] Notwithstanding, the CJEU starting with the *Case of Age Concern* has provided that since Article 6(1) constitutes an exception to the principle prohibiting discrimination on account of age it must be strictly interpreted.[158] This level of scrutiny has been consistently applied by that tribunal in subsequent cases.[159] Moreover, the CJEU has applied a high level of scrutiny when interpreting other provisions of Directive 2000/78 which authorize exceptions to the principle of non-discrimination on account of age, including Articles 2(5),[160] 4(1)[161] and Article 6(2).[162]

In conclusion, European human rights treaties and EU Directive 2000/78 provide solid protection to older persons regarding the right to equality and non-discrimination on account of age. However, to fully comprehend the span of such protection, it is important to note certain distinctions regarding the scope of application of the provisions of the human rights treaties and the EU Directive. First, it is worth mentioning that while Article 14 of the *European Convention* applies mostly to civil and political rights, Protocol 12 enshrines a free-standing right to non-discrimination that ensures protection to a broader range of rights, including economic, social, and cultural rights. At the same time, the *European Convention* imposes upon states the obligation to respect and ensure the rights of individuals within their jurisdiction, regardless of their nationality. Finally, even though the protection of non-discrimination on account of age falls within the scope of "other status" under Article 14 of the *European Convention* and 1 of Protocol 12, as such "age" is not considered as a "suspect ground" by the ECtHR for purposes of scrutiny.

On the other hand, Article E of the *Revised Social Charter* offers protection against non-discrimination on account of age in the exercise of the rights protected therein. Moreover, according to the interpretation of the ESCR, under Article 23 of the *Revised Social Charter*, states have a duty to combat age discrimination in the access to goods, facilities and services, but not in the area of employment, which is covered by Article 1(2) and 24 of that treaty. In contrast to the *European Convention*, the *Revised Social Charter* has a more restrictive scope of personal application. According to its Appendix, with few exceptions the rights protected by this treaty apply only to foreigners who are nationals of other state parties to the

157 *Id.*

158 Case C-388/07, *Age Concern England v. Secretary of State for Business, Enterprise and Regulatory Reform*, *supra* note 142, 62.

159 See, *inter alia*, Case C-159/10, *Gerhard Fuchs and Peter Köhler v. Land Hessen*, *supra* note 146, 62, and Case C-499/08, *Ingeniørfoeningen i Danmark v. Region Syddanmark*, *supra* note 145, 33.

160 Case C-341/08, *Domnica Petersen v. Berufungsausschuss für Zahnärtzte für den Bezirk westfalen-Lippe*, *supra* note 148, 60.

161 Case C-447/09, *Prigge v. Deutsche Lufthansa AG*, *supra* note 131, 72.

162 Case C-546/11, *Dansk Jurist- og Økonomforbund v. Indenrigs- og Sundhedsministeriet*, September 23, 2013, 41.

Revised Social Charter and who are either lawful residents or working regularly within the territory of the state party concerned.[163] Like the ECtHR, the case-law and practice of ECSR does not consider age as a "suspect ground" requiring a higher standard of scrutiny.

Directive 2000/78/EC and Article 21 of the *Charter on Fundamental Rights* expressly protect against discrimination on the basis of age. While Directive 2000/78/EC applies only in the context of employment and occupation, it is not clear whether Article 21 will apply beyond that limitation since there is not sufficient practice of the CJEU on this issue. As stated before, so far the CJEU has provided that Directive 2000/78 gives expression to the principle of non-discrimination enshrined in Article 21, but has never applied the last provision by itself.[164] Furthermore, in contrast to the *European Convention*, Directive 2000/78/EC includes some limitations regarding its application to nationals of non-EU states. Even if the language of Article 3 of the Directive poses some doubts as to the application of the Directive to third country nationals and stateless persons, it seems that the general interpretation and practice has been that the prohibition of discrimination also applies to those nationals, with the exception of treatment based on nationality and without prejudice to provisions governing entry and residence.[165] Finally, given that the protection of non-discrimination on the basis of age is expressly recognized in Directive 2000/78 and Article 21 of the *Charter on Fundamental Rights*, distinctions on account of such grounds seemed to have been subject to a high level of scrutiny by the CJEU. Whether this more rigorous approach allows for considering age as a suspect ground is still a subject of debate in legal academia and practice.[166]

[163]Specifically, the Appendix to the *Revised European Social Charter* provides: 1. Without prejudice to Article 12, paragraph 4, and Article 13, paragraph 4, the persons covered by Articles 1–17 and 20–31 include foreigners only in so far as they are nationals of other Parties lawfully resident or working regularly within the territory of the Party concerned, subject to the understanding that these articles are to be interpreted in the light of the provisions of Articles 18 and 19. This interpretation would not prejudice the extension of similar facilities to other persons by any of the Parties. *Appendix to the Revised European Social Charter, supra* note 24.

[164]Case C-476/11, *HK Danmark v. Experian A/S*, *supra* note 132, 31.

[165]European Commission on Human Rights, *Report from the Commission to the European Parliament and the Council*, Joint Report on the application of Council Directive 2000/43/EC of 29 June 2000 implementing the principle of equal treatment between persons irrespective of racial or ethnic origin ('Racial Equality Directive') and of Council Directive 2000/78/EC of 27 November 2000 establishing a general framework for equal treatment in employment and occupation ('Employment Equality Directive').

[166]For a discussion on the matter see, Colm O'Cinneide, *Age Discrimination and the European Court of Justice: EU Equality Law Comes to Age*, supra note 156.

3. The Right to Freedom from Cruel, Inhuman or Degrading Treatment or Punishment

The prohibition of torture, inhuman and degrading treatment or punishment is embedded in Article 3 of the *European Convention*, which states that: "No-one shall be subjected to torture or to inhuman or degrading treatment or punishment." The ECtHR has consistently stated that such prohibition is absolute and does not admit of any limitations, even in situations of emergency[167] or as a result of the victim's behaviour.[168] Moreover, it has provided that the distinction between torture and other forms of ill-treatment lays on the severity of the treatment or punishment experienced by the victim.[169] The ECtHR has indicated that the following aspects must be taken into account when assessing the severity of the treatment or punishment: its duration, its physical or mental effects, and the victim's sex, age, and state of health.[170] The ECtHR has characterized certain treatment as inhuman when "it was premeditated, was applied for hours at a stretch and caused either actual bodily injury or intense physical and mental suffering."[171] On the other hand, "where treatment humiliates or debases an individual, showing a lack of respect for or diminishing his or her human dignity, or arouses feelings of fear, anguish or inferiority capable of breaking an individual's moral and physical resistance, it may be characterised as degrading."[172] It is important to underscore that what is prohibited as inhuman and degrading is treatment that is not connected with some form of legitimate treatment or punishment.[173]

Violations of this provision that are of particular relevance to the rights of older persons derive mostly from violence, abuse and intentional or unintentional neglect.[174] Elder maltreatment has been defined as "a single or repeated act or lack of appropriate action occurring within any relationship where there is an expectation of trust which causes harm or distress to an older person. This type of violence constitutes a violation of human rights and includes physical, sexual, psychological, emotional; financial and material abuse; abandonment; neglect; and

[167]See, *inter alia*, ECtHR, *Case of Selmouni v. France*, 25803/94, Judgment (Merits and Just Satisfaction), July 28, 1999, 95.

[168]ECtHR, *Case of Labita v. Italy*, 26772/95, Judgment (Merits and Just Satisfaction), April 6, 2000, 119.

[169]*Id.* 100.

[170]*Id.*

[171]Id., 120.

[172]ECtHR, *Case of Semikhvostov v. Russia*, 2689/12, Judgment (Merits and Just Satisfaction), February 6, 2014, 70.

[173]ECtHR, *Case of Labita v. Italy*, 26772/95, *supra* note 168, 120.

[174]*Explanatory Memorandum to the Recommendation on the promotion of the human rights of older persons*, *supra* note 63, 27.

serious loss of dignity and respect."[175] Elder abuse or neglect may occur in state institutions or at home. The *Recommendation on the human rights of older persons* provides that states should protect older persons from such abuse and neglect.[176] *The Explanatory Memorandum to the Recommendation on the human rights of older persons* indicates that the protection against that abuse and neglect derives from Article 3 of the *European Convention*.[177]

With respect to abuse or neglect of older persons by state authorities, Article 3 of the *European Convention* has been litigated through the ECtHR mostly in relation to imprisonment. The question of keeping individuals in criminal detention in poor health, old age, or frailty has arisen in several cases. The ECtHR has reiterated that old age per se is not an impediment for imprisonment, but that age together with other factors, including the state of health, may be taken into account when assessing whether the deprivation of liberty may create an exceptional hardship to the person concerned.[178] Moreover, the ECtHR has provided that detention of an older person for a lengthy sentence may raise issues under Article 3 of the *European Convention*.[179]

States have a positive obligation to ensure that detention conditions are compatible with human dignity and that prisoners are not exposed to an exceeding level of hardship or suffering than that arising from the inherent nature of imprisonment.[180] In this vein, lack of appropriate medical attention to an older person who is sick may involve a level of suffering that falls within the notion of ill-treatment prohibited by European human rights standards.[181] Even if there is no duty under the *European Convention* for states to release a prisoner who is sick or transfer him or her to a civil hospital, states have an obligation to guarantee that their health and well-being are adequately protected through the provision of the medical assistance needed.[182] According to the ECtHR, such assessment should be done in light of the following elements: (a) the condition of the prisoner; (b) the quality of the medical care; (c) a determination on whether the prisoner should continue in detention in view of his or her state of health.[183] In the *Farbtuhs* case, the ECtHR found that the age of the complainant, who was 84 at the time of his incarceration, and the fact that he was paraplegic, invalid and suffered from other

---

[175]World Health Organization, Elder abuse, Fact sheet N°357, December 2014, available at http://www.who.int/mediacentre/factsheets/fs357/en/ (last visited Dec. 29, 2014).

[176]*Recommendation CM/Rec(2014)2 of the Committee of Ministers to member states on the promotion of the human rights of older persons, supra* note 15, Article IV.16.

[177]*Explanatory Memorandum to the Recommendation on the promotion of the human rights of older persons, supra* note 63, 29.

[178]ECtHR, *Case of Papon v. France*, 64666/01, Decision of Inadmissibility, July 6, 2001.

[179]ECtHR, *Case of Haidn v. Germany*, 6587/04, Judgment (Merits and Just Satisfaction), January 13, 2011, 106.

[180]ECtHR, *Case of Farbtuhs v. Latvia*, 4672/02, Judgment, June 6, 2005, 51.

[181]*Id.*

[182]*Id.*, 52.

[183]ECtHR, *Case of Enea v. Italy*, 74912/01, Judgment, September 17, 2009, 59.

related illnesses, required that in order for the state to keep him in prison, it had to provide the adequate medical care to avoid exposing him to a humiliating treatment. Since the state was unable to meet the standard of treatment needed, the ECtHR concluded that by not releasing him, the state exposed him to degrading treatment in violation of Article 3 of the *European Convention.*[184] In a later case, *Enea v. Italy*, the ECtHR found that even if the complainant had advanced age, used a wheelchair and underwent two major operations requiring special medical treatment, the state fulfilled its obligation to monitor his health closely and provided appropriate medical treatment.[185] Thus, in that case, it found that the complainant had not endured treatment exceeding the unavoidable level of suffering inherent in detention.[186]

The prison conditions endured by an older person who is sick are also taken into account when assessing whether the overall treatment received in prison attains a level of suffering falling within the definition of ill-treatment provided by Article 3 of the *European Convention.*[187] Lack of intention on the part of state authorities to humiliate or degrade the prisoner does not prevent a finding of ill-treatment, since a violation of the prohibition established in Article 3 of the *European Convention* may occur as a result of a failure to act with due diligence.[188] For example, in *Hénaf v. France*, the ECtHR found that shackling an older person to his hospital bed for security reasons before undergoing an operation amounted to inhuman treatment.[189] The prisoner's age, state of health and the absence of any security risk showed that the measure was disproportionate and unnecessary.[190] Additionally, in two cases regarding prisoners with physical disabilities, the ECtHR found that the treatment received by the complainants amounted to inhuman and degrading treatment as a result of the state's failure to ensure that the premises of the detaining institution allowed them to have independent access to sanitation facilities, the place where food was served, and to be able to move freely without assistance from others.[191] Moreover, in both cases the prison authorities relied on other prisoners' voluntary services to assist the disabled prisoners with their basic needs, thereby creating a situation in which they were exposed to additional abuse and neglect.[192] Even if these cases did not

---

[184]ECtHR, *Case of Farbtuhs v. Latvia*, *supra* note 180, 61.

[185]ECtHR, *Case of Enea v, Italy*, *supra* note 183, 62.

[186]*Id.*, 66–7.

[187]*Id.*, 57.

[188]*ECtHR, Case of Farbtuhs v. Latvia*, *supra* note 180, 58.

[189]ECtHR, *Case of Hénaf v. France*, 65436/01, Judgment, February 27, 2004, 56.

[190]*Id.*

[191]ECtHR, *Case of Grimailovs v. Latvia*, 6087/03, Judgment (Merits and Just Satisfaction), June 25, 2013, 162*; Case of Semikhvostov v. Russia*, *supra* note 172, 86.

[192]ECtHR, *Case of Grimailovs v. Latvia*, 6087/03, Judgment (Merits and Just Satisfaction), *supra*, 160–162; *Case of Case of Semikhvostov v. Russia*, *supra*, 83–4.

involve older prisoners, the standards developed are very relevant to advocate for the protection of the rights of older disabled prisoners within states and in future international complaints.

Beyond imprisonment, the ECtHR has found that Article 3 applies in the context of individuals placed in social care homes or hospitals run by state authorities.[193] In *Stanev v. Bulgaria*, the ECTHR found that the applicant's placement at a social care home amounted to deprivation of liberty and that Article 3 applied in full force as in other situation of detention.[194] Ultimately, it held that the conditions to which the applicant was exposed during seven years amounted to degrading treatment in violation of that provision.[195] Likewise, the ECtHR concluded that Article 3 applied in the context of a hospital where the applicant had received compulsory medical treatment due to a mental illness.[196] However, it held that there was no sufficient evidence that such treatment amounted to a degree of suffering beyond that prohibited by the European standards.[197] Again, although these cases did not relate to older persons, the rulings of the ECtHR on treatment provided in state-run social home care and hospitals are essential for discussing the standard of protection that these institutions must afford to older individuals placed in their care.

With respect to abuse or neglect occurring in private institutions or at home, there is no case yet decided by the ECtHR addressing specific situations where the victims have been older persons. However, there are judgments regarding ill-treatment perpetrated by private individuals either at institutions or at home that could be of relevance to address future cases where the complainants are older individuals. First, the ECtHR has consistently stated that the prohibition of torture and other forms of ill-treatment imposes upon states a positive obligation to ensure that individuals under their jurisdiction are not subject to torture, inhuman or degrading treatment, even when such actions are perpetrated by private actors.[198] This positive obligation, however, must be interpreted in a way that does not place upon states an excessive burden, given that authorities do not always have the power to prevent all forms of violence perpetrated within a community.[199] Consequently, the ECtHR has framed this positive duty as requiring states to provide effective protection to children and other vulnerable groups and take "reasonable steps to

---

[193]ECtHR, *Case of Stanev v. Bulgaria*, 36760/06, Judgment (Merits and Just Satisfaction), January 12, 2012; *Case of Case of Shtukaturov v. Russia*, *supra* note 53.

[194]ECtHR, *Case of Stanev v. Bulgaria*, 36760/06, Judgment (Merits and Just Satisfaction), *supra*, note 193, 186.

[195]*Id.*, 209–12.

[196]ECtHR, *Case of Case of Shtukaturov v. Russia*, supra note 53, 126.

[197]*Id.*, 128–9.

[198]ECtHR, *Case of O'Keeffe v. Ireland*, 35810/09, Judgment (Merits and Just Satisfaction), January 28, 2014, 144.

[199]*Id.*

prevent ill-treatment of which the authorities had or ought to have had knowledge."[200]

The ECtHR found a violation of Article 3 of the *European Convention* on the basis of this positive obligation, for example, in the *O'Keeffe* case regarding ill-treatment suffered by children in non-state managed schools in Ireland. In that case, which referred to the perpetration of sexual assaults in schools mostly run by the Catholic Church, the ECtHR found that the state should have been aware of the risk run by students in those settings, and failed to take any preventive action to protect the integrity of the minors through effective mechanisms of control.[201] The fact that the victims were children in a primary education context required the state to take its duty to prevent ill-treatment to a higher level of scrutiny given the "acute importance" of this obligation.[202]

Moreover, the ECtHR has consistently found that violence perpetrated within the context of the family may amount to ill-treatment prohibited by Article 3 of the *European Convention*.[203] In *Eremia v. the Republic of Moldova*, the ECtHR held that for a state to meet its positive obligation under that provision it had to adopt "a legislative framework aimed at preventing and punishing ill-treatment by private individuals and, on the other hand, when aware of an imminent risk of ill-treatment of an identified individual or when ill-treatment has already occurred, to apply the relevant laws in practice, thus affording protection to the victims and punishing those responsible for ill-treatment."[204] First, in that case the ECtHR acknowledged that the state set up the necessary laws punishing domestic violence.[205] However, the state failed to do enough to prevent the incidents of violence perpetrated against the victim, though the authorities were aware of the risk that she faced.[206] Finally, the state suspended the investigation of prior acts of violence, which in practice shielded the perpetrator from punishment rather than deterring him from committing other acts of violence against the complainant.[207] The vulnerability of the victim in this case was also taken into account when deciding whether or not the state had fulfilled its positive obligation.[208]

Both cases and the standards therein developed could be directly applicable in cases of abuse of older persons either at institutional settings or within the home.

---

[200] *Id.*

[201] *Id.*, 168–9.

[202] *Id.*, 168.

[203] See, *inter alia*, *Case of Opuz v. Turkey*, 33401/02, Judgment (Merits and Just Satisfaction), June 9, 2009, 176.

[204] ECtHR, *Case of Eremia v. the Republic of Moldova*, 3564/11, Judgment (Merits and Just Satisfaction), May, 28, 2013, 56; *Case of T.M. and C.M. v. The Republic of Moldova*, 26608/11, Judgment (Merits and Just Satisfaction), January 1, 2014, 43.

[205] *Id.*, 57.

[206] *Id.*, 60, 66.

[207] *Id.*, 64–66.

[208] *Id.*, 61.

First, older persons fall within the notion of vulnerable group requiring additional efforts by state authorities to ensure that their personal integrity is protected. Second, to guarantee such heightened level of protection, states must devise mechanisms to facilitate that older persons facing abuse have access to authorities to denounce any forms of ill-treatment that they have suffered. To address that issue, the *Recommendation on the human rights of older persons* encourages states to adopt a number of measures, including those geared towards raising awareness among medical staff, care workers, informal workers and other individuals who provide care to older persons to detect abuse in private and public settings; training personnel on the courses of action if such abuse is identified; and encouraging them to report abuse to authorities.[209] Some states have already implemented mechanisms to report the abuse of older persons such as helplines, protocols that can be filled out by providers of services interacting with older persons in residential care, and direct phone lines to police stations for older persons living alone, among others.[210] Also, the Commissioner of Human Rights at the Council of Europe has recommended the adoption of independent complaints and inspection systems to ensure that older persons residing in institutions are not subject to abuse or neglect.[211] Additional recommendations have been made in the Accompanying Guide to the *European Charter of the Rights and Responsibilities of Older Persons in Need of Long-Term Care and Assistance (European Charter of Older Persons in Long-Term Care)*, adopted by a network of older person rights organizations in Europe.[212]

Furthermore, the *Recommendation on the human rights of older persons* provides that states "should take measures to protect persons reporting abuses from any form of retaliation."[213] This Recommendation extends protection to staff or other personnel that might otherwise face dismissal or other consequences when reporting the existence of abuse or neglect to the authorities. The ECtHR addressed a situation of retaliation in *Heinisch v. Germany*, where a nurse at an institution that provided assistance to highly dependent older persons was dismissed from her job, after she complained publicly of the deficiencies in the services provided.[214] The ECtHR concluded that, under certain circumstances,

[209]*Recommendation CM/Rec(2014)2 of the Committee of Ministers to member states on the promotion of the human rights of older persons*, *supra* note 15, Article IV.18.

[210]*Id*, section on Best Practices, at 14–15.

[211]Commissioner for Human Rights, Viewpoint of the Commissioner for Human Rights of 28 April 2008: "*Aged people are too often ignored and denied their full human rights*," CDDH-AGE (2012)02, at 48.

[212]Accompanying Guide, *European Charter of the Rights and Responsibilities of Older Persons in Need of Long-Term Care and Assistance*, 2010, available at http://www.age-platform.eu/age-work/age-projects/health-and-long-term-care/659-daphne?showall=&limitstart= (last visited Dec. 7, 2014).

[213]*Recommendation CM/Rec(2014)2 of the Committee of Ministers to member states on the promotion of the human rights of older persons*, *supra* note 15, Article IV.18.

[214]ECtHR, *Case of Heinisch v. Germany*, 28274/08, Judgment (Merits and Just Satisfaction), July 21, 2011.

whistle-blowers may be protected under the right to freedom of expression when reporting to the public illegal conduct or wrongdoing at their work place.[215] This conclusion is essential to ensure the protection of service providers who report elder abuse; their reporting ensures that the ill-treatment suffered by older persons does not remain hidden.

Finally, states must adopt special legislative frameworks to ensure that incidents of abuse in private settings by non-state actors are properly investigated and punished. The *Recommendation on the human rights of older persons* provides that states should carry out effective investigations into alleged incidents of elder abuse or neglect. Moreover, it also recommends states to afford the victims of abuse an effective remedy and appropriate redress within a reasonable time. Likewise, the *European Charter of Older Persons in Long-Term Care* recognizes a right to redress for older persons in cases of mistreatment, abuse, or neglect.[216] The state obligation to investigate alleged torture or other forms of ill-treatment is also embodied in Article 3 of the *European Convention*, in relation to Article 1 of that instrument.[217] The ECtHR has consistently found that state's failure to investigate acts of domestic violence that can be characterized as ill-treatment constitutes a violation of the duty to prevent torture, cruel, inhuman and degrading treatment under European standards.[218] Moreover, the right to an effective remedy and to adequate redress in cases of ill-treatment is enshrined in Article 13 of the *European Convention*. Under that provision, states must provide a remedy for establishing the responsibility of state and non-state actors for acts or omissions that violate the rights protected by the *European Convention*.[219] Adequate redress in the form of pecuniary or non-pecuniary damages falls within the notion of an effective remedy.[220]

The prohibition of torture and inhuman or degrading treatment or punishment is also established in Article 4 of the *Charter of Fundamental Rights*. Additionally, Article 3 of that instrument protects the right to personal integrity. According to the latter, states must ensure the free and informed consent of persons in the fields of medicine and biology. This provision complements the *Recommendation on the human rights of older persons*, which provides that older persons should receive medical treatment only upon their free and informed consent, and may withdraw such consent at any time.[221] This Recommendation also states that in cases in

---

[215] *Id.*, 63.

[216] See Article 9, *European Charter of the Rights and Responsibilities of Older Persons in Need of Long-Term Care and Assistance*, 2010, available at http://www.age-platform.eu/age-work/age-projects/health-and-long-term-care/659-daphne?showall=&limitstart= (last visited Dec. 7, 2014).

[217] ECtHR, *Case of T.M. and C.M. v. the Republic of Moldova*, *supra* note 204, 38.

[218] *Id*, 46–49.

[219] ECtHR, *Case of O'Keeffe v. Ireland*, *supra* note 198, 176.

[220] *Id.*

[221] *Recommendation CM/Rec(2014)2 of the Committee of Ministers to member states on the promotion of the human rights of older persons*, *supra* note 15, Article VI.36.

which older persons do not have the capacity to consent, the authorization of a representative, authority, or a person or body designated according to the law is required for a medical procedure to proceed.[222] In cases of emergency, when the appropriate consent cannot be obtained, older persons must receive the necessary medical care to ensure the protection of their health. In both cases, either lack of capacity to provide consent or emergency, effective safeguards must be provided to prevent abuse.[223]

4. The Right to Life

The *European Convention* guarantees the right to life as protected under Article 2 of the *European Convention*, which in paragraph 1 states that: "Everyone's right to life shall be protected by law. No one shall be deprived of his life intentionally save in the execution of a sentence of a court following his conviction of a crime for which this penalty is provided by law." The ECtHR has consistently interpreted this provision as imposing both negative and positive obligations upon state parties. The negative obligation requires states not to intentionally and unlawfully deprive individuals of life.[224] The positive obligation imposes upon states the duty "to take appropriate steps to safeguard the lives of those within its jurisdiction."[225] In cases of older persons, the positive duty to protect has been central for the protection of the right to life of this group. As part of the positive duty arising out of Article 2 of the *European Convention*, states must adopt regulations compelling public and private hospitals to implement appropriate measures to ensure the protection of patients' lives while in the care of the institution, as well as effective judicial remedies to determine the cause of patients' death and the liability of those responsible for such deprivation of life. With respect to the first prong, state regulation of nursing homes is essential for the protection of the right to life of older persons living in such institutions.[226] As to judicial remedies, criminal, civil or disciplinary remedies must be established to investigate a death of a patient, depending on whether or not the deprivation of life was intentional or the result of medical or institutional negligence. In cases of negligence, the ECtHR has found that the duty to provide remedies may be satisfied by ensuring civil remedies alone, or in combination with criminal or disciplinary measures, as long as that those actions enable the victims to establish the liability of the medical or other

---

[222] *Id.*

[223] *Id.*

[224] ECtHR, *Case of Osman v. The United Kingdom*, 23452/94, Judgment (Merits and Just Satisfaction), October 28, 1998, 115.

[225] *Id.*

[226] ECtHR, *Case of Dodov v. Bulgaria*, 59548/00, Judgment (Merits and Just Satisfaction), January 1, 2008, 80.

staff responsible for the death of a patient.[227] In *Dodov v. Bulgaria*, for example, the ECtHR examined the case of the disappearance of an older woman with Alzheimer's disease from the medical wing of a state-run nursing home as a result of staff negligence who failed to supervise her closely and left her alone in the backyard.[228] The state did not provide remedies capable of establishing the facts and securing the accountability of those who had placed the patient's life in danger.[229] The ECtHR found that Article 2 was applicable and held that there had been a violation of the right to life by the state's failure to comply with its obligation to provide appropriate criminal, civil or disciplinary remedies.[230]

The ECtHR has also consistently established that, under certain circumstances, the positive obligation under Article 2 of the *European Convention* requires states parties to take preventative operational measures to protect the life of an individual who is at risk.[231] The measures must be adopted in regard to activities involving private or public action. However, for a state to be responsible, "it must be established that the authorities knew or ought to have known at the time of the existence of a real and immediate risk to the life of an identified individual and that they failed to take measures within the scope of their powers which, judged reasonably, might have been expected to avoid that risk."[232] Overall, a positive obligation must not be interpreted as to impose an impossible burden upon the state, particularly taking into account that operational choices are made in light of existing priorities and resources.[233]

In *Watts v. United Kingdom*, the ECtHR found that a badly managed transfer of older residents from a care home involved the application of Article 2 of the *European Convention*, because it could have a negative impact on the life expectancy of the older persons affected, due to their general frailty and resistance to change.[234] However, given the careful planning of the transfer implemented by the responsible institution in this case, which adopted a number of measures to minimize the risk of the older persons affected, including full consideration of the residents and their families' views, transfer together with friends and to a place close by, the ECtHR concluded that the measures adopted by the state to handle the transfer met the positive obligations arising out of Article 2 of the *European Convention*.[235] Even if the ECtHR acknowledged that the measures tailored by the

---

[227] *Id.*, 87.

[228] *Id.*, 6–12.

[229] *Id.*, 89–96.

[230] *Id.*, 97–98.

[231] ECtHR, *Case of Osman v. The United Kingdom*, *supra* note 224, 115.

[232] *Id.*, 116.

[233] *Id.*

[234] ECtHR, *Case of Watts v. the United Kingdom*, 53586/09, Decision of Admissibility, May 4, 2010, 88.

[235] *Id.*, 92.

state did not completely ensure that the older residents' life expectancy would not be affected,[236] they stroke a proportional balance between the rights of the applicants and the operational and resource constraints faced by the state authorities.[237] In coming to this conclusion, the ECtHR took into account that the closure of the care home was decided in light of the need to rationalize existing resources and due to the need of placing older individuals in residential homes that met new standards for space and facilities.[238] Ultimately, the ECtHR found that the life of the older persons living in the institution was not disrupted, and requesting the state to keep the residential home open would have imposed an impossible or disproportionate burden on the state.[239]

Another important issue related to the right to life of older persons involves the discussion on whether or not Article 2 of the *European Convention* protects the right to end life in a dignified manner. In *Pretty v. United Kingdom*, the ECtHR analyzed the claim of a woman who was paralyzed and suffered from a degenerative and incurable illness.[240] She could not commit suicide, an action lawful in the UK, because her paralysis prevented her from taking such steps without assistance. Since assisted suicide was a crime in the UK, she had requested the Director of the Public Prosecutors Office to give an undertaking not to prosecute her husband so that he could help her to end her life in a dignified manner.[241] Upon rejection of her request, she brought her claim before the ECtHR arguing that such denial violated her right to choose whether or not to go on living under Article 2 of the *European Convention*.[242] She claimed that Article 2 encompassed protection of the right to die to avoid suffering and indignity, as a corollary of the right to life.[243] The ECtHR rejected this interpretation and concluded that Article 2 does not protect the right to die.[244] Even if other provisions of the *European Convention* such as Article 11 involve both the right to join and the right not to be forced to join an association, Article 2 does not involve a negative aspect.[245] Thus, issues related to the quality of living or related to the autonomy to decide whether to live or die fall outside of the scope of that provision.[246]

On the other hand, the ECtHR in *Pretty* acknowledged that the right to self-determination to put an end to one's life is protected under Article 8 of the

[236] *Id.*, 86.

[237] *Id.*, 89.

[238] *Id.* 90.

[239] *Id.*

[240] ECtHR, *Case of Pretty v. United Kingdom*, 2346/02, Judgment (Merits), April 29, 2002.

[241] *Id.*, 9–10.

[242] *Id.*, 35.

[243] *Id.*

[244] *Id.*, 39–40.

[245] *Id.*, 39.

[246] *Id.*

European Convention.[247] That provision, which ensures the right to private life, encompasses protection of the notion of personal autonomy in a broad sense.[248] Thus, under Article 8, for example, individuals "may claim to exercise a choice to die by declining to consent to treatment which may have the effect of prolonging [their] life."[249] However, with respect to other forms of ending life, such as through assisted suicide, states have a certain margin of discretion to establish limitations without violating the *European Convention*.[250] For example, in *Pretty*, the ECtHR found that even if the prohibition of assisted suicide constituted an interference with the victim's right to self-determination, it was not disproportionate because it was put in place to protect the right to life of those who were not in condition to take informed decisions regarding acts intended to end life.[251] Moreover, the ECtHR acknowledged that this prohibition was balanced through a system of enforcement and adjudication that took into account the particular circumstances of each case before convicting for assisted suicide.[252]

The ECtHR followed a similar approach in subsequent cases regarding the right to end life in a dignified manner. In *Hass v. Switzerland*, the ECtHR reiterated its interpretation that Article 8 protects the right to self-determination to decide "by what means and at what point his or her life will end, provided he or she is capable of freely reaching a decision on this question and acting in consequence."[253] In *Hass*, the state imposed a restriction on access to sodium pentobarbital upon submission of a medical prescription based on thorough psychiatric assessment. The claimant had argued that access to that substance was essential to ensure his right to commit suicide in a safe and dignified manner and that the existence of such restriction constituted a disproportionate interference with his right.[254] As in *Pretty*, the ECtHR found that the legal restriction imposed on access to sodium pentobarbital was justified to protect the life of those who could not make an informed decision regarding suicide and to avoid abuse.[255] In *Gross v. Switzerland*, the ECtHR initially requested that the state provide clear guidelines on when doctors may issue the medical prescription to obtain sodium pentobarbital, particularly with regard to

[247] *Id.*, 61.

[248] *Id.*

[249] *Id.*, 63.

[250] *Id.*, 74.

[251] *Id.*, 76.

[252] *Id.*

[253] ECtHr, *Case of Haas v. Switzerland*, 31322/07, Judgment (Merits and Just Satisfaction), January 20, 2011, 51.

[254] *Id.*, 32.

[255] *Id.*, 58.

cases where the persons wishing to end their life did not suffer a terminal illness.[256] However, this case was later rejected due to an abuse of the right of petition.[257]

In sum, although with the exception of *Gross*, the claimants in the cases above were not older persons; issues regarding the end of life at older age due to frailty or illness have become more prominent in recent years. As admitted by the ECtHR in its cases, the domestic laws of the European states have not reached consensus as to the way to regulate this matter. As a result, issues of this nature and complexity will continue to be decided by the ECtHR in highly contextualized cases without affording clear and unique standards to older persons who may be willing to exercise their right to self-determination to end life in a dignified manner. At least, the ECtHR has acknowledged that the right to decline to consent to treatment, which may have the effect of prolonging life, is protected under Article 8 of the *European Convention*.

5. The Right to Liberty and Security

The right to liberty is protected by Article 5 of the *European Convention*, which states in paragraph 1 that: "Everyone has the right to liberty and security of person." Moreover, no one will be deprived of his or her liberty except in certain specific circumstances and provided that the detention is carried out in accordance with procedures established by law. The ECtHR has interpreted Article 5(1) as containing an exhaustive list of permissible grounds to justify deprivation of liberty; consequently, a measure depriving somebody of liberty would not be lawful unless that it falls within one of those grounds.[258] For Article 5(1) to apply, the first element that needs to be established is whether a restriction of liberty can be characterized as deprivation of liberty. This provision does not apply to other restrictions to freedom of movement, which is protected and regulated under Article 2 of Protocol 4 to the *European Convention*.

For older persons, admission procedures in hospitals or institutional care settings may risk violating their liberty and security when presenting themselves for treatment. The necessity of informed consent and shared decision making, even in situations of diminished capacity is integral to participation and realization of the right to liberty. In *H.M. v. Switzerland*, an older woman was placed by the state authorities in a nursing home on account of neglect.[259] Even though initially the placement was carried out against her will, later on she admitted in a hearing that she had no reason to be unhappy with the nursing home. The state authorities confirmed that since the applicant had agreed to stay in the nursing home, they had

[256] ECtHR, *Case of Gross v. Switzerland*, Judgment (Merits and Just Satisfaction), May 14, 2013, 66–9.

[257] ECtHR, *Case of Gross v. Switzerland*, 67810/10, Judgment (Merits), September 30, 2014.

[258] ECtHR, *Case of Stanev v. Bulgaria*, 36760/06, *supra* note 193, 144.

[259] ECtHR, *Case of H.M. v. Switzerland*, 39187/98, Judgment (Merits and Just Satisfaction), February 26, 2002.

lifted the order for the applicant's compulsory placement. The ECtHR found that the determination on whether the applicant was deprived of liberty involved considering at the starting-point the specific situation of the applicant together with "a whole range of factors such as the type, duration, effects and manner of implementation of the measure in question."[260] Moreover, "[t]he distinction between a deprivation of and restriction upon liberty is merely one of degree or intensity, and not one of nature or substance."[261] Ultimately, the ECtHR concluded that the placement of the applicant did not involve a deprivation of liberty because it was "a responsible measure taken by the competent authorities" in her interest.[262] Moreover, the ECtHR weighed in the fact that the applicant had agreed to stay in the nursing home.[263]

In *Storck v. Germany*,[264] the ECtHR further stated that deprivation of liberty involves not only "the objective element of a person's confinement in a particular restricted space for a not negligible length of time," but also the subjective element of having provided consent for such confinement.[265] The existence of consent to be placed in an institution or residential home has also been considered an essential element for the determination of the existence of deprivation of liberty in more recent cases, including in *Shtukaturov v. Russia*[266] and *Stanev v. Bulgaria*.[267] In *Shtukaturov*, the applicant lacked *de jure* legal capacity and was represented by her mother. He was placed in a psychiatric hospital upon his mother's request. Similarly, in *Stanev*, the applicant was placed under partial guardianship and a state official was appointed as his guardian. The state official negotiated an agreement with a social care home for his placement without consulting the applicant. The ECtHR found that lack of full legal capacity did not imply that the applicants in both cases were unable to understand their situation.[268] In contrast, the ECtHR found that the applicants had challenged their guardians' decisions, which showed that they were placed in an institution against their will.[269] Thus, the ECtHR concluded in both cases that the applicants had not consented to their placement and were consequently deprived of their liberty under Article 5.1 of the

---

[260] *Id.*, 42.

[261] *Id.*

[262] *Id.*, 48.

[263] *Id.*, 47.

[264] ECtHR, *Case of Storck v. Germany*, 61603/00, Judgment (Merits and Just Satisfaction), June 16, 2005. *Storck* involved the confinement of the applicant by her father, in several private psychiatric hospitals for different periods of time, without a court order or the applicant's consent.

[265] *Id.*, 74.

[266] ECtHR, *Case of Shtukaturov v. Russia*, *supra* note 53.

[267] ECtHR, Case of Stanev v. Bulgaria, *supra* note 193.

[268] ECtHR, *Case of Shtukaturov v. Russia*, *supra* note 53, 108; *Case of Stanev v. Bulgaria*, *supra note* 193, 130.

[269] ECtHR, *Case of Shtukaturov v. Russia*, *supra* note 53, 108; *Case of Stanev v. Bulgaria*, *supra note* 193, 130.

European Convention.[270] It is important to note, however, that in *Stanev*, the ECtHR acknowledged that there might be situations in which a person may not be able to express his or her wishes due to an impairment of his or her mental faculties, in which case the person's interests may be validly represented by another individual acting in the context of a protective measure.[271]

Also, the ECtHR has found that states have a positive obligation to protect vulnerable persons against deprivations of liberty perpetrated by a private party without state intervention, as long as the state knew or ought to have known of such situation.[272] In *Storck*, for example, the applicant was placed in a private psychiatric hospital and medically treated on several occasions upon her father's request. There was never a court order instructing the applicant's placement and, at least in one occasion in which she tried to escape, she was returned back to the institution by the police. The ECtHR found that the state had violated its positive obligation to protect the applicant against an unjustified deprivation of liberty, because it failed to assess whether the law authorized her placement in a private institution and supervise the lawfulness of her detention for approximately twenty months.[273]

Even if these cases do not relate to older persons specifically, the standards developed by these judgments are of relevance to assess the situation of older people placed in public or private nursing institutions or residential settings. What transpires from the existing case law is that the notion of deprivation of liberty is broad and comprises not only the physical "detention," but also the desire or acceptance of the person concerned to be placed in an institution. The inability to leave, the location of the institution, and management's access to the identity papers and the finances of the person placed in the institution are elements to assess if significant restriction to the physical right to liberty exists.[274] Moreover, the lack of consent is essential to conclude under Article 5(1) of the *European Convention* that there is a deprivation of liberty. In the same vein, the *Recommendation on the human rights of older persons* requires that, in principle, older persons should only be placed in residential, institutional, or psychiatric care with their free and informed consent.[275]

Once a deprivation of liberty is confirmed, Article 5(1) specifies the situations in which such deprivation of liberty may be justified under the *European Convention*. First, the deprivation of liberty must be lawful. The notion of lawfulness requires observance of procedures prescribed by domestic law, but also that

---

[270] ECtHR, *Case of Shtukaturov v. Russia*, *supra* note 53, 109; *Case of Stanev v. Bulgaria*, *supra* note 193, 131.

[271] ECtHR, *Case of Stanev v. Bulgaria*, *supra* note 193, 130.

[272] ECtHR, *Case of Storck v. Germany*, *supra* note 264, 102.

[273] *Id*, 104–8.

[274] ECtHR, *Case of Stanev v. Bulgaria*, *supra* note 193, 124–128.

[275] *Recommendation CM/Rec(2014)2 of the Committee of Ministers to member states on the promotion of the human rights of older persons*, *supra* note 15, Article VI.C.43.

such procedures ensure that deprivation of liberty is not arbitrary.[276] Moreover, if less severe measures than deprivation of liberty exist to safeguard the individual or public interest, states should choose such measures before proceeding to detention.[277] In other words, "it does not suffice that the deprivation of liberty is in conformity with national law; it must also be necessary in the circumstances."[278] Finally, as stated above, deprivation of liberty will not be lawful unless it falls within one of the exhaustive list of permissible grounds provided in Article 5(1) of the *European Convention*.

With respect to older persons placed in nursing care or other similar institutions, the only ground from those listed in Article 5(1) that would justify deprivation of liberty appears to be the lawful detention of "persons of unsound mind," contemplated in subparagraph (1)(e) of that provision.[279] To justify the deprivation of liberty of an individual for being of "unsound mind", three requirements must be satisfied: (1) he or she must be reliably shown to be of unsound mind before a competent authority on the basis of objective medical expertise; (2) the mental disorder must be of a nature or degree requiring compulsory confinement; and (3) the lawfulness of the continued confinement will depend upon the persistence of the mental disorder.[280] As to the second condition, the need for compulsory confinement will apply in cases where the individual needs medication and/or treatment, but also where there is a need to control a person to prevent him or her from hurting himself or herself or others.[281] If these requirements are not met, the deprivation of liberty will be considered unlawful and in violation of Article 5(1) of the *European Convention*.[282]

Applying the above principles to older persons admitted into nursing or residential homes may present a challenge in cases in which the underlying reason for

---

[276]ECtHR, *Case of Shtukaturov v. Russia*, *supra* note 53, 111.

[277]ECtHR, *Case of Stanev v. Bulgaria*, *supra* note 193, 143.

[278]*Id.*

[279]Article 5(1)(e) reads in full: the lawful detention of persons for the prevention of the spreading of infectious diseases, of persons of unsound mind, alcoholics or drug addicts or vagrants. The other permissible grounds are: (a) the lawful detention of a person after conviction by a competent court; (b) the lawful arrest or detention of a person for noncompliance with the lawful order of a court or in order to secure the fulfilment of any obligation prescribed by law; (c) the lawful arrest or detention of a person effected for the purpose of bringing him before the competent legal authority on reasonable suspicion of having committed an offence or when it is reasonably considered necessary to prevent his committing an offence or fleeing after having done so; (d) the detention of a minor by lawful order for the purpose of educational supervision or his lawful detention for the purpose of bringing him before the competent legal authority; (e) the lawful arrest or detention of a person to prevent his effecting an unauthorised entry into the country or of a person against whom action is being taken with a view to deportation or extradition.

[280]ECtHR, *Case of Shtukaturov v. Russia*, *supra* note 53, 114.

[281]ECtHR, *Case of Stanev v. Bulgaria*, *supra* note 193, 153.

[282]See, *inter alia*, ECtHR, *Case of Shtukaturov v. Russia*, *supra* note 53, 115–116; *Case of Stanev v. Bulgaria*, *supra* note 193, 156–160.

the placement resides not in the existence of a mental disorder, but on the basis of frailty, physical illness or neglect, as presented in the case of *H.M. v. Switzerland*. In those situations, if the older person concerned does not consent to the institutional placement, it would be possible to conclude that there is an unlawful deprivation of liberty because the "detention" does not fall within one of the permissible grounds listed under Article 5(1) of the *European Convention*. This strict interpretation would require states to adopt less stringent measures than placement in a nursing institution in those situations, saved for the existence of consent by the older person concerned.[283] Alternatively, situations of frailty, physical illness, or neglect could fall within a more extensive interpretation of the notion of "persons of unsound mind" for purposes of applying Article 5(1)(e).[284] Even if the former interpretation can be rejected for not pursuing the most protective approach for older persons, it is expected that under certain extreme situations the placement of older persons in institutions will be justified on the basis of such argument.

In fact, placement in institutional care for older persons for reasons beyond the existence of a mental illness is contemplated, for example, in the *Explanatory Memorandum to the Recommendation on the human rights of older persons*. The Memorandum acknowledges that there might be situations in which state authorities believe that placement in an institution is for the older person's own benefit,[285] particularly when "the person is becoming a danger to him/herself or third persons, is incapable of seeing to his or her basic personal needs, or is becoming a threat to law and order."[286] In those situations, the *Recommendation on the human rights of older persons* and its *Explanatory Memorandum* acknowledge that the placement of an individual in an institution is acceptable, provided that such measure is a proportionate response to a risk of potential harm.[287]

In any event, any deprivation of liberty entitles the person affected to challenge the lawfulness of the detention under Article 5(4) of the *European Convention*.[288] In cases of persons deprived of liberty on the grounds of being of "unsound mind," unless there is automatic review of a judicial nature, they have a right to take proceedings at reasonable intervals before a court to challenge the lawfulness of their detention.[289] The form of the judicial remedy, however, "may vary from one

[283] See for example the analysis of Judge Loucaides in his dissenting opinion in the ECtHR, *Case of H.M. v. Switzerland*, 39187/98, *supra* note 257.

[284] See, for example the analysis of Judge Gaukur Jörundsson in his concurring opinion in the ECtHR, Case of H.M. v. Switzerland, 39187/98, *supra* note 259.

[285] *Explanatory Memorandum to the Recommendation on the promotion of the human rights of older persons*, *supra* note 63, 66.

[286] *Id.*

[287] *Id.*

[288] Article 5(4) states: Everyone who is deprived of his liberty by arrest or detention shall be entitled to take proceedings by which the lawfulness of his detention shall be decided speedily by a court and his release ordered if the detention is not lawful.

[289] ECtHR, *Case of Shtukaturov v. Russia*, *supra* note 53, 121.

domain to another, and depend on the type of deprivation of liberty in issue."[290] Consequently, states have a margin of discretion to design judicial remedies to challenge the lawfulness of detention, including setting up a system of automatic review or ensuring other forms of remedies that the persons affected may access periodically. Those remedies must ensure procedural safeguards "to protect the interests of persons, who on account of their mental disabilities, are not fully capable of acting for themselves."[291]

In cases in which an older person provides consent to be placed in institutional care or that consent is duly substituted based on his or her inability to provide it, the *European Convention* affords protection to the person concerned under the right to freedom of movement enshrined in Article 2 of Protocol 4 to the *European Convention*. Even if there is no related European case law regarding restrictions of freedom of movement of older persons placed in institutional care, overall the approach of the ECtHR has been to accept the restriction of this right only under certain circumstances. In particular, for a restriction to be permissible, it must be established by law; pursue a legitimate aim, namely: national security or public safety, maintenance of public order, prevention of crime, protection of health or morals, or protection of the rights and freedoms of others; and be proportional to the aims sought.[292] The *Recommendation on the human rights of older persons* recognizes the right of older persons in institutional care to freedom of movement and provides, in line with the jurisprudence of the ECtHR, that any restrictions to this right must be lawful, necessary, and proportional.[293] Also this Recommendation provides that any restriction to the right to freedom of movement should be implemented with the free and informed consent of the older person concerned, or as a proportionate response to a risk of harm.[294] Furthermore, it establishes that restrictions to the freedom of movement of older persons in institutional care should be subject to regular monitoring and review.[295] Similarly, the *European Charter of Older Persons in Long-Term Care* recognizes that physical restraint may be exceptionally acceptable as long as it is a proportionate response to a risk of potential harm.[296] The standard that this Charter requires for the determination to restrain an older person is supposedly based on his or her best interest, and is to be made through a transparent and independently verifiable process that can be reversed.[297]

---

[290] *Id.*, at para. 123.

[291] ECtHR, *Case of Stanev v. Bulgaria*, *supra* note 193, 170.

[292] See, *inter alia*, ECtHR, *Case of Rosengren v. Romania*, 70786/01, Judgment (Merits and Just Satisfaction), April 24, 2008, 33.

[293] *Recommendation CM/Rec(2014)2 of the Committee of Ministers to member states on the promotion of the human rights of older persons*, *supra* note 15, Article VI.41.

[294] *Id.*, Article VI.43.

[295] *Id.*, Article VI.42.

[296] See Article 2.8, *European Charter of the Rights and Responsibilities of Older Persons in Need of Long-Term Care and Assistance*, 2010, *supra* note 216.

[297] *Id.*

In sum, the right to liberty and security is meant to ensure protection of older persons in cases of placement in institutional care. First, deprivation of liberty exists when there is physical detention, understood broadly, and the person concerned has not provided consent to be placed. Second, for such deprivation of liberty to be lawful in terms of Article 5(1) of the *European Convention*, it must not be arbitrary, and must be provided for in law. It must fall within one of the permissible grounds listed in Article 5(1). In principle, the only ground that would justify deprivation of liberty in cases of older persons placed in care institutions is the lawful detention of "persons of unsound mind," contemplated in Article 5(1)(e). The *Recommendation on the human rights of older persons* acknowledges the possibility of placing older people in nursing homes on grounds different from mental illness, including in situations of neglect and when the person concerned may hurt himself or herself, or others. Undoubtedly, this is an area in which further exploration and interpretation by the ECtHR is required. In any event, Article 5(4) of the *European Convention* requires states to ensure effective remedy to challenge any form of detention on the basis of the lawfulness of the measure. Access to such remedy must be provided by the state automatically or the persons affected must be able to seek judicial review at reasonable intervals. Finally, if an older person consents to be placed at a care institution, the *European Convention* still affords protection to the person concerned under the right to freedom of movement enshrined in Article 2 of Protocol 4. Any constraints imposed upon such right must be lawful, necessary, and proportional. Moreover, any restriction to this right should be implemented with the free and informed consent of the older person concerned, or as a proportionate response to a risk of harm. Finally, states must ensure that measures subjecting older persons to restrictions of their right to freedom of movement are periodically monitored and can be challenged by the persons affected through appropriate procedures.

6. The Right to Private Life

The right to respect for private life is protected under Article 8 of the *European Convention*, which states that: "Everyone has the right to respect for his [or her] private and family life, his [or her] home and his [or her] correspondence" and establishes the permissible limitations applicable to its exercise. For Article 8 to apply there needs to be an interference with the exercise of the right to private life. Once such interference is confirmed, Article 8(2) provides the grounds to assess whether it is based on a permissible ground, pursues a legitimate aim and is necessary in a democratic society.[298] The last element of the analysis involves an

[298]Article 8(2) provides: There shall be no interference by a public authority with the exercise of this right except such as is in accordance with the law and is necessary in a democratic society in the interests of national security, public safety or the economic wellbeing of the country, for the prevention of disorder or crime, for the protection of health or morals, or for the protection of the rights and freedoms of others.

assessment on the proportionality between the goals pursued by the measure and the means used to achieve them. Furthermore, Article 7 of the *Charter of Fundamental Rights* protects the right to respect for private and family life.

The ECtHR has consistently stated that "the concept of 'private life' is a very broad term not susceptible to exhaustive definition."[299] Among others, this notion encompasses elements such as name, gender identification, sexual orientation and sexual life, personal development and the right to establish and develop relationships with other persons.[300] In *Schlumpf v. Switzerland*, the ECtHR reaffirmed that issues related to gender identity constituted one of the most essential aspects of the notion of private life.[301] In the case, the applicant aged 67 years was rejected reimbursement for the health cost involved in a sex conversion surgery because two years had not elapsed since the determination of her transsexualism, as required by the Swiss Federal Insurance Court. The ECtHR found that this decision constituted a disproportionate measure because it did not balance out properly the applicant's interest and those of the insurance company's. Even if the insurance policy could apply in other cases, given the age of the applicant, waiting for two years was likely to influence her decision to proceed with the operation, thereby impairing her freedom to determine her gender identity and, consequently, her right to private life.[302]

Also, the ECtHR has found that Article 8 protects the right to personal autonomy[303] and the rights to a person's physical and psychological integrity.[304] These two notions are essential for the protection of the right to privacy of older persons. The notion of personal autonomy was developed by the ECtHR in *Pretty vs. the United Kingdom*.[305] In that case, the ECtHR concluded that personal autonomy relates to the principle of self-determination which could be defined as "the ability to conduct one's life in a manner of one's own choosing, [including] the opportunity to pursue activities perceived to be of a physically or morally harmful or dangerous nature for the individual concerned."[306] As stated before in the section on the right to life, in cases of older persons this could involve decisions regarding refusal to accept medical treatment, as well as other measures related to ending life in a dignified manner.[307] Most recently, in *McDonald v. United Kingdom*, the

[299]ECtHR, *Case of Haas v. Switzerland*, 31322/07, *supra* note 253, 50.

[300]*Id.*

[301]ECtHR, *Case of Schlumpf v. Switzerland*, 29002/06, Judgment (Merits and Just Satisfaction), January 8, 2009.

[302]*Id.*, 110–15.

[303]ECtHR, *Case of Pretty v. United Kingdom*, 2346/02, *supra* note 240, 61.

[304]ECtHR, *Case of McDonald v. The United Kingdom*, 4241/12, Judgment (Merits and Just Satisfaction), 46.

[305]ECtHR, *Case of Pretty v. United Kingdom*, *supra* note 240, 61.

[306]*Id.*, 62.

[307]*Id.*, 63, 65.

ECtHR reviewed the right of private life of an older woman who alleged that the state authorities' decision to withdraw night-time care from her and forced her to use incontinence pads, even though she was not incontinent, conflicted with her sense of dignity and self-identity.[308] The applicant had suffered a stroke that had left her with severely limited mobility, a situation that was followed by several falls that added to her inability to access the toilet unaided. Given her situation, state authorities had provided her with night-time care services. After a year of services, the state authorities decided to reduce the funds allocated to her care because it was decided that the applicant would be provided with incontinence pads in lieu of night-time care. She challenged the decision before domestic courts and later before the ECtHR arguing that the measure violated her right to private life because it affected her personal and psychological integrity.[309] Even though the ECtHR concluded that the decision to withdraw the night-time care and forced her to use incontinence pads constituted an interference with her right to private life,[310] the measure was not disproportionate because it stroke a proper balance between the applicant's needs and the state authorities' duty to allocate scarce resources to serve the community.[311] Therefore, the ECtHR did not find a breach of Article 8 of the *European Convention*.

The right of older persons to autonomy and personal dignity is also acknowledged in the *Recommendation on the human rights of older persons*. Article III on Autonomy and Participation provides that older persons "are entitled to lead their lives independently, in a self-determined and autonomous manner."[312] This right encompasses taking independent decisions on property, income, finances, place of residence, health, medical treatment or care, and funeral arrangements.[313] Moreover, the Article states that this right may be subject to limitations that are proportional to the specific situation, but states must afford appropriate and effective safeguards to prevent abuse and discrimination.[314] Furthermore, the Article ensures the right to private and family life of older persons, including the right to respect for their sexual intimacy.[315]

Nursing or residential institutions occupy a special position of power to control the privacy of older individuals in their care. Article 23 of the *Revised Social Charter* compels states to adopt special measures to protect the privacy of older persons living in institutions and to guarantee their participation in decisions

[308]ECtHR, *Case of McDonald v. the United Kingdom*, *supra* note 304.

[309]*Id.*, 38–42.

[310]*Id.*, 47.

[311]*Id.*, 57–8.

[312]*Recommendation CM/Rec(2014)2 of the Committee of Ministers to member states on the promotion of the human rights of older persons*, *supra* note 15, Article III.9.

[313]*Id.*

[314]*Id.*

[315]*Id.*, Article III.11.

concerning living conditions in those institutions. According to the Explanatory Report of that provision, "[r]espect for privacy is mentioned only in relation to elderly persons living in institutions, a situation warranting special mention. Everyone in all circumstances is naturally entitled to respect for his or her private life as guaranteed by the *European Convention*."[316] In the conclusions regarding national reports, the ECSR has insisted that states provide information on legal and administrative measures protecting that right, though it has not yet received adequate responses to develop standards.[317] The issue of protecting respect for private and family life is also relevant for older persons in a broader range of situations than the Explanatory Report explores, such as in relation to visits with family or visitation with grandchildren.

In the same vein, Articles 2 and 3 of the *European Charter of Older Persons in Long-Term Care* ensure the rights to self-determination and privacy, respectively.[318] Article 2 ensures the right of older persons in long-term care to make their own life choices and respect for their own free will.[319] Article 3, on the other hand, protects the right to intimacy and underlines that protection of privacy is essential for maintaining the self-esteem and well-being of older persons.[320]

Article 25 of the *Charter of Fundamental Rights* protects the right of older people to lead a life of dignity and independence.[321] The explanatory report regarding the adoption of the Charter specifies that the language of this provision was drawn from Article 23 of the *Revised Social Charter*[322]; therefore it is expected that this provision is interpreted in accordance with the understandings adopted by the ECSR, including the protection of the right to private life. Likewise, Article 8(2) of the *Charter of Fundamental Rights* protects personal data. In relevant part, Article 8(2) states that personal date "must be processed fairly for specified purposes and on the basis of the consent of the person concerned or some other legitimate basis laid down by law."[323] In the same vein, Article VI on Care of the *Recommendation on the human rights of older persons* provides that "care providers should treat any sensitive personal data of older persons confidentially and carefully in accordance

[316]See *Explanatory Report to the Additional Protocol to the European Social Charter*, E.T.S. No. 128, (1988), 59, available at http://conventions.coe.int/Treaty/en/Reports/Html/128.htm (last visited Dec. 29, 2014).

[317]See, *inter alia*, ECSR, Conclusions 2003, France, at 93; Conclusions 2003, Denmark, at 49; Conclusions 2003, Italy, p. 76; Conclusions 2005, Norway, at 49.

[318]*European Charter of the Rights and Responsibilities of Older Persons in Need of Long-Term Care and Assistance*, 2010, *supra* note 216.

[319]*Id.*

[320]*Id.*

[321]*Charter of Fundamental Rights of the European Union*, *supra* note 38.

[322]Text of the explanations relating to the complete text of the Charter as set out in CHARTE 4487/00 CONVENT 50, *supra* note 42.

[323]*European Charter of the Rights and Responsibilities of Older Persons in Need of Long-Term Care and Assistance*, 2010, *supra* note 216.

with their right to privacy."[324] Also, Article 3 of the *European Charter of Older Persons in Long-Term Care* provides that the right to privacy encompasses the obligation to handle personal data with confidentiality.[325]

In sum, the right to private life is essential to protect the autonomy, self-determination, physical and psychological integrity as well as the intimacy of older persons. The protections afforded by the *European Convention*, the *Revised Social Charter* and the *Charter of Fundamental Rights* of the European Union serve as a departure point from which to articulate more specific standards applicable to older persons. Additional provisions enshrined in the *Recommendation on the human rights of older persons* and the *European Charter of Older Persons in Long-Term Care* are essential contributions for the design of those specific standards because they were drafted to address the issues that affect older persons the most.

7. The Right to Information, Association and Participation

Article 10 of the *European Convention* states that "[e]veryone has the right to freedom of expression," which includes the freedom to hold opinions, receive information and ideas, and impart information and ideas. Article 11 of the *Charter of Fundamental Rights* also enshrines the right to freedom of expression and information. Overall these notions have been broadly interpreted by the ECtHR in order to ensure enhanced protection to a right that it is considered fundamental for the proper functioning of a democratic society, and "one of the basic conditions for its progress and for each individual's self-fulfillment."[326] The right to freedom of expression may be subjected to restrictions as provided in Article 10(2) of the *European Convention*.[327] For a restriction to be permissible it must be prescribed by law, pursue a legitimate aim, including the protection of national security, territorial integrity or public safety, the prevention of disorder or crime, the protection

[324] *Recommendation CM/Rec(2014)2 of the Committee of Ministers to member states on the promotion of the human rights of older persons, supra* note 15, Article VI.32.

[325] *European Charter of the Rights and Responsibilities of Older Persons in Need of Long-Term Care and Assistance*, 2010, *supra* note 216.

[326] ECtHR, *Case of Tešić v. Serbia*, 4678/07 and 50591/12,Judgment(Merits and Just Satisfaction), February 11, 2014, 61.

[327] Article 10(2) provides: The exercise of these freedoms, since it carries with it duties and responsibilities, may be subject to such formalities, conditions, restrictions or penalties as are prescribed by law and are necessary in a democratic society, in the interests of national security, territorial integrity or public safety, for the prevention of disorder or crime, for the protection of health or morals, for the protection of the reputation or rights of others, for preventing the disclosure of information received in confidence, or for maintaining the authority and impartiality of the judiciary.

of health or morals, the protection of the reputation or rights of others, for preventing the disclosure of information received in confidence, or for maintaining the authority and impartiality of the judiciary, and be necessary in a democratic society. The assessment of the necessity is determined on the basis of a proportionality test between the goals sought with the measure and the means employed to achieve them. In *Tešić v. Serbia*, the only case decided where the right to freedom of expression of an older person was at stake, the ECtHR found that a civil judgment and its enforcement against the applicant in defamation proceedings initiated by a lawyer as a result of allegedly offensive comments made by her in a published article constituted an interference with her right to freedom of expression.[328] The applicant, who was 74 years old at the time the enforcement judgment was issued, was obliged to pay the damages out of the pension she received. Such payments constituted two-thirds of the total amount of her pension, a situation that forced the applicant to sustain her basic needs with the remaining one third.[329] Her dire financial situation was also compounded by poor health requiring costly medications and other medical treatments.[330] The ECtHR found that even if the measure was prescribed by law and pursue a legitimate aim, namely the protection of the reputation of others, the way the judgment was enforced was disproportionate given the precarious situation faced by an older woman suffering from a number of serious diseases.[331] Therefore, the ECtHR found a violation of Article 10 of the *European Convention* in this case.

Access to appropriate information is essential to enable older persons "to remain full members of society for as long as possible." First, Article I.4 of the *Recommendation on the human rights of older persons* provides that older persons are entitled to have access to sufficient information regarding their rights. Also, Article 23 of the *Revised Social Charter* establishes that states must provide, directly or in co-operation with public or private organizations, "information about services and facilities available for elderly persons and their opportunities to make use of them."[332] In *Central Association of Carers in Finland v. Finland*, the ECSR found that lack of upper limits in pricing established in law for the provision of 24-h assistance in service housing for older persons deprived this group from their right to information "about services and facilities available for elderly persons and their opportunities to make use of them," as provided by Article 23b of the *Revised Social Charter*.[333]

Older persons have the right to be informed about their state of health as well as existing treatment and care options available to address their medical needs. Dissemination of information regarding issues of relevance to older persons

[328]ECtHR, *Case of Tešić v. Serbia*, *supra* note 326, 64.

[329]*Id.*, 26, 28.

[330]*Id.*, 30–31.

[331]*Id*, 67–68.

[332]For a full text of Article 23, see *supra* note 25.

[333]ECSR, *The Case of the Central Association of Carers in Finland v. Finland*, *supra* note 33, 53.

contributes to enhancing a positive perception of the role of older persons in society.[334] The media has a crucial role in disseminating such information as well as in producing educational programs regarding older persons.[335]

Another facet of life for older persons in Europe which affects their right to freedom of information is the difficulty some experience with new technologies. Access to the internet and computer literacy can have dramatic positive effects on the right to exercise access to information, to associate freely, and to participate in society. Article 6 of the *European Charter of Older Persons in Long-Term Care* acknowledges the right of older persons to equal access to new technologies and training to learn how to use them.[336] In the last decade, the EU has sponsored a number of programs and research protocols for "effective and efficient use of ICT [information and communication technology] to deliver public services and policy outcomes that improve the lives of citizens, through greater choice and personalization."[337] These policies supporting the integration of older people in the technological world represent positive implementation of the right to information. Within the EU, access to the Internet for people with disabilities and for older persons was an objective for Europe 2002.[338] It aimed to ensure that these groups of people have access to public websites and their contents as a way to improve their participation in society.[339] In 2006, the Riga Ministerial Declaration on e-Inclusion set out as a specific objective to "halve the gap in internet usage by 2010 for groups at risk of exclusion, such as older people, people with disabilities, and unemployed persons."[340] The EU has continued to work on web-accessibility, especially for

---

[334] *Appendix to Recommendation No. R(94)9, Guiding principles to be taken into account when taking measures concerning elderly people*, adopted by the Committee of Ministers on October 10, 1994. See also, Council of Europe, Parliamentary Assembly, *Recommendation 1428 (1999)1, The future of senior citizens: protection, participation and promotion*, September 1999, 17–8.

[335] *Id.*

[336] *European Charter of the Rights and Responsibilities of Older Persons in Need of Long-Term Care and Assistance*, 2010, *supra* note 216.

[337] Fernando, S., et al., *Transformational government and improving services for older adults through the DIADEM assistive technology*, Transforming Government: People, Process and Policy, 2010, Vol. 4 No. 1, at 55.

[338] eEurope 2002: Accessibility of Public Web Sites and their Content, European Parliament resolution on the Commission communication eEurope 2002: Accessibility of Public Web Sites and their Content (COM(2001) 529—C5-0074/2002-2002/2032(COS)), available at http://www.europarl.europa.eu/omk/omnsapir.so/pv2?PRG=DOCPV&APP=PV2&LANGUE=EN&SDOCTA=18&TXTLST=1&POS=1&Type_Doc=RESOL&TPV=DEF&DATE=130602&PrgPrev=PRG@TITRE|APP@PV2|TYPEF@TITRE|YEAR@02|Find@%2577%2565%2562%2520|FILE@BIBLIO02|PLAGE@1&TYPEF=TITRE&NUMB=1&DATEF=020613 (last visited Dec. 9, 2014).

[339] *Id.*, consideration E.

[340] European Commission, Press Release Database, Internet for all: EU ministers commit to an inclusive and barrier-free information society, June 12, 2006, available at http://europa.eu/rapid/press-release_IP-06-769_en.htm (last visited Dec. 9, 2014).

persons with disabilities, and this work has gained new momentum upon ratification of the United Nations Convention on the Rights of Persons with Disabilities (UNCRPD) by the EU in 2010.[341] Ostensibly, improved website accessibility would facilitate the participation of groups such as older persons that require special accommodation to access information.

The CoE has also fostered access to technology by older persons as a way to ensure social inclusion and engagement in active ageing. In *Recommendation No. R(94)9*, the Committee of Ministers encouraged public authorities to promote the use of new forms of technology by older persons in order to contribute to maintaining their autonomy. Moreover, in *Recommendation 1428*, the Parliamentary Assembly suggested that states should bring older persons in closer contact with Internet and other new forms of communication.[342] Additionally, in *Recommendation CM/Rec (2009)6*, the Committee of Ministers acknowledged the need to enhance availability of free and accessible access to modern information systems and counselling, including via Internet, to promote the autonomy and independent and active life of older persons.[343] The Committee of Ministers has also noted the importance of information and communication technologies (ICT) to the democratic process, and that lack of access can "deprive individuals of their right to exercise fully their human rights."[344]

Long term care institutions have a tremendous amount of control over access to information for residents. Human rights protections around older persons' access to mail, telephones, internet services, and visitors are vital to the protection of their right to freedom of information. Article 3 of the *European Charter of Older Persons in Long-Term Care* expressly acknowledges the right of older persons to the privacy of correspondence, including letters, electronic mail as well as ensures the right to make telephone calls in private.[345] While in residential or care institutions, older persons have the right to express their opinions, wishes and choices and must have their values, beliefs and life philosophy respected, regardless of whether those values or beliefs are shared by others or their caregivers.[346]

Another relevant issue regarding long term care and freedom of expression is the need to protect persons reporting elderly abuse from retaliation. In *Heinisch v.*

---

[341]For more information, see http://ec.europa.eu/ipg/standards/accessibility/eu_policy/index_en.htm (last visited Dec. 9, 2014).

[342]Parliamentary Assembly, *Recommendation 1428 (1999)1, The future of senior citizens: protection, participation and promotion*, *supra* note 334, 16.2.

[343]Committee of Ministers, *Recommendation CM/Rec (2009)6, on ageing and disability in the 21st century: sustainable frameworks to enable greater quality of life in an inclusive society*, *supra* note 14, 1.12.

[344]Council of Europe, *Declaration of the Committee of Ministers on Human Rights and Rule of Law in the Information Society*, CM(2005)56, May 13, 2005, 3.

[345]*European Charter of the Rights and Responsibilities of Older Persons in Need of Long-Term Care and Assistance*, 2010, *supra* note 216, Article 3.

[346]*Id*, Articles 2.2 and 7.1.

Germany, the ECtHR reviewed the dismissal of a worker who publicly complained of the deficiencies in the services provided to people in a care facility.[347] Before denouncing the deficiencies publicly in a criminal complaint, the applicant had informed her superiors of her concerns.[348] Moreover, the information she released was not factually incorrect or frivolous[349] and she acted in good faith.[350] The ECtHR held that the information released by the nurse was of public interest because it related to a particularly vulnerable group of older persons who were unable to report by themselves the shortcomings in care and hygiene that they were receiving.[351]

The right to freedom of expression also involves the right to express opinions and participate in matters that affect one's personal life. Article I.5 of the *Recommendation on the human rights of older persons* provides that older persons should be consulted prior to the adoption of measures that affect the enjoyment of their human rights.[352] Also, Article 2 of the *European Charter of Older Persons in Long-Term Care* states the right of older persons to be consulted and participate in the decision-making process regarding measures that have an impact on their lives.[353] Furthermore, according to *Recommendation CM/Rec (2009)6* of the Committee of Ministers of the Council of Europe, participation in the design, implementation and evaluation of services to older persons require their full and direct involvement.[354]

The right to participation of older persons also encompasses involvement in political life and the right to participate in elections. The right to vote and participate in elections is protected in Article 3 of Protocol 1 to the *European Convention*.[355] Also, Article 25 of the *Charter of Fundamental Rights* covers the right of older persons to political participation.[356] In the same vein, Article 6 of the *European Charter of Older Persons in Long-Term Care* acknowledges the right of older persons to exercise their civic rights, including the right to vote and

[347]ECtHR, *Case of Heinisch v. Germany*, *supra* note 214.

[348]*Id*, 76.

[349]*Id*, 81.

[350]*Id*., 87.

[351]*Id*, 71.

[352]*Recommendation CM/Rec(2014)2 of the Committee of Ministers to member states on the promotion of the human rights of older persons*, *supra* note 15, Article I.5.

[353]*European Charter of the Rights and Responsibilities of Older Persons in Need of Long-Term Care and Assistance*, 2010, *supra* note 216, Article 2.2.

[354]Council of Europe, Committee of Ministers, *Recommendation CM/Rec (2009)6, on ageing and disability in the 21st century: sustainable frameworks to enable greater quality of life in an inclusive society*, *supra* note 14, 2.2.

[355]Article 3 of Protocol 1 states: The High Contracting Parties undertake to hold free elections at reasonable intervals by secret ballot, under conditions which will ensure the free expression of the opinion of the people in the choice of the legislature.

[356]Text of the explanations relating to the complete text of the Charter as set out in CHARTE 4487/00 CONVENT 50, supra note 42.

participate in elections.[357] If the exercise of this right requires assistance by a third person, this individual must maintain confidentiality as to the choice made by the older person.[358]

The right to association is protected in Article 11 of the *European Convention*, which provides that "Everyone has the right to freedom of peaceful assembly and to freedom of association with others, including the right to form and to join trade unions for the protection of his interests." Also, Article 12 of the *Charter of Fundamental Rights* protects freedom of assembly and association. Like the right to freedom of expression, the right to freedom of association may be subject to permissible restrictions, provided that they are prescribed by law, pursue a legitimate aim and are necessary in a democratic society.[359] In a similar vein, and specifically tailored for older persons, the *European Charter of Older Persons in Long-Term Care* ensure the right to establish an association, join a group or refuse to associate or join.[360] This right applies to older persons who are living in long term care institutions or outside of them.

The right to information, association and participation is also connected to the right to freedom of thought, conscience and religion under Article 9 of the *European Convention,* which states that: "Everyone has the right to freedom of thought, conscience and religion." In a Eurostat survey, 63 % of adults over 55 stated that they believed in the existence of a God, the highest of any age group.[361] This is an aspect of the current population of older persons that must be kept in mind in terms of their participation in society, access to information they may need, and associations they may wish to keep that differ from the values of their younger counterparts. The right to freedom of thought, conscience and religion is further enshrined in the *Charter of Fundamental Rights of the European Union's* Article 10(1), which declares that "Everyone has the right to freedom of thought, conscience and religion."[362] Additionally, Article 7 of the *European Charter of*

---

[357] *European Charter of the Rights and Responsibilities of Older Persons in Need of Long-Term Care and Assistance*, 2010, supra note 216, Article 6.5.

[358] *Id.*

[359] Article 11(2) of the *European Convention* states: No restrictions shall be placed on the exercise of these rights other than such as are prescribed by law and are necessary in a democratic society in the interests of national security or public safety, for the prevention of disorder or crime, for the protection of health or morals or for the protection of the rights and freedoms of others. This Article shall not prevent the imposition of lawful restrictions on the exercise of these rights by members of the armed forces, of the police or of the administration of the State.

[360] *European Charter of the Rights and Responsibilities of Older Persons in Need of Long-Term Care and Assistance*, 2010, supra note 216, Article 7.4.

[361] European Commission, Eurostat, *Social Values, Science and Technology*, 2005, at 10, available at http://ec.europa.eu/public_opinion/archives/ebs/ebs_225_report_en.pdf (last visited Dec. 29, 2014).

[362] *Charter of Fundamental Rights of the European Union, supra* note 38, Article 10(1).

Older Persons in Long-Term Care notes the right of older persons to practice and observe religious or spiritual beliefs, or to refuse to do so if the persons concerned do not share any religious or spiritual beliefs.[363]

In sum, the rights to freedom of expression, association and participation are fundamental for the continuous engagement of older persons in the social and cultural life of the communities where they live. Existing provisions in the *European Convention* and the *Charter of Fundamental Rights* constitute the baseline from which specifically standards tailored to the needs of older persons must be developed. Non-binding instruments such as the *Recommendation on the human rights of older persons* and the *European Charter of Older Persons in Long-Term Care* may provide additional sources to draw from when interpreting the rights of freedom of expression, association and participation under the binding treaties applicable to the European states.

D. Economic, Social and Cultural Rights of Older Persons in Europe

I. The Right to Culture and Education

Article 23 of the *Revised Social Charter* provides that states undertake to adopt measures to enable elderly persons to remain full members of society for as long as possible, *inter alia*, by enabling them to play an active part in public, social and cultural life.[364] The Explanatory Report of the Additional Protocol to the European Social Charter states that the notion "full members" implies that older persons "must suffer no ostracism on account of their age, since the right to take part in society's various fields of activity is not granted or refused depending on whether an elderly person has retired or is still vocationally active or whether such a person is still of full legal capacity or is subject to some restrictions in this respect (diminutio capitis)."[365] States may implement this obligation directly or through cooperation with private or public institutions. Likewise, Article III.10 of the *Recommendation on the human rights of older persons* provides that older persons should interact with other members of society and fully participate in social, cultural and education activities.[366] These measures will improve social cohesion and

[363]*European Charter of the Rights and Responsibilities of Older Persons in Need of Long-Term Care and Assistance*, 2010, *supra* note 216, Article 7.3.

[364]For a full text of Article 23, see *supra* note 25.

[365]See *Explanatory Report to the Additional Protocol to the European Social Charter*, *supra* note 316, 54.

[366]*Recommendation CM/Rec(2014)2 of the Committee of Ministers to member states on the promotion of the human rights of older persons*, *supra* note 15, Article III.10.

active citizenship in Europe; while at the same time contribute significantly to the intellectual, social and physical state of the older persons in general.[367] Older persons are also entitled to be given the opportunity to participate in social life, on the basis of their interests and abilities and in the spirit of solidarity between generations.[368] In the same vein, *Recommendation CM/Rec(2009)6* encouraged states to ensure opportunities for active participation in all areas of life of older persons where their knowledge and life experience can be recognized.[369] Opportunities of this nature will also contribute to eliminate situations of isolation or loneliness among older persons, a situation that affects this group immensely.[370] The promotion of inter-generational exchanges of experience and transmission of knowledge and know-how from older persons to younger generations has also been recognized by the Parliamentary Assembly of the Council of Europe in *Recommendation 1428 (1999)*.[371]

In general terms, the right to education is protected under Article 2, Protocol 1 of the *European Convention*, which states that: "No-one shall be denied the right to education." This right, however, is not absolute but may be subjected to permissible restrictions.[372] In contrast to other provisions, such as Articles 8–11, the *European Convention* does not provide for clear guidelines as to the grounds on which the right to education may be restricted.[373] The ECtHR has stated that for a limitation to the right to education to be permissible, it must be provided by law, pursue a legitimate aim and strike a reasonable relationship of proportionality between the means employed and the aim sought to be achieved.[374] Also, the right to education is protected by Article 14 of the *Charter of Fundamental Rights*, which additionally ensures access to vocational and continuing training. Furthermore, Article 10 of the *Revised Social Charter* protects the right to vocational training. According to the ECSR, this right is applicable to all members of the population on the basis of need.[375] Furthermore, paragraph 3.a of Article 10 establishes that states undertake to provide or promote adequate and readily available training for adult workers. It would be interested to explore in future

---

[367] *Explanatory Memorandum of Recommendation CM/Rec(2014)2 of the Committee of Ministers to member states on the promotion of human rights of older persons*, *supra* note 63, 21.

[368] *European Charter of the Rights and Responsibilities of Older Persons in Need of Long-Term Care and Assistance, 2010*, *supra* note 216, Article 6.1.

[369] Recommendation CM/Rec (2009)6, on ageing and disability in the 21st century: sustainable frameworks to enable greater quality of life in an inclusive society, *supra* note 14, Appendix to Recommendation, 1.8.

[370] *Id.*, 1.9.

[371] Council of Europe, Parliamentary Assembly, *Recommendation 1428 (1999)1, The future of senior citizens: protection, participation and promotion*, *supra* note 334, at para. 16.8.

[372] ECtHR, *Case of Velyo Velev v. Bulgaria*, 16032/07, Judgment (Merits and Just Satisfaction), May 27, 2014, 32.

[373] *Id.*

[374] *Id.*

[375] ECSR, European Action of the Disabled (AEH) v. France, *supra* note 121, 25.

interpretations whether this provision could be used to justify the right of older persons to education and training.

The ECtHR has acknowledged that the right to education is one of the rights where the existence of resources is a matter to be assessed and considered.[376] Thus, "in deciding how to regulate access to education, a state must strike a balance between, on the one hand, the educational needs of those under its jurisdiction and, on the other, its limited capacity to accommodate them."[377] Notwithstanding this constraint, education is the only public service that is afforded direct protection by the *European Convention*. Moreover, it is "a very particular type of public service, which not only directly benefits those using it but also serves broader societal functions."[378] In this vein, the *Recommendation on the human rights of older persons* calls states to ensure older persons' access to education and training activities and opportunities.[379] Moreover, *Recommendation CM/Rec(2009)6* encourages states to apply the principle of life-long learning, regardless of the age of the individuals.[380] The continuous development of skills contributes to ensuring autonomy and participation in society, particularly for older persons.[381] Similarly, the *European Charter of Older Persons in Long-Term Care* also acknowledges the right of older persons to take part in lifelong learning opportunities.[382]

One area in which older persons may face limitations to the exercise of their right to education involves physical barriers, including commuting and transit issues, lack of access due to the inexistence of physical accessibility and/or a lack of alternative arrangements.[383] Moreover, many state scholarships are only open to people under a certain age.[384] Interference with the right to education of older persons may have grave implications on other areas of life as well. Where older persons' employment and promotion in the workplace are dependent on access to training within the workplace, older persons are consistently excluded. Older

---

[376] *Id.*, 33.

[377] *Id.*

[378] *Id.*

[379] *Recommendation CM/Rec(2014)2 of the Committee of Ministers to member states on the promotion of the human rights of older persons*, *supra* note 15, Article III.10.

[380] *Recommendation CM/Rec (2009)6*, on ageing and disability in the 21st century: sustainable frameworks to enable greater quality of life in an inclusive society, *supra* note 14, 1.7.

[381] *Id.*

[382] *European Charter of the Rights and Responsibilities of Older Persons in Need of Long-Term Care and Assistance*, 2010, *supra* note 216, Article 6.1.

[383] American Council on Education, *Framing New Terrain: Older Adults and Higher Education*, 2007, at 19.

[384] Van Bueren, *The Protection of the Human Rights of Older Persons in Europe: A Legal Perspective*, International Symposium on the Rights of Older People Hosted by Age Concern England in collaboration with HelpAge International and the International Federation on Ageing, January 2009, at 2.

workers may face stereotypes of poor adaptability, including the ability to utilize education opportunities to learn new skills and technology.[385]

Overall ensuring access of older persons to culture and education will promote social cohesion and active citizenship in Europe, whereas at the same time significantly contributing to the intellectual, social and physical state of older persons in general. Given the resources required for the full implementation of the right to education and the need to strike a balance among the competing groups entitled to such right, there is a risk that older persons may be left behind when scarce resources are allocated by state authorities. More work needs to be carried out to establish priorities and articulate the notion that older persons' right to education needs to be satisfied on an equal basis with other groups.

2. The Right to Work

The right to work is protected under Article 1 of the *Revised Social Charter*. Also, Article 15 of the *Charter of Fundamental Rights* enshrines the freedom to choose an occupation and the right to engage in work.

Since access to work by older persons may become challenging, one of the most important protections afforded to elderly workers is the right not to be subject to discrimination on account of age in employment relations. As stated in the section on the right of older persons to equality and non-discrimination, the principle of non-discrimination is protected by Articles 14 of the *European Convention* and Article 1 of Protocol 12 to that treaty.[386] Moreover, the right not to be discriminated on the basis of age in the area of employment is protected by Article 1(2) and 24 of the *Revised Social Charter*.[387]

Under Article 1, para. 2 of the *Revised Charter*, states undertake to "protect effectively the right of the worker to earn his living in an occupation freely entered upon." On the other hand, Article 24 provides for protection in cases of termination of employment. The ECSR has concluded that under Article 1, para. 2, states must eliminate "all forms of discrimination in employment, whatever the legal nature of the professional relationship."[388] Moreover, that provision prohibits discrimination at least on account of the following grounds: sex, race, ethnic origin, religion, disability, age, sexual orientation and political opinion.[389] The ECSR follows a similar

[385]K. Smith, L. Webber, *Promotion: Age matters, International Congress Series*, 2005, vol. 1280, at 403.

[386]*Id.* See section on the right of older persons to equality and non-discrimination.

[387]Article E of the *Revised Social Charter* ensures the right to non-discrimination in relation to the other rights protected by that treaty. See, also, ECSR, *Fellesforbundet for Sjøfolk (FFFS) v. Norway*, Complaint No. 74/2011, July 2, 2013, 48.

[388]*Id.*, 104.

[389]See, *inter alia*, ECSR, Conclusions XVIII-1, Iceland, at 423–424.

approach to the determination of what constitutes discrimination under this provision as it does under Article E of the Revised Charter. Thus, for discrimination to exist in employment there must be a difference in treatment between people in comparable situations that fails to pursue a legitimate aim and is disproportionate in relation to the ends that it seeks to achieve.[390] In *Fellesforbundet for Sjøfolk (FFFS) v. Norway*, the ECSR found a violation of Article 1, para. 2, concluding that the claimants in the case were discriminated on account of age when they were given notice of termination solely on the basis that they were 62 years old.[391] Domestic law in Norway setting out that age was found disproportionate because it only affected seamen, while other workers in the country were allowed to work until the age of 70.[392] The ECSR held that under Article 1, para. 2 older persons may not be deprived of their right to earn their living in an occupation freely entered upon by them.[393] Moreover, protecting the right to work of older persons is "consistent with one of the primary objectives of Article 23 [of the Revised Charter], which is to enable elderly persons to remain full members of society."[394]

In the same vein, the ECSR found that the Norwegian restriction breached Article 24 of the Revised Charter because the seamen could be dismissed by their employers for having reached the age of 62, without a need to properly justify termination on valid reasons established in that provision, including those related to capacity or conduct of the employee or based on the operational requirements of the employer's business.[395]

Furthermore, as extensively discussed in the section on the right of older persons to equality and non-discrimination, EU Directive 2000/78/EC establishes a general framework to ensure equal treatment in employment and occupation.[396] Article 2 of the Directive provides that the principle of equal treatment means that direct and indirect discrimination are prohibited on the basis of Article 1, which specifically includes age.[397] This Directive, coupled with the prohibition on discrimination on the basis of age established in Article 21 of the *Charter of Fundamental Rights*,[398] constitutes an important shield against discrimination of

[390] *Id.*, 107.

[391] *Id.*, 114, 116–117.

[392] *Id.*, 109–110.

[393] *Id.*, 115.

[394] *Id.*, 116.

[395] *Id.*, 97.

[396] Council Directive 2000/78/EC of 27 November 2000 establishing a general framework for equal treatment in employment and occupation, Official Journal L 303, *supra* note 47.

[397] Article 1 of Council Directive 2000/78/EC states: The purpose of this Directive is to lay down a general framework for combating discrimination on the grounds of religion or belief, disability, age or sexual orientation as regards employment and occupation, with a view to putting into effect in the Member States the principle of equal treatment.; Article 2(1) provides: For the purposes of this Directive, the "principle of equal treatment" shall mean that there shall be no direct or indirect discrimination whatsoever on any of the grounds referred to in Article 1.

[398] For a full text of Article 21 of the *Charter of Fundamental Rights*, see *supra* note 130.

older persons in employment relations on the basis of age. Since the Directive required states to transpose the prohibition on age discrimination in employment to domestic law, a wealth of cases have been decided both at the local and European level.

One area that has been addressed on several occasions by the CJEU when applying the principle of non-discrimination on account of age, and that remains of extreme relevance for the protection of the rights of older persons, is the issue of mandatory retirement age. The case-law of the CJEU on this matter shows that a wide margin of discretion is provided to states to establish a mandatory retirement age.[399] Overall, the approach of the CJEU has been to place upon states the authority to fix, in light of political, economic, social demographic and budgetary considerations, the age limit to retire.[400] However, in making such assessment states must not "go beyond what is appropriate and necessary to achieve the legitimate aim pursued."[401] One consideration that has been very relevant in making such analysis on proportionality involves what has been called "inter-generation equality."[402] This notion requires states to strike a balance between the interests of older and younger workers when fixing age retirements, particularly in certain sectors.[403]

In addition, the *Recommendation on the human rights of older persons* specifically establishes that states should ensure that older persons do not face discrimination in employment on the basis of age, in both the private and public sectors. Areas of employment that older persons should enjoy on an equal basis with other workers include recruitment, vocational and continuous training, working conditions such as those related to dismissal and remuneration, membership in trade unions and retirement.[404] Moreover, this Recommendation encourages states "to include the promotion of participation of older persons in the labour market in their employment policies."[405] Specific attention should be provided in state policies to the issue of safety and health problems of older workers.[406] In *Resolution 1793(2011)*, the Parliamentary Assembly of the CoE encouraged states to adopt

[399]European Commission, *The Evolution and Impact of the Case-Law of the Court of Justice of the European Union on Directives 2000/43/EC and 2000/78/EC, European Network of Legal Experts in the non-discrimination field*, November 2012, Part III.

[400]Case C-45/09, *Gisela Rosenbladt v. Oellerking Gebäudereinigungsges.mbH., supra* note 140, 44.

[401]Case C-159/10, *Gerhard Fuchs and Peter Köhler v. Land Hessen, supra* note 146, at para. 65.

[402]European Commission, *The Evolution and Impact of the Case-Law of the Court of Justice of the European Union on Directives 2000/43/EC and 2000/78/EC, supra* note 399, Part III.

[403]See, *inter alia*, Joined Cases C-250/09 and C-268/09, *Vasil Ivanov Georgiev v Tehnicheski universited-Sofia, filial Plovdiv, supra* note 142, 51.

[404]*Recommendation CM/Rec(2014)2 of the Committee of Ministers to member states on the promotion of the human rights of older persons, supra* note 15, Article V.26.

[405]*Id.*, Article V.27.

[406]*Id.*, Article V.28.

legislation prohibiting discrimination, removing labor market barriers and empowering older persons to "enter, remain in, or return to the labour market in accordance with their capabilities and willingness to work."[407]

Finally, it is important to note that Articles 2, 3 and 4 of the *Revised Social Charter* protect other important aspects of labor rights that are essential for older workers, including the right to just conditions of work, the right to safe and healthy working conditions and the right to a fair remuneration. Moreover, the *Community Charter of the Fundamental Social Rights of Workers* ensures several labor rights, including those to which older persons are entitled, including retirement, medical and social assistance.[408] Additionally, Articles 27, 28, 29, 30 and 31 of the *Charter of Fundamental Rights* completes the labor protection afforded under European law.[409] As with the protections provided by the *Revised Social Charter*, some of these rights are essential for ensuring the right of older persons to work in fair conditions and on equal basis with other workers.

3. The Right to an Adequate Standard of Living and Housing

Article 23 of the *Revised European Social Charter* requires states to adopt specific measures towards enabling older persons to remain full members of society for as long as possible, including through the provision of the services and facilities needed to afford the elderly an adequate standard of living.[410] Even if Article 23 only makes reference to the state duty to provide information about services and facilities, the ECSR has interpreted that provision broadly.[411] The ECSR has concluded that states have an obligation to provide those services and facilities[412] and has also framed this duty as a right of older persons to claim certain services and facilities from the state.[413] On this matter, Article 23 is complemented by Article 14 of the *Revised Social Charter* which protects the right to benefit from social

[407]Parliamentary Assembly, *Resolution 1793(2011), Promoting active ageing – capitalising on older people's working potential*, January 28, 2011, 6.1.

[408]See Articles 24 and 25 of the Community Charter of the Fundamental Social Rights of Workers.

[409]Articles 27–32 protect the following labor rights: workers' right to information and consultation within the undertaking; right of collective bargaining and action; right of access to placement services; protection in the event of unjustified dismissal; and fair and just working conditions.

[410]See Article 23(1)(b), *supra* note 25.

[411]See, *inter alia*, ECSR, Conclusions 2003, France, at 90.

[412]*Id.*, where the Committee states that Article 23(1)(b) presupposes the existence of those services and facilities and consequently, when monitoring compliance with this duty it will inquire about the services and facilities themselves.

[413]*Id.*, where the Committee considers that in the case of older persons, the right to benefit from social welfare services protected in Article 14 of the *Revised Social Charter* will be monitored under Article 23.

welfare services. According to the ECSR, Article 14 enshrines the right to equal and effective access to those services, understood as access that is guaranteed in law and in practice and is capable of keeping pace with user's needs.[414] Those social services may include counselling, advice, rehabilitation and other forms of support from social workers, home help services (assistance in the running of the home, personal hygiene, social support, delivery of meals), residential care, and social emergency care (shelters).[415]

With respect to older persons specifically, the ECSR consistently monitors the provision of the following services: the existence, extent and cost of home help services or community services and how the provision of those services is monitored by the state[416]; the existence of specific care centers for persons with dementia and other related conditions[417]; and the training and financial assistance available for persons caring for an older person.[418] Another important issue that the ECSR monitors is whether states provide access to complaint procedures to claim the adequate provision of those services.[419] Access to such procedures relates to the existence of an available effective remedy, which is protected by Article 13 of the *European Convention*. Even if that provision applies in principle only with regard to the protection of the substantive rights enshrined in the treaty, the ECtHR has interpreted the scope of some of those rights, in particular private life, broadly enough to cover access to services to older persons under that notion.[420] The ECSR has acknowledged under Article 14 that an effective implementation of the right to benefit from social welfare services requires states to provide effective remedies to ensure equal and effective access of the beneficiaries to those services.[421]

The right of older persons to receive adequate facilities and services enabling them to enjoy and adequate standard of living is recognized by Article V.21 of the *Recommendation on the human rights of older persons*.[422] Among those services, states should take measures to facilitate mobility of older persons and ensure access to transport and other infrastructure.[423] Additionally, states should provide directly or in cooperation with private organizations other supplementary services such as adult day care, nursing care or preparation of meals.[424] All these measures

[414]ECSR, *International Federation for Human Rights (FIDH) v. Belgium*, *supra* note 122, at 108.

[415]ECSR, Digest of the Case Law of the European Committee of Social Rights, September 1, 2008, at 107.

[416]See, *inter alia*, ECSR, Conclusions 2013, Andorra, at 38.

[417]See, *inter alia*, ECSR, Conclusions 2005, France, at 29.

[418]*Id.*

[419]See, *inter alia*, ECSR, Conclusions 2013, Spain, at 35.

[420]See, *inter alia*, ECtHR, *Case of McDonald v. The United Kingdom*, *supra* note 304.

[421]ECSR, *International Federation for Human Rights (FIDH) v. Belgium*, supra note 122, at 108.

[422]*Recommendation CM/Rec(2014)2 of the Committee of Ministers to member states on the promotion of the human rights of older persons*, supra note 15, Article Principle V.21.

[423]*Id.*, Article V.22.

[424]*Id.*, Article V.24.

are intended to help older persons to remain living in their homes for as long as they choose, while ensuring that basic services are accessible to them.[425]

The *European Charter of Older Persons in Long-Term Care* acknowledges the right of older persons to quality of life.[426] This notion involves, *inter alia*, the right to be cared by persons with the skills and support needed to provide adequate care and treatment.[427] These individuals, either family members or professionals must receive specialized training and guidance as well appropriate technical, structural and financial resources.[428] In addition, caregivers must be provided with time breaks to ensure their own quality of life.[429] Another important issue contemplated within the notion of quality of life includes the right of older persons to receive timely support for their daily needs.[430] Also, the *European Charter of Older Persons in Long-Term Care* acknowledges the tailored care required by older persons who live in long-term residences,[431] including the support needed to ensure their mobility.[432]

With respect to housing, Article 23 establishes a duty of states to provide housing suited to the needs and health of older persons or adequate support to adapt elderly's housing to those needs. The underlying principle is that by ensuring access to appropriate housing, states will fulfil the overarching goal of enabling older persons "to choose their life-style freely and lead independent lives in their familiar surroundings for as long as they wish and are able."[433] Article 23 is complemented by Article 31 of the Revised European Social Charter which protects the right to housing. According to the ECSR "the notion of an adequate house implies a dwelling which is safe from a sanitary and health point of view …This means that dwellings must have access to natural and common resources, namely safe drinking water, electricity, sanitation facilities and waste disposal."[434] These requirements, which are applicable to everyone, must be adapted to ensure adequate housing to older persons given their particular vulnerability. To ensure that states implement that additional obligation, the ECSR constantly monitors how local authorities execute their duty to provide adequate housing to the elderly,[435]

[425] *Explanatory Memorandum of Recommendation CM/Rec(2014) of the Committee of Ministers to Member states on the promotion of human rights of older persons*, *supra* note 63, 44.

[426] *European Charter of the Rights and Responsibilities of Older Persons in Need of Long-Term Care and Assistance*, 2010, *supra* note 216, Article 4.

[427] *Id.*, Article 4.2.

[428] *Id.*

[429] *Id.*, Article 4.3.

[430] *Id.*, Article 4.6.

[431] *Id.*, Article 4.7.

[432] *Id.*, Article 4.8.

[433] Article 23 of the Revised Social Charter.

[434] ECSR, *European Roma Rights Centre v. Portugal*, Complaint No. 61/2010, 31.

[435] See, *inter alia*, ECSR, Conclusions 2003, France, at 92; Conclusions 2005, France, at 30.

the rules on security of tenure,[436] and if the supply of housing for older persons is adequate.[437] Also, it supervises whether public finance mechanisms are available to older persons for home renovation or adjustment.[438]

Article V.23 of the *Recommendation on the human rights of older persons* also acknowledges that states should provide support to "enable older persons to have housing adapted to their current and future needs."[439] Similar principles are promoted in other Resolutions of the Committee of Ministers and the Parliamentary Assembly.[440]

The effectiveness of the right to adequate housing requires access to affordable and impartial remedies, either judicial or administrate.[441] In the case of older persons, the age of an applicant should be considered a factor when analyzing the reasonableness of the length of proceedings under Article 6 of the *European Convention*,[442] and this includes administrative review of housing proceedings.[443] In *Spadea & Scalabrino v. Italy*, applicants brought the complaint during a subsidized housing shortage where evictions were suspended in Milan.[444] The landlord applicants claimed the Italian government had violated their right to property by allowing the older women to remain in the property for years after the end of their lease.[445] The older women were awaiting housing assignment from Milan's Housing Council.[446] The ECtHR ruled that the state legislation protected tenants during a serious housing shortage of subsidized housing, and that therefore the legislation did not violate the *European Convention*.[447] The ECmHR report, annexed to the ECtHR decision, stated that because the tenants were elderly and had low

---

[436]See, *inter alia*, ECSR, Conclusions 2003, France, at 92.

[437]*Id.*

[438]See, *inter alia*, ECSR, Conclusions 2013, Andorra, at 38.

[439]*Recommendation CM/Rec(2014)2 of the Committee of Ministers to member states on the promotion of the human rights of older persons*, supra note 15, Article V.23.

[440]See, *inter alia*, *Recommendation CM/Rec(2009)6 of the Committee of Ministers to member states on ageing and disability in the 21st century: sustainable frameworks to enable greater quality of life in an inclusive society*, Appendix to Recommendation CM/Rec(2009)6, para. 1.13; Parliamentary Assembly, *Resolution 1008 (1993), Social Policies for elderly persons and their self-reliance*, 5.c.

[441]ECSR, Conclusions 2003, France, at 128.

[442]See ECtHR, *Case of Süßmann v. Germany*, 20024/92, Judgment (Merits and Just Satisfaction), September 16, 1996, paras. 73–84 (noting the factors to be taken under consideration in determining the length of proceedings under Article 6 of the European Convention).

[443]See ECtHR, *Case Spadea & Scalabrino v. Italy*, 12868/87, Judgment (Merits), September 28, 1995, 31.

[444]ECtHR, *Case of Spadea & Scalabrino v. Italy*, *supra*, 8–16.

[445]*Id.*, 26.

[446]*Id.*, 39.

[447]*Id.*, 41.

income, they "were in fact entitled to special social protection."[448] The ECmHR also ruled that the protection afforded to the tenants was not disproportionate to the aim of social justice.[449] In contrast, in *A.O. v. Italy,* although it acknowledged that the tenant was an older sick person, the ECtHR still found that eviction proceedings should have been carried out more speedily in order not to breach the right of the owner to the enjoyment of his property.[450] It is important to note that in this case, however, the ECtHR took into account that the state authorities were able to find a subsidized apartment to the tenant immediately before the eviction was carried out.[451] Therefore, it concluded that such action could have taken place before, saved for the negligence of the state to act diligently.[452]

Another right ensured by Article 23 regarding services and housing involves the duty of states to guarantee older persons living in institutions appropriate support. Institutions pose a particular challenge for protecting older persons' rights; older persons who require institutionalization are often more vulnerable than older persons not requiring institutionalization. Compounding this dependence are heightened responsibilities to ensure the protection of all human rights while in the care of the state. This responsibility has been consistently upheld by the ECtHR with respect to other vulnerable persons, including persons with disability.[453]

This heightened duty of states to persons in care extends to institutions when provided by private contractors.[454] Under the *Revised Social Charter*, states have a duty to supervise that the rights protected by that treaty are protected by other institutions beyond the state, including private parties.[455] In its supervisory role, the ECSR has consistently requested states more details about the type of accommodations available as well as whether the supply of institutional facilities for older persons is sufficient.[456] Also, it has inquired whether older persons may be compulsorily placed in institutional care.[457] Other issues monitored included training and wage levels of staff of care institutions,[458] affordability of care[459] and

---

[448]See ECmHR, *Case of Giovanni Spadea et Michelangela Scalabrino v. Italy*, 12868/87, Report (Merits), 45.

[449]*Id.*, 46.

[450]ECtHR, *Case of A.O. v. Italy*, 22534/93, Judgment (Merits), May 30, 2000, 29.

[451]*Id.*

[452]*Id.*

[453]See, *inter alia*, ECtHR, *Case of Ketreb v. France*, 38447/09, Judgment (Merits and Just Satisfaction, July 19, 2012, 70–74.

[454]See, *inter alia*, ECSR, *European Roma Rights Center v. Italy*, Complaint n°27/2004, December 7, 2005, 26.

[455]See, *inter alia*, ECSR, *ERRC v. Greece*, Complaint No. 15/2003, December 8, 2004, 29.

[456]See, *inter alia*, ECSR, Conclusions 2009, Spain, at p. 31; Conclusions 2013, Ireland, at 26.

[457]See, *inter alia*, ECSR, Conclusions 2003, France, at p. 93; Conclusions 2005, France, at 31.

[458]*Id.*

[459]See, *inter alia*, ECSR, Conclusions 2013, The Netherlands, at 39.

whether all institutions are properly licensed.[460] Most importantly, the ECSR has consistently emphasized the duty of states to provide for an independent inspection system of public and private residential care services[461] and for a complaint mechanism to challenge the adequacy and availability of services.[462]

Finally, it is important to note that access to adequate social services and housing under the *Revised Social Charter* are not rights of immediate application, but of progressive implementation. With respect to the implementation of Article 14, the ECSR has concluded that "when the implementation of one of the rights guaranteed by the Charter is exceptionally complex and expensive, the measures taken by the state to achieve the Charter's aims must fulfil the following three criteria: (i) a reasonable timeframe, (ii) measurable progress and (iii) financing consistent with the maximum use of available resources."[463] With respect to the right to housing protected by Article 31, the ECSR acknowledged that this is not a right of "results", but of "means."[464] However, it noted that "the rights recognized in the Social Charter must take a practical and effective, rather than purely theoretical, form."[465] Thus, for states to be in compliance with the right to housing under the *Revised Social Charter*, they must: "a. adopt the necessary legal, financial and operational means of ensuring steady progress towards achieving the goals laid down by the Charter; b. maintain meaningful statistics on needs, resources and results; c. undertake regular reviews of the impact of the strategies adopted; d. establish a timetable and not defer indefinitely the deadline for achieving the objectives of each stage; e. pay close attention to the impact of the policies adopted on each of the categories of persons concerned, particularly the most vulnerable."[466] In consequence, applying these interpretations to the rights of older persons under Article 23 we must conclude that the protections regarding social services and facilities as well as housing are similarly to be considered of progressive implementation. Other obligations such as the principle of equality in access to services and housing and the right to an effective remedy must be considered as of immediate application. The obligation to create an independent system to supervise institutional residences for purposes of complying with existing regulations must be characterized as of immediate implementation.

---

[460]See, *inter alia*, ECSR, Conclusions 2013, Spain, at 36.

[461]See, *inter alia*, ECSR, Conclusions 2013, Andorra, at 39; Conclusions 2013, Spain, at 36.

[462]See, *inter alia*, ECSR, Conclusions 2005, France, at 31.

[463]ECSR, *International Federation for Human Rights (FIDH) v. Belgium*, *supra* note 122, 147.

[464]ECSR, *European Federation of National Organisations working with the Homeless (FEANTSA) v. France*, Complaint No. 39/2006, December 5, 2007, 53.

[465]*Id.*

[466]*Id.*, 54.

4. The Right to Social Protection and Security

Social protection encompasses cash-transfer and in-kind support. The area is exceedingly thoroughly litigated in Europe. The social protection scheme is woven into the fabric of European cultural identity. Older persons may face the urgency of dire circumstances, without the capability -or in some states the legal right- to work in formal or informal employment in old age. Where older persons' options have been curtailed by legal, physical, or economic barriers, the immediacy with which social protection is linked to the right to health, food, and housing is compounded.

There are numerous treaty provisions, enforcement from the ECSR and the ECtHR, and a rich body of case law regarding social protection in Europe. The following section will examine the law, its enforcement mechanisms, and the variety of ways social protection is analyzed in European jurisprudence.

a. European Treaty Law and Other Instruments Regarding Social Protection

European law protecting social security includes Article 34 of the *Charter of Fundamental Rights*. This article recognizes the right to social security in old age, and notes that "*Everyone* residing and moving legally within the European Union is entitled to social security benefits and social advantages in accordance with Community law and national laws and practices."[467] This provision draws, among others, from the corresponding provisions of the *European Charter for Social Rights* and the *Community Charter of the Fundamental Social Rights of Workers*.[468] It links social security to a decent existence and the fight against social exclusion and poverty.[469]

The European human rights system regularly refers to the treaties adopted within the CoE, including the *Revised Social Charter and* the *1988 Additional Protocol to* the *European Social Charter*, which articulates the right to social protection for older persons in Article 4. The language of the *1988 Additional Protocol* as well as its *Explanatory Report*, were then incorporated verbatim into Article 23 of the *1996 Revised Charter* and accompanying *Explanatory Report*.[470] The language of Article 23 establishing the right of older persons to social

[467]*Charter of Fundamental Rights of the European Union*, *supra* note 38, Article 34(2).

[468]Including Articles 12 and 13 of the *Revised Social Charter* and Articles and Articles 24 and 25 of the Community Charter of the Fundamental Social Rights of Workers. See, Text of the explanations relating to the complete text of the Charter as set out in CHARTE 4487/00 CONVENT 50, *supra* note 42.

[469]*Charter of Fundamental Rights of the European Union*, *supra* note 38, Article 34(3).

[470]See *Explanatory Report to the Additional Protocol to the European Social Charter*, *supra* note 316, 81–82.

protection provides that states undertake to ensure the elderly "with adequate resources enabling them to lead a decent life and play an active part in public, social and cultural life."[471] The ESCR stated that the underlying objective of this provision is to "enable elderly persons to play an active part and have some influence in society, to guarantee them sufficient resources to live independently, to provide housing and an environment suited to their needs, and to guarantee adequate health care and social services."[472] Article 23 is complemented by the protection afforded by Articles 12 and 13 of the *Revised Social Charter* which protect the rights to social security and social and medical assistance, respectively.

Though the *Revised Charter* names the right of older persons to social protection as one of the aims which state parties are to pursue by all appropriate means,[473] the protection of older persons has been revealingly limited.[474] This is one of the few areas where older persons are explicitly referenced in international European law; however, agreement to be bound on the right to social protection was not necessary for ratification of the *1996 Revised Charter*.[475] There has been speculation that the cost of social security systems' development, combined with robust compliance mechanisms, has discouraged the embrace of true consensus on international social security.[476]

Additionally, the *Recommendation on the human rights of older persons* provides that older persons should receive appropriate resources enabling them access to an adequate standard of living and participation in public, economic, social and

---

[471]For a full text of Article 23, see *supra* note 25.

[472]ECSR, Conclusions XV-2 2001, Finland, at pp. 149, 182 (responding to Finland's report on the developments in the protections of the social rights of the elderly).

[473]See Article A of the *Revised Social Charter* (stating that states undertake the consideration of particular aims declared in Part I of the Charter). Part I provides that "every elderly person has the right to social protection." See Part I, 23.

[474]See Chart of Signatures and Ratifications of the European Social Charter (Revised) http://www.coe.int/t/dghl/monitoring/socialcharter/Presentation/SignaturesRatifications_en.pdf; see also List of Declarations Made with Respect to Treaty no. 163, available at http://conventions.coe.int/Treaty/commun/ListeDeclarations.asp?NT=163&CM=10&DF=27/10/2011&CL=ENG&VL=1 (last visited Dec. 14/2014). As of March 2013 only 17 out of the 33 states parties to the *Revised Social Charter* had accepted to be bound by Article 23.

[475]*Revised Social Charter*, Article A, Undertakings. Upon ratification, states parties undertake to be bound by six of the following nine rights: the right to work (Article 1); the right to organize (Article 5); the right to bargain collectively (Article 6); the right of children and young persons to protection (Article 7); the right to social security (Article 12); the right to social and medical assistance (Article 13); the right of the family to social, legal, and economic protection (Article 16); the right of migrant workers and their families to protection and assistance (Article 19); and the right to equal opportunities and equal treatment in matters of employment and occupation without discrimination on the grounds of sex (Article 20). States may decide to undertake to be bound by other provisions, including Article 23.

[476]See, e.g. Diego Rodriguez-Pinzon and Claudia Martin, *The International Human Rights Status of Elderly Persons*, 18 Am. U. Int'l L. Rev. 915, 991, 1000–1004 (2003).

cultural life.[477] Moreover, the Parliamentary Assembly of the Council of Europe called the Committee of Ministers to urge Member states to preserve the right to pensions and its adequacy and sustainability.[478]

b. European Committee of Social Rights

As stated before, the ECSR monitors whether states party are in conformity in law and in practice with the provisions of the *Revised Social Charter*. The *Revised Charter* does not require states to follow a particular pension scheme, but the ECSR adopts Conclusions in response to national reports and decisions for collective complaints assessing whether state parties to the treaty are ensuring the rights protected under Article 23 of the *Revised Social Charter*.

The primary focus of the right of older persons to adequate resources is on ensuring their access to pensions, contributory or non-contributory, and other social benefits enabling them to have a decent life.[479] To achieve that goal, the ECSR examines the amount of pensions and other complementary cash benefits available to elder persons and compares them with the "median equivalised income" of the country.[480] Moreover, the ECSR also takes into account other "relevant indicators relating to risk-of-poverty for persons aged 65 and over."[481] On those bases, it assesses, if pensions or other social benefits are sufficient to meet the standard of "adequate resources" under Article 23 of the *Revised Social Charter*.[482] Additionally, the ECSR monitors national qualification conditions of beneficiaries such as minimum age, nationality and residence requirements.[483] As part of supervising the standard of living for the elderly, it also examines the cost of care, specifically home care and pharmaceutical products, to ensure that the costs do not negatively affect their status as full members of society by creating too heavy a financial burden.[484]

---

[477] *Recommendation CM/Rec(2014)2 of the Committee of Ministers to member states on the promotion of the human rights of older persons*, *supra* note 15, Article V.21.

[478] Parliamentary Assembly, *Recommendation 2000(2012), Decent pensions for all*, adopted on April 27, 2012.

[479] See, *inter alia*, ECSR, Conclusions 2013, France, at 43; Conclusions 2013, Finland, at 31.

[480] *Id.* "The equivalised disposable income is the total income of a household, after tax and other deductions, that is available for spending or saving, divided by the number of household members converted into equalised adults; household members are equalised or made equivalent by weighting each according to their age, using the so-called modified OECD equivalence scale," available at http://ec.europa.eu/eurostat/statistics-explained/index.php/Glossary:Equivalised_disposable_income (last visited Dec. 14, 2014).

[481] *Id.*

[482] ECSR, Conclusions 2009, Spain, at 30; Conclusions 2013, Ireland, at 24.

[483] See, *inter alia*, ECSR, Conclusions 2003, Italy, at 74; Conclusions 2009, Spain, at 30.

[484] See, inter alia, ECSR, Conclusions XV-2, Finland, at 183–84 (assessing the amount of expenditures Finland devotes to the elderly in subsidizing the costs of home care and pharmaceuticals).

Additionally, the *Explanatory Report* to the *Revised Social Charter* states that the phrase "adequate resources" (enabling them to lead a decent life) in Article 23 is to be interpreted in light of Article 13 (Social and Medical Assistance),[485] and if necessary, Article 12 (Social Security)[486] of the *1996 Revised Charter*. The *Explanatory Report* states that "social protection" in Article 23 and "social assistance" in Article 13 are compatible,[487] though not equivalent. The aim of Article 23 is to ensure participation in the community as full members, and the aim of Article 13 is to reduce poverty, which may require filling a gap left when a pension does not cover an individual's basic necessities.[488]

According to the ECSR, under Article 13 states must provide social and medical assistance to those in need as a matter of right,[489] in other words assistance may not depend only on the discretion of public authorities. This right must be prescribed by law and claimants must be ensured a right to appeal if assistance is denied.[490] The law must provide for objective and precise terms.[491] In particular, "it must define the elements taken into account in order to assess the state of need and make the criteria for assessment of that need clear, as well as the procedure for determining whether a person lacks adequate resources, including the methods to investigate resources and needs."[492] With respect to the right to appeal, the body in charge of deciding can be a court or administrative agency, but must be independent and hand down binding decisions.[493] Unfavorable decisions from that body must be subject to appeal.[494] The ECSR has also inquired whether legal aid is provided to those who would like to resort to available remedies, but do not have the resources to afford the litigation.[495]

The *Revised Charter*'s Article 13(1) provides that "any person who is without adequate resources" is entitled to adequate social and medical assistant. The ECSR

---

[485]*Revised Social Charter*, Article 13 (setting forth the obligations to provide social and medical assistance).

[486]See *id.* Article 12 (setting forth the obligations to provide social security).

[487]*Explanatory Report to the Additional Protocol to the European Social Charter, supra* note 316, (noting that the language expressed in Article 13 and Article 23 does not render the provisions incompatible).

[488]ECSR, Malta, Conclusions XV-1 at pp. 395, 411 (2001) (stating that a goal of the Charter is for states to have a strategy to assist individuals and families in overcoming poverty).

[489]ECSR, *European Roma Rights Centre (ERRC) v. Bulgaria*, Complaint No. 48/2008, February 18, 2009, 37.

[490]ECSR, *Digest of the Case Law of the European Committee of Social Rights*, *supra* note 415, at 100.

[491]*Id.*

[492]*Id.*

[493]*Id.*

[494]See, *inter alia*, ECSR, Conclusions 2013, Hungary, at p. 24; Conclusions, 2013, at 32.

[495]ECSR, Conclusions 2006, Bulgaria, section on Article 13; Conclusions 2009, Finland, at 23; Conclusions 2013, Hungary, at 24; Conclusions 2013, Turkey, at 30.

has interpreted "any person" under this paragraph to include, in addition to nationals and foreigners legally residing or lawfully working, other foreign migrants who are in an irregular situation of stay in the territory of the state. State parties to the *Revised Social Charter* must provide the latter with urgent medical assistance and such basic social assistance as is necessary to cope with an immediate state of need (accommodation, food, emergency care and clothing).[496] On the other hand, Article 13(4) provides that states must provide social and medical assistance "on an equal footing with their own nationals to nationals of other [state] Parties who are lawfully resident or working regularly within their territory, or otherwise lawfully present within their territories."[497] Narrowing of criteria for eligibility, including length of residency or age, may be in violation of the protections afforded by the *Revised Social Charter*.[498]

Need must be the sole criterion for social assistance, illness must be the sole criterion for medical assistance. Need may be defined as the inability to obtain by his or her own efforts or from other sources, including benefits from a social security scheme, adequate resources to live a decent life and meet basic needs.[499] The ECSR monitors the adequacy of the assistance provided by the state in case of need by taking into account the amount perceived by a single person alone and assessing whether such amount is not manifestly below the poverty threshold.[500] As Europeans remain mobile into older age, this provision may become increasingly important. Older persons may wish to retire in a different country from the country where they spent their working years; older persons may move to be with adult children who have immigrated or older persons may seek to live where home assistance is relatively inexpensive.

On the other hand, Article 12 guarantees the right to social security by establishing a social security system (paragraph 1), defining a minimum level for this system (paragraph 2) describing its progressive implementation (paragraph 3) and, finally, by encouraging measures to ensure equality of treatment.[501] The *Revised Social Charter* does not impose a common model of social security system or harmonization on social security legislation, but it establishes minimum common standards that all state parties must abide by.[502] Thus, in practice, states are free to establish their own models of social security systems.[503]

---

[496]ECSR, Conclusions (2013), General Introduction, at 8.

[497]*Id.*, at 7–8.

[498]ECSR, Conclusions XVII-1, 2004, (Spain), section on Article 13; Conclusions 2009, Spain, at p. 23; Conclusions 2013, Spain, at 25.

[499]ECSR, *Digest of the Case Law of the European Committee of Social Rights*, *supra* note 415, at 98.

[500]See, *inter alia*, ECSR, Conclusions 2009, France, at p. 32; Conclusions 2013, Hungary, at 23.

[501]Lenia Samuel, *Fundamental Social Rights, Case Law of the European Social Charter*, 2nd ed., (2002), at 283.

[502]ECSR, *Digest of the Case Law of the European Committee of Social Rights*, *supra* note 415, at 89.

[503]*Id.*

According to the ECSR, a social security system exists when the following standards are met: (1) the system affords medical care, sickness benefit, unemployment benefit, old age benefit, employment injury benefit, family benefit and maternity benefit; (2) it is collectively funded through contributions made by workers, employers and the state; and (3) it covers a significantly percentage of the population and must be awarded on the basis of the principle of non-discrimination.[504] Furthermore, the benefits must be adequate to ensure the right to social security. [505]As with the obligation to provide adequate resources to older persons, the ECSR examines adequacy of pensions in general on the basis of the notion of "median equivalised income."[506]

Next, in light of paragraph 2 of Article 12, to ensure the effective right to social security, states must at least establish and maintain a system equal to that required by the *European Code of Social Security*. Consequently, for states that have ratified the former instrument, the ECSR monitors compliance with Article 12(2) in light of the obligations arising out of that treaty.[507] In relation to old-age benefits, for example, the Code regulates the age of qualification for a pension,[508] the persons comprised by this benefit,[509] and even the procedure by which the amount of the benefits should be calculated.[510]

Thirdly, under Article 12(3) states undertake to "raise progressively the system of social security to a higher level." The application of this provision has proven to be controversial in times of financial crisis requiring drastic reductions of public budgets. In *PanHellenic Federation of Public Service Pensioners v. Greece*, the ECSR assessed whether the reductions of pensions resulting from the financial and economic crisis in Greece violated the state's duties under Article 12(3) of the *Revised Social Charter*. First, the ECSR reaffirmed that reductions to benefits available in the social security system do not per se constitute a violation of that provision.[511] However, when economic reasons are at the basis of a state's decision to modify the benefits so far provided, it is necessary "for that state to maintain the social security system on a satisfactory level that takes into account the legitimate expectations of beneficiaries of the system and the right of all persons to

---

[504]See, *inter alia*, ECSR, Conclusions 2009, Belgium, at 20; ECSR, *Digest of the Case Law of the European Committee of Social Rights*, *supra* note 415, at 89.

[505]*Id.*

[506]*Id.*

[507]ECSR, Conclusions 2013, Ireland, at 143.

[508]See *id.* Article 26 (codifying that the age at which benefits are first granted cannot be over sixty-five, unless appropriate circumstances, such as economic or social criteria, justify a higher age).

[509]See *id.* Article 28 (setting out the qualifications of individuals who would receive old-age benefits).

[510]See *id.* Articles 29, 71–72 (describing the process of calculating old-age benefits).

[511]ECSR, *PanHellenic Federation of Public Service Pensioners (POPS) v. Greece*, Complaint No. 77/2012, December 7, 2012, 63.

effective enjoyment of the right to social security."[512] Second, for a restriction of the right to social security to be compatible with Article 12(3) it must be prescribed by law, pursue a legitimate aim such as the protection of the rights and freedoms of others, the protection of public interest, national security, public health or morals.[513] When assessing the proportionality of restrictions to the right to social security due to economic and financial considerations, the ECSR has taken into account four factors, namely: "a. the nature of the changes (field of application, conditions for granting allowances, amounts of allowance, lengths, etc.); b. the reasons given for the changes and the framework of social and economic policy in which they arise; the extent of the changes introduced (categories and numbers of people concerned, levels of allowances before and after alteration); c. the necessity of the reform, and its adequacy in the situation which gave rise to these changes (the aims pursued); d. the existence of measures of social assistance for those who find themselves in a situation of need as a result of the changes made; and e. the results obtained by such changes."[514]

In *PanHellenic Federation of Public Service Pensioners* and in *Federation of Employed Pensioners of Greece*, the ECSR reviewed the measures adopted by the state in light of those standards. It concluded that even if some of the reductions in themselves did not violated Article 12(3),[515] the cumulative effect of the restrictions, particularly as they affected the most vulnerable groups of society, failed to comply with the obligations stated in the treaty.[516] Interestingly, in both cases, the ECSR took into account the obligations arising out of Article 4 of the Optional Protocol, similar to Article 23 of the Revised Charter, to analyze the measures adopted by Greece since an important group of the beneficiaries affected included older persons whose pensions or similar benefits were reduced as a result of the measures of reduction passed by the state.[517]

In addition to complement the interpretation of Article 23, Articles 12 and 13 are essential to monitor the right to social security and the right to social and medical services of older persons in cases in which states have not accepted to be bound by Article 23.[518]

---

[512] *Id.*

[513] *Id.*, 65.

[514] *Id.*, 67.

[515] ECSR, *PanHellenic Federation of Public Service Pensioners (POPS) v. Greece*, *supra* note 511, 72; *Federation of Employed Pensioners of Greece (IKA-ETAM) v. Greece*, Complaint No. 76/2012, December 7, 2012, 77.

[516] *Id.*, 73–78 and 78–83, respectively.

[517] *Id.* 71, 76 and 76, 81, respectively.

[518] ECSR, Conclusions 2013, Bulgaria, at 25 and 28–9, respectively; Conclusions 2013, Austria, at 21–22 and 31–32, respectively.

c. *European Court of Human Rights*

Even if the ECtHR does not provide explicitly for the right to social security, both the ECmHR and ECtHR have applied certain civil and political rights recognized in the *European Convention* to protect the rights of older persons to social security. In particular, issues related to social security and other social benefits affecting the elderly have been litigated before the ECtHR on the basis of the rights to obtain a decision within a reasonable time, the right to property and the right to equal protection and non-discrimination.

i. ECtHR on Immediacy

The evolution of the case law has gradually taken more and more notice of the special situation of older persons, depending on social security or pension payments to obtain a judgment in a dispute within a reasonable amount of time. In the *Deumeland case*,[519] the applicant applied for a widow's supplementary pension when industrial accident caused her husband's death.[520] She began proceedings in 1970, and they were not finalized until 1981, when the case was rejected.[521] The applicant claimed that the German courts had not given the case a fair hearing within a reasonable time, which violated Article 6 paragraph 1 of the *European Convention*.[522] The ECtHR case law dictated an analysis of the degree of complexity of the case, the behavior of the applicant, and the conduct of the competent courts.[523] The ECtHR found a violation of the right to a hearing within a reasonable time, noting that social security cases required "particular diligence."[524] This case represents the beginnings of an understanding of the immediacy of the right to social security for older persons in Europe within the ECtHR.

In *Süßmann*,[525] another pension case, the ECtHR assessed the reasonableness of the length of the proceedings, the complexity of the case, the behavior of the parties, and the significance of what was at risk for the applicant in the litigation.[526] Elaborating upon the right to a fair trial further, the ECtHR in the *Süßmann* case examined whether the right to an expeditious pension trial was contravened by a 3-year delay in a verdict. In such situations, the Court established that the

---

[519]ECtHR, *Case of Deumeland v. Germany*, 9384/81, Judgment (Merits and Just Satisfaction), May 29, 1986.

[520]*Id*, 10.

[521]*Id*, 11, 45.

[522]*Id.*, 58 (quoting the *European Convention on Human Rights*, Article 6, paragraph 1, in support of the claim that plaintiff had not received a fair hearing within a reasonable time).

[523]*Id.*, 78–89.

[524]*Id.* 90 (noting that the litigation at issue spanned approximately an eleven year period).

[525]ECtHR, *Case of Süßmann v. Germany*, 20024/92, *supra* note 442.

[526]*Id.* 48.

administrative and judicial authorities are obligated under Article 6 paragraph 1 to act with "exceptional diligence."[527] While carefully circumscribing the ruling within the historical deluge of pension cases in Germany at that time, and to distinct standards of constitutional courts, *Süßmann* court found no violation of the right to a hearing within a reasonable time.[528] Importantly, the ECtHR analyzed that given the age of the applicant, the risk in litigation in domestic courts were "of undeniable importance to him."[529]

In *Solodyuk v. Russia,* the ECtHR addressed the issue of immediacy with respect to the right to social security payments.[530] The Court found that the applicants' entitlement to receive their pension in the month for which it was due was established by law and had been indirectly confirmed by court decisions.[531] However, during an eleven-month period, payment of the pensions had been delayed for up to four months, at a time when inflation was very unstable and resulted in a significant loss of purchasing power, and in the Court's view that had imposed an individual and excessive burden on the applicants.[532] The ECtHR concluded that this situation disproportionately affected the right of the applicants to their right to property under Article 1 of Protocol 1 to the *European Convention.*[533]

ii. ECtHR on Property Rights to Pensions

The ECtHR and the former ECmHR, have created a robust body of rulings recognizing the right to benefits in connection with the payment of contributions to the social insurance system.[534] Individuals can assert these rights in accordance with the right to peaceful enjoyment of possessions protected by Article 1 of the *European Convention's* Protocol 1.[535] Even if this provision does not impose upon

---

[527]See, *inter alia, A. v. Denmark*, 20826/92, Judgment (Merits and Just Satisfaction), February 2, 1996, 78.

[528]ECtHR, *Case of Süßmann v. Germany, supra* note 442, 62.

[529]*Id.* 61.

[530]ECtHR, *Case of Solodyuk v. Russia*, 67099/01, Judgment (Merits and Just Satisfaction), July 12, 2005.

[531]*Id.*, 27.

[532]*Id.*, 35–6.

[533]*Id.*, 36.

[534]See, *inter alia*, ECmHR, *Mueller v. Austria*, App. No. 5849/72, 3 Eur. Comm'n H.R., Dec. & Rep. 25 (1975) (deciding pension rights based on contributions to a fund may fall within the protection of Article 1).

[535]Article 1 of Protocol 1 states: Every natural or legal person is entitled to the peaceful enjoyment of his possessions. No one shall be deprived of his possessions except in the public interest and subject to the conditions provided for by law and by the general principles of international law. The preceding provisions shall not, however, in any way impair the right of a State to enforce such laws as it deems necessary to control the use of property in accordance with the general interest or to secure the payment of taxes or other contributions or penalties.

the states a duty to put in place a social security system, once that system is created the beneficiary is entitled to a proprietary interest falling under the scope of that Article.[536] This does not insulate individuals from state interference with pension amounts. However, for an interference to be compatible with Article 1 of Protocol 1 it must be lawful, pursue a legitimate public or general interest and be reasonably proportionate to the aim that it intends to achieve.[537] States have a wide margin of appreciation to assess what falls within the notion of public interest, given the fact that the enactment of laws regarding social security benefits commonly involves consideration of social and economic policies.[538] At the same time, the ECtHR still retains the power to oversee whether the measures concerned impose upon persons an individual and excessive burden. The ECtHR assesses in each case on the basis of a "fair balance" standard[539] and within the context in which the facts transpired, whether the interference at issue imposes such an excessive burden.[540]

Many cases have been litigated before the ECtHR regarding reduction or termination of social security or related benefits. Overall, the ECtHR has found that reduction or termination measures may constitute an interference with the right to property under Article 1 of Protocol 1.[541] However, since this provision does not guarantee per se a right to a pension of a particular amount,[542] measures of reduction or termination need to be assessed on its own merits to determine whether or not they breached the right to property.[543] In the *Case of Janković v. Croatia*, the applicant served as an officer in the former Yugoslav People's Army and retired as such before the dissolution of that state.[544] Upon dissolution, his social security payments stopped, but were later resumed by the Croatian Social Security Fund.[545] As a consequence, the amount of his pension was reduced to about 63 % of what he had received before.[546] The ECtHR reviewed the case and concluded that in the balance between his right to a pension and the general interest of the

---

536 See, *inter alia*, ECtHR, *Case of Rasmussen v. Poland*, 38886/05, Judgment (Merits and Just Satisfaction), April 28, 2009, 71.

537 See, *inter alia*, ECtHR, *Case of Stefanetti and Others v. Italy*, 21838/10, 21849/10, 21852/10, 21855/10, 21860/10, 21863/10, 21869/10 and 21870/10), Judgment (Merits), April 15, 2014, 52.

538 See, *inter alia*, ECtHR, *Case of Iwaszkiewicz v. Poland*, 30614/06, Judgment (Merits and Just Satisfaction), July 26, 2011, 43.

539 The fair balance must be struck between the demands of the general interest of the community and the requirements of the protection of the individual's fundamental rights. Id., at para. 44.

540 See, *inter alia*, ECtHR, *Case of Wieczorek v. Poland*, 18176/05, Judgment (Merits and Just Satisfaction), December 8, 2009, 64.

541 See, *inter alia*, ECtHR, *Case of Iwaszkiewicz v. Poland*, *supra* note 538, 41; ECtHR, *Case of Rasmussen v. Poland*, *supra* note 536, 71.

542 See, *inter alia*, ECtHR, ECtHr, *Case of Iwaszkiewicz v. Poland*, *supra* note 538, 41.

543 See, *inter alia*, ECtHR, *Case of Stefanetti and Others v. Italy*, *supra* note 537, 59–60.

544 ECtHR, *Case of Jancović v. Croatia*, 43440/98, Decision on Admissibility, October 12, 2010.

545 *Id.*, at 2.

546 *Id.*

community the measure was not disproportionate.[547] Even if he lost a portion of his pension, the essence of the applicant's pension rights was not impaired.[548]

Even if the ECtHR has been willing to accept reductions of pensions based on social and economic policies as compatible with Article 1 of Protocol 1, under certain circumstances the deprivation of the entirety of a pension has been considered a violation of the right to property.[549] In the *Kjartan Ásmundsson* case, the applicant lost his disability pension due to law changes in the parameters used to assess disability. At the time the pension payments started, the law only required inability to work as a seaman, a condition that the applicant fulfilled.[550] Later on, the law was changed to address the financial problems of the disability pension fund.[551] Under the new law, disability pensions would be paid only to workers who had become incapacitated to perform any work.[552] Since the applicant was considered 25 % incapacitated to work and had found employment as an office assistant, he did not fulfill the requirements of the new law.[553] Thus, the disability pension that he perceived until then was discontinued.[554] The ECtHR found that the measure failed to strike a proper balance between the applicant's right to property and the general public interest in addressing the financial issues of the disability fund.[555] Consequently, the termination of her pension was found disproportionate and an excessive burden to the applicant, in violation of Article 1 of Protocol 1 to the *European Convention*.[556]

iii. ECtHR on Discrimination

The issue of discrimination in pension cases has been reviewed by the ECtHR on several occasions.[557] As provided in the section of the right of older persons to equality and non-discrimination, the test for establishing discrimination under Article 14 of the *European Convention* involves a measure that treats people in similar or analogous situation differently, does not pursue a legitimate aim and/or

---

547 *Id.*, at 5.

548 *Id.*

549 See, *inter alia*, ECtHR, *Case of Kjartan Ásmundsson v. Iceland*, 60669/00, Judgment (Merits and Just Satisfaction), October 12, 2004. The ECtHR has also found that substantial reductions to the pension payments may constitute a violation of the right to property. See, ECtHR, *Case of Stefanetti and Others v. Italy*, *supra* note 537, 66–7.

550 ECtHR, *Case of Kjartan Ásmundsson v. Iceland*, *supra* note 549, 8.

551 *Id.*, 10.

552 *Id.*

553 *Id.*, 12.

554 *Id.*

555 *Id.*, 44–45.

556 *Id.*

557 See, *inter alia*, ECtHR, *Case of Stec and Others v. United Kingdom*, 65731/01 and 65900/01, Judgment (Merits), April 12, 2006; *Case of Andrle v. the Czech Republic*, *supra* note 102.

fails to strike a reasonable proportionality between the means employed and the aim sought to be realized.[558] Even though the ECtHR has strictly scrutinized measures establishing distinctions on the bases of some so called "suspect grounds," including sex, sexual orientation, race and ethnic origin and nationality,[559] at the same time it has shown a high degree of deference for the state role in setting social policy, especially in the context of establishing social security schemes.[560] National authorities are better place to appreciate what is involved in the public interest of their society when it comes to social or economic policies.[561] Thus, the ECtHR "will generally respect the State's policy choice unless that it is "manifestly without reasonable foundation."[562]

In *Stec and Others v. United Kingdom*, the applicants argued that the conditions governing entitlement to the non-contributory benefits in issue in their cases were discriminatory on grounds of sex since the right to continue to receive the benefits was reliant upon the different minimum age for men and women used by the old-age pension laws in the United Kingdom, namely 65 for men and 60 for women.[563] To rely on the non-discrimination Article 14 of the *European Convention*, the applicants persuaded the ECtHR that, notwithstanding the non-contributory nature of the benefits, they had a proprietary interest to which Article 1 of Protocol No. 1 applied.[564] The ECtHR reiterated in this connection that this provision does not require states to set up a particular form of social security system or a specific amount in the benefits that it provides.[565] However, if a Contracting state has in force legislation providing for the payment as of right of a welfare benefit, such entitlement must be regarded as generating proprietary interest falling within the ambit of Article 1 of Protocol No. 1.[566] Moreover, states must ensure that, regardless of the form or scope of the benefits or pension scheme, access to the system is provided in full compliance with the principle of equality.[567] This does not impact the right to social security directly, only the right to non-discrimination in the application of a social security scheme, regardless of whether it is contributory or non-contributory.

---

[558]See, *inter alia*, ECtHR, *Case of Stec and Others v. United Kingdom*, *supra*, 51.

[559]European Commission Justice, *The Prohibition of Discrimination under European Human Rights Law. Relevance for the EU non-discrimination directives – an update*, *supra* note 103. Also, ECtHR, *Case of Stec and Others v. United Kingdom*, *supra* note 557, 52.

[560]See, *inter alia*, ECtHR, *Case of Stec and Others v. United Kingdom*, *supra* note 557, 52; *Case of Andrle v. the Czech Republic*, *supra* note 102, 50.

[561]See, *inter alia*, ECtHr, *Case of Stummer v. Austria*, 37452/02, Judgment (Merits), July 7, 2011, at para. 89; *Case of Andrle v. the Czech Republic*, *supra* note 102, 50.

[562]*Id.*

[563]ECtHR, *Case of Stec and Others v. United Kingdom*, *supra* note 557, 15–25.

[564]*Id.*, 53.

[565]*Id.*

[566]*Id.*

[567]*Id.*

The ECtHR found that there had been no violation of Article 14 of the *European Convention* taken in conjunction with Article 1 of Protocol No. 1 "because the difference in pensionable able between men and women was originally intended to correct the disadvantaged economic position of women."[568] Even if any distinction on the basis of sex required very weighty reasons to be compatible with Article 14, the state had a wide margin of discretion to implement measures of economic or social policy. In using that discretion, the state had assessed the need to make gradual changes to equalize the pensionable ages after it was demonstrated that social conditions have changed and women were not affected because of a shorter working life.[569] Those changes had been implemented slowly and in stages to reflect the far-reaching consequences for women and the economy in general. Thus, in this particular case, the state had not overstepped its margin of discretion to establish justified distinctions within the framework of its social security scheme.[570] Also, the ECtHR concluded that linking the eligibility of the non-contributory benefit at issue with the pensionable able was not discriminatory either because such benefit was similarly intended to compensate for reduced earning capacity during one's working life.[571]

A similar discrimination argument was presented in *Andrle v. the Czech Republic*, where the applicant claimed that since he had cared for two minor children after his divorce he could retire at the age of 57, the pensionable age set for women who took care of children.[572] When his petition was denied by the domestic courts, he filed a complaint with the ECtHR arguing a violation of his right to enjoy property without discrimination.[573] Again in this case, the ECtHR found that this measure was rooted on the particular circumstances of protecting women who had to face "factual inequality and hardship arising out of the combination of the traditional mothering role of women and the social expectation of their involvement in work on a full-time basis."[574] As with the situation in the UK, the Czech Republic had already moved towards pension equalization at the time the case was litigated.[575] The ECtHR concluded that the distinction in the pensionable age of women and men intended to address past inequalities and its modification involved a complex process requiring financial commitments and long-term planning.[576] Thus, the progressive implementation of measures toward equalization fell within

---

568 *Id.*, 66.

569 *Id*, 62–64.

570 *Id.*, 65.

571 *Id.* 66.

572 ECtHR, *Case of Andrle v. the Czech Republic*, *supra* note 102, 5–14.

573 *Id.*, 26.

574 *Id.*, 54, 56.

575 *Id.*, 57.

576 *Id.*, 58.

the margin of appreciation enjoyed by the state and did not violate Article 14 in connection to Article 1 of Protocol 1 to the *European Convention*.[577]

In *Andrejeva v. Latvia*, the issue at stake was whether a distinction on the basis of nationality used to calculate the pension rights of the applicant involved a violation of Article 14 in connection to Article 1 of Protocol 1.[578] Ms. Andrejeva was born in Kazakhstan and came to Latvia at the age of 12 when it was one of the republics of the Soviet Union.[579] She resided permanently in Latvia since then.[580] In 1991, after Latvia's independence and the cessation of the Soviet Union as a state, the applicant became stateless.[581] To address the situation of individuals in the same position as Mr. Andrejeva, Latvia passed a law awarding former Soviet Union citizens without Latvian or other citizenship the status of "permanent resident non-citizen."[582] The applicant worked for an enterprise that was initially under the authority of the Soviet Union and later under autonomous management until 1993 when she was fired in staff cutbacks.[583] She worked for another employer in Latvia until her retirement in 1997.[584] At the time of claiming her pension she was informed that, as a stateless person, only the years worked in Latvia would be counted towards her pension; the years she worked for the Soviet company were considered as performed outside Latvia because she was "employed by entities based in Kyiv and Moscow."[585] Citizens of Latvia placed in a similar position that hers, however, were acknowledged the years of work towards the calculation of their pensions.[586] The ECtHR found that the distinction based on nationality in this case was not reasonably proportional to justify the measure.[587] Even if the state had a margin of discretion, in this particular case, the state had not overcome the weighty reasons required to justify the difference of treatment between the applicant and the Latvian citizens.[588] Thus, the ECtHR found a violation of Article 14 in connection to Article 1 of Protocol 1 of the *European Convention*.[589]

Additionally, the ECtHR has considered cases on alleged discrimination in pension rights in which applicants claimed that the difference in treatment was based

---

[577]*Id.*, 60–61.

[578]ECtHR, *Case of Andrejeva v. Latvia*, *supra* note 90.

[579]*Id.*, 10.

[580]*Id.*

[581]*Id.*, 13.

[582]*Id.*

[583]*Id.*, 11, 14.

[584]*Id.*, 14.

[585]*Id.*, 15.

[586]*Id.*, 87.

[587]*Id.*, 87–88.

[588]*Id.*, 89.

[589]*Id.*

on their place of residence,[590] their status as a married person,[591] or their status as prisoners.[592] In all those cases, the ECtHR followed a similar approach of balancing out whether the reasons for treating individuals differently could be justified in light of the wide margin of appreciation enjoyed by states when dealing with social security matters. Interestingly to note in the *Stummer* case, where the applicant claimed discrimination because his status as a prisoner prevented him from being affiliated with the old-pension system even if he worked for lengthy periods, the ECtHR followed an approach similar to the cases alleging sex discrimination. The ECtHR found no discrimination, but suggested that the state should keep this issue under review due to the evolving trends regarding incorporation of prisoners into the state pension scheme.[593] Since the notion of what constitutes discrimination is an evolving concept, it is clear that some cases in which the ECtHR ruled in favor to the state on the basis of the general public interest may not in the future be found to be compatible with the *European Convention*, particularly those in which sex has been the ground for distinction.

5. The Right to Health

Article 23(b) of the *Revised European Social Charter* establishes that states must adopt measures to provide older persons with the services necessitated by their state of health. In this regard, "services" includes admission to specialized institutions for older persons when appropriate.[594] The requirement of the provision of these services assumes the existence of an adequate number of institutions that the state can establish care directly or in cooperation with other relevant public or private organizations.[595] Additionally, states must provide special training for individuals caring for the elderly[596] and must improve accessibility and quality of long term and geriatric care.[597] However, when supervising compliance with this provision, the ECSR has interpreted the scope of that Article much more broadly, encompassing other duties applicable to older persons beyond health care services at nursing homes. In particular, the ESCR has consistently required states to establish health care and services, especially primary health care services, aimed at the

---

[590]ECtHR, *Case of Carson and Others v. United Kingdom*, *supra* note 100.

[591]ECtHR, *Case of Zubczewski v. Sweden*, Decision of Admissibility, January 12, 2010.

[592]ECtHR, *Case of Stummer v. Austria*, *supra* note 561.

[593]*Id.*, 105–106, 110.

[594]*Explanatory Report to the Additional Protocol to the European Social Charter*, *supra* note 316, 57.

[595]*Id.*,

[596]See, *inter alia*, ECSR, Conclusions 2009, France, at 41; Conclusions 2009, Ireland, at 30.

[597]See, *inter alia*, ECSR, Conclusions 2009, Finland, at 33; Conclusions 2013, Andorra, at 39.

elderly, as well adopt guidelines on health care for older persons.[598] Furthermore, the ECSR has asked states to set forth mental health programs to address the psychological problems that are faced by the elderly, particularly with respect to dementia.[599] Also, the ESCR has underlined the need to adopt policies regarding health promotion and maintenance of older persons who decide to stay home, including preventive home visits, regular check-ups and self-help groups.[600]

Article 23 of the Revised Charter is complemented by Article 11 of the same instrument that ensures the right to protection of health.[601] The ECSR has extensively monitored compliance with this particular right noting in its first conclusions regarding Article 11 that a state party will be deemed in satisfaction of its duties under this article if its health system comprises certain specific protections.[602] One of these protections, according to the ECSR, includes special measures to protect the health of older persons.[603] Another important issue raised under Article 11 that is essential to older persons relates to the cost of healthcare. The ECSR has underlined that the right to access to health care requires that the cost of providing such service must be borne by the community as a whole.[604] In the same vein, the ECSR has required states to provide information on the costs paid by older persons when obtaining healthcare services as a measure to assess access under Article 23 of the *Revised Social Charter*.[605] Moreover, under Article 11 the ESCR has consistently stated that access to the healthcare system must be ensured to everyone, including members of disadvantaged and vulnerable groups. [606]

Additionally, the ECSR has stated that the protection afforded by Article 11 is linked to the positive obligations embodied in Articles 2 and 3 of the *European Convention*.[607] Also, the right to health care is protected by Article 35 of the *Charter of Fundamental Rights*, which ensures "access to preventive health care

---

[598]See, *inter alia*, ECSR, Conclusions 2009, Spain, at 31; Conclusions 2013, Greece, at 42.

[599]See, *inter alia*, ECSR, Conclusions 2003, France, at 92; Conclusions 2013, Spain, at 35–36.

[600]See, *inter alia*, ECSR, Conclusions 2009, France, at 41.

[601]*Revised European Social Charter*, Article 11 states:

With a view to ensuring the effective exercise of the right to protection of health, the Parties undertake, either directly or in co-operation with public or private organisations, to take appropriate measures designed inter alia:

1. to remove as far as possible the causes of ill-health;
2. to provide advisory and educational facilities for the promotion of health and the encouragement of individual responsibility in matters of health;
3. to prevent as far as possible epidemic, endemic and other diseases, as well as accidents.

[602]See, Lenia Samuel, *Fundamental Social Rights, Case Law of the European Social Charter, supra* note 501, at 263.

[603]*Id.*

[604]ECSR, *Médecins du Monde- International v. France*, Complaint No. 67/2011, September 11, 2012, 139.

[605]See, *inter alia*, ECSR, Conclusion 2003, France, at 91.

[606]ECSR, *Médecins du Monde- International v. France*, *supra* note 604, 140.

[607]See, *inter alia*, ECSR, Conclusions 2005, General Introduction, at 10.

and medical treatment under the conditions established by national laws and practices." Article 35 draws from the obligations arising out of Article 11 of the *Revised Social Charter*. Consequently, the combination of the provisions of these three European instruments has created a complementary set of obligations that when jointly applied and implemented may provide older persons with an extensive protection to their right to health in today's Europe.

In addition to these obligations, other non-binding instruments enshrine further guidance for the protection of the rights to health of older persons. In particular, Article VI, Care of the *Recommendation on the human rights of older persons* acknowledges that states should take all necessary and appropriate measures to promote, maintain and improve the health care and well-being of older persons.[608] Health care should be provided for those who stay in their homes as well as those who are placed in long-term care institutions.[609] In both cases, health care should be available, accessible and of good quality.[610] This Recommendation also provides that caregivers should be provided training and health and social services, through the adoption of a multi-dimension approach to serve the elderly.[611] Likewise, the *European Charter of Older Persons in Long-Term Care* emphasizes the right of older persons "to receive high quality, timely and affordable health and long-term care services that are adapted to [their] individual needs and wishes and without discrimination of any kind."[612] Moreover, the Charter ensures protection against medical and pharmaceutical abuse, including unnecessary or excessive medical treatment or drug use, as well as denial of treatment.[613]

Another very relevant issue regarding the right to health of older persons includes the provision of palliative services to people who are terminally ill and dying. Palliative care has been defined as "an approach that improves the quality of life of patients and their families facing the problem associated with life-threatening illness, through the prevention and relief of suffering by means of elderly identification and impeccable assessment and treatment of pain and other problems, physical, psychological and spiritual."[614] The Committee of Ministers of the Council of Europe in a specific recommendation asking states to regulate palliative care has stated that such care "is a vital and integral part of health services. Provisions for its development and functional integration should be incorporated into national health

[608] *Recommendation CM/Rec(2014)2 of the Committee of Ministers to member States on the promotion of the human rights of older persons*, *supra* note 15, Article VI.29.

[609] *Id.*, Article VI.29 and 30.

[610] *Id.*, Article VI. 29, 30 and 33.

[611] *Id.* Article VI.31.

[612] *European Charter of the Rights and Responsibilities of Older Persons in Need of Long-Term Care and Assistance*, 2010, *supra* note 216, Article 4.1.

[613] *Id.*, Article 1.2.9.

[614] *Explanatory Memorandum of Recommendation CM/Rec(2014)2 of the Committee of Ministers to member States on the promotion of the human rights of older persons*, *supra* note 63, 73.

strategies."[615] Such resolution provides detailed guiding principles on the development and implementation of palliative care services in Europe that are applicable to all persons, including older persons. With respect to older persons, those guiding principles have been specifically tailored in Article VI of the *Recommendation on the human rights of older persons* and in the *European Charter of Older Persons in Long-Term Care*. From these two instruments, it is possible to draw the following principles: (1) the principle of dignity is at the core of the provision of palliative care services and must be ensured to older persons in the process of dying[616]; (2) access to palliative care services must be ensured without undue delay, at home or in long-term care settings[617]; (3) health care professionals involved in palliative care should respect the rights of patients, including respect for and observance of their religious beliefs[618]; (4) palliate-care medicines should be adequate, available and accessible[619]; (5) advanced instructions should be respected regarding the extent of treatment and other end-of-life care[620]; and (6) trained specialists in the field of palliative care should be available to carry out educational programs and research in this subject.[621] Given the core importance of availability of palliative care services for older persons, the ECSR has insisted upon the states' duty to provide such services when monitoring compliance with the obligations arising out of Article 23 of the *Revised Social Charter*.[622]

As a final note, it is important to consider that the ECSR has found that the right to health, as the right to an adequate standard of living and housing, may involve progressive implementation given the need of extensive resources to operate a health care system.[623] In that regard, it has reiterated that when the

---

[615]Committee of Ministers, *Recommendation Rec(2003)24 of the Committee of Ministers to member states on the organization of palliative care*, adopted on November 12, 2003, Appendix, Guiding Principles I.1.

[616]*Recommendation CM/Rec(2014)2 of the Committee of Ministers to member states on the promotion of the human rights of older persons*, *supra* note 15, Article VI.44; *European Charter of the Rights and Responsibilities of Older Persons in Need of Long-Term Care and Assistance*, 2010, *supra* note 216, Article 8.1 and 8.2.

[617]*Recommendation CM/Rec(2014)2 of the Committee of Ministers to member states on the promotion of the human rights of older persons*, *supra* note 15, Article VI.45.

[618]*Recommendation CM/Rec(2014)2 of the Committee of Ministers to member states on the promotion of the human rights of older persons*, *supra* note 15, Article VI.47; *European Charter of the Rights and Responsibilities of Older Persons in Need of Long-Term Care and Assistance*, 2010, *supra* note 216, Article 8.7.

[619]*Recommendation CM/Rec(2014)2 of the Committee of Ministers to member states on the promotion of the human rights of older persons*, *supra* note 15, Article VI.49.

[620]*European Charter of the Rights and Responsibilities of Older Persons in Need of Long-Term Care and Assistance*, 2010, *supra* note 216, Article 8.4, 8.5 and 8.6.

[621]*Recommendation CM/Rec(2014)2 of the Committee of Ministers to member states on the promotion of the human rights of older persons*, *supra* note 15, Article VI.48.

[622]See, *inter alia*, ECSR, Conclusions 2013, Spain, at 36; Conclusions 2013, Ireland, at 25.

[623]See, *inter alia*, ECSR, *International Federation for Human Rights (FIDH) v. Greece*, Complaint No. 72/2011, 129; Conclusions 2013, Turkey, at 16.

implementation of a right under the *Revised Social Charter* is "exceptionally complex and particularly expensive to resolve, a state Party must take measures that allows it to achieve the objectives of the Charter within a reasonable time, with measurable progress and to an extent consistent with the maximum use of available resources. States Parties must be particularly mindful of the impact that their choices will have for groups with heightened vulnerabilities as well as for others persons affected."[624] The achievement of the goals set by the authorities, however, cannot be deferred indefinitely.[625] Notwithstanding the principle of progressive implementation, the ESCR has stated clearly that access to existing health care services must be provided without discrimination in light of Article E of the *Revised Social Charter*.[626]

E. The Protection of Other Vulnerable Groups and the Notion of "Multiple Discrimination"

The notion of "multiple discrimination" has been debated extensively in regards to the principle of equality and the protection afforded to individuals who may be subjected to unjustified distinctions on the basis of more than one reasons or grounds. The *Recommendation on the human rights of older persons* acknowledges the importance of this notion for the enjoyment of rights with equality by the elderly. Article II. 8 specifically provides that states "should take effective measures to prevent multiple discrimination of older persons."[627] The *Explanatory Memorandum to the Recommendation on the human rights of older persons* defines multiple discrimination as "as a situation in which a person is being discriminated against for several reasons."[628] In particular, states should be aware of the situation and needs of older women, older migrants, older person with disabilities and older persons with a different sexual orientation.[629]

Recommendation 1796 (2007) of the Parliamentary Assembly on the situation of elderly persons in Europe specifically contemplates the situation of older women and acknowledges that they are a particularly vulnerable group that is subject to

---

[624] *Id.*

[625] *Id.*

[626] See, *inter alia*, ECSR, Conclusions 2005, *supra* note 615, at 10–11.

[627] *Recommendation CM/Rec(2014)2 of the Committee of Ministers to member states on the promotion of the human rights of older persons*, *supra* note 15, Article II.8.

[628] *Explanatory Memorandum to the Recommendation on the promotion of the human rights of older persons*, *supra* note 63, 15.

[629] *Id.*

discrimination in many ways.[630] First, older women face insecurity due to inadequate social support and low income.[631] Second, since women are often responsible for caring for family members, they may stop working and face ensuing consequences for their income levels and pensions.[632] Moreover, both the Parliamentary Assembly of the Council of Europe[633] and the European Parliament[634] have stressed the impact of the recent economic and financial crisis for women in general, and older women in particular. Since women in poverty or at risk of poverty are more likely to work in low-paid, precarious and informal jobs, their contributions to the social security system are insufficient to lift them out of poverty in old age.[635] This assertion is confirmed by Eurostat's statistics according to which single women over 65 are at substantially higher risk of poverty than single men of the same age.[636]

Also, *Recommendation 1796 (2007)* draws attention to the potential multiple discrimination faced by older migrants. In particular, it states that older migrant persons "face a particular risk of double or even triple discrimination and requires specific policies and culturally sensitive services."[637] In the same vein, *Recommendation 1619 (2003)* provides guidelines to states as to the measures required to ensure the social inclusion of older migrants and their access to basic services, including social welfare, health care and active participation in society.[638] Most recently, *Recommendation CM/Rec (2011) of the Committee of Ministers on reducing the risks of vulnerability of elderly migrants and improving their welfare* adopted a comprehensive set of principles encouraging states to take appropriate action at the different levels of their administration to promote the welfare of older migrant persons, including the prevention of abuse, the reduction of their vulnerability and their social isolation and marginalization.[639] This Recommendation specifically acknowledges the risks for triple discrimination faced by elderly migrant women.[640] These Recommendations are complemented by Articles 19 of the *Revised Social Charter*

---

[630]Parliamentary Assembly, *Recommendation 1796 (2007), The situation of elderly persons in Europe*, *supra* note 18, 9.

[631]*Id.*

[632]*Id.*

[633]See, in general, Resolution 1719 (2010), Women and the economic and financial crisis, adopted on April 28, 2010.

[634]See, in general, European Parliament resolution of 12 March 2013 on the impact of the economic crisis on gender equality and women's rights (2012/2301(INI)).

[635]See, *inter alia*, Council of Europe, Commissioner for Human Rights, Safeguarding human rights in times of economic crisis, January 2014, at 24.

[636]Europe 2020 Targets: Poverty and Social Exclusion Active Inclusion Strategies, p. 4, visited 12/15/2014, at http://ec.europa.eu/europe2020/pdf/themes/33_poverty_and_social_inclusion.pdf.

[637]Recommendation 1796 (2007), The situation of elderly persons in Europe, *supra* note 18, 10.

[638]Parliamentary Assembly, *Recommendation 1619 (2003), Rights of elderly migrants*, 4.

[639]Recommendation CM/Rec(2011)5 of the Committee of Ministers on reducing the risk of vulnerability of elderly migrants and improving their welfare, adopted on May 25, 2011.

[640]Id. Appendix to Recommendation CM/Rec (2011)5, 8–9.

which ensures the rights of migrant workers and their families.[641] Although this Article applies in principle to lawful migrants from other state parties,[642] in some cases the ESCR has considered that certain basic rights such as medical assistance must be provided to foreign nationals, even if they reside in the country concerned in an irregular manner, on the basis of the respect of human dignity, an underlying principle of the *Revised Social Charter*.[643] Likewise, the ESCR has recently concluded that "the restriction of the personal scope of the Charter included in its Appendix should not be read in such a way as to deprive migrants in an irregular situation of

[641]Article 19 of the *Revised Social Charter* provides: With a view to ensuring the effective exercise of the right of migrant workers and their families to protection and assistance in the territory of any other Party, the Parties undertake:

to maintain or to satisfy themselves that there are maintained adequate and free services to assist such workers, particularly in obtaining accurate information, and to take all appropriate steps, so far as national laws and regulations permit, against misleading propaganda relating to emigration and immigration;

to adopt appropriate measures within their own jurisdiction to facilitate the departure, journey and reception of such workers and their families, and to provide, within their own jurisdiction, appropriate services for health, medical attention and good hygienic conditions during the journey;

to promote co-operation, as appropriate, between social services, public and private, in emigration and immigration countries;

to secure for such workers lawfully within their territories, insofar as such matters are regulated by law or regulations or are subject to the control of administrative authorities, treatment not less favourable than that of their own nationals in respect of the following matters:

remuneration and other employment and working conditions;

membership of trade unions and enjoyment of the benefits of collective bargaining;

accommodation;

to secure for such workers lawfully within their territories treatment not less favourable than that of their own nationals with regard to employment taxes, dues or contributions payable in respect of employed persons;

to facilitate as far as possible the reunion of the family of a foreign worker permitted to establish himself in the territory;

to secure for such workers lawfully within their territories treatment not less favourable than that of their own nationals in respect of legal proceedings relating to matters referred to in this article;

to secure that such workers lawfully residing within their territories are not expelled unless they endanger national security or offend against public interest or morality;

to permit, within legal limits, the transfer of such parts of the earnings and savings of such workers as they may desire;

to extend the protection and assistance provided for in this article to self-employed migrants insofar as such measures apply;

to promote and facilitate the teaching of the national language of the receiving state or, if there are several, one of these languages, to migrant workers and members of their families;

to promote and facilitate, as far as practicable, the teaching of the migrant worker's mother tongue to the children of the migrant worker.

[642]The Appendix to the European Social Charter (Revised) Scope of the Revised European Social Charter in terms of persons protected states: Without prejudice to Article 12, paragraph 4, and Article 13, paragraph 4, the persons covered by Articles 1 to 17 and 20 to 31 include foreigners only in so far as they are nationals of other Parties lawfully resident or working regularly within the territory of the Party concerned, subject to the understanding that these articles are to be interpreted in the light of the provisions of Articles 18 and 19.

[643]ECSR, *International Federation of Human Rights Leagues (FIDH) v. France*, Complaint No. 14/2003, September 8, 2014, 30–32.

the protection of the most basic rights enshrined in the Charter, or to impair their fundamental rights, such as the right to life or to physical integrity or to human dignity."[644] Consequently, at least with respect to access to basic services and benefits, older migrant persons even in situation of irregularity may benefit from the protections afforded by the *Revised Social Charter*.

With respect to older persons with disabilities, *Recommendation CM/Rec (2009)6 of the Committee of Ministers on ageing and disability in the 21st Century: sustainable frameworks to enable greater quality of life in an inclusive society* stresses the need to adopt measures tailored to address the needs of ageing people with disabilities and older persons with disabilities.[645] This Recommendation defines ageing people with disabilities as persons "who grow old having experienced a disability for much of their lives, sometimes from birth" whereas it refers to older persons with disabilities as those for whom "disability is first experienced at a relatively advanced age."[646] Also, it encourages states to create frameworks to enable greater quality of life in an inclusive society for the members of these two groups. Towards that end, it promotes the respect of the principles of autonomy and an independent active life, the enhancing of quality of services as well as the equal access to social and legal services.[647] The principles of this Recommendation are complemented by Articles 15 of the *Revised Social Charter* protecting the right of persons with disabilities to independence, social integration and participation in the life of the community[648] and 26 of the *Charter of Fundamental Rights* regarding integration of persons with disabilities.[649]

---

[644]ECSR, *European Federation of National Organisations working with the Homeless (FEANTSA) v. the Netherlands*, Complaint No. 86/2012, July 2, 2014, 58.

[645]*Recommendation CM/Rec (2009)6 of the Committee of Ministers on ageing and disability in the 21st Century: sustainable frameworks to enable greater quality of life in an inclusive society*, *supra* note 15.

[646]*Id., Appendix to Recommendation CM/Rec(2009)6, supra* note 15, Introduction.

[647]See, in general, *Appendix to Recommendation CM/Rec(2009)6, supra* note 15.

[648]Article 15 of the *Revised Social Charter* provides: With a view to ensuring to persons with disabilities, irrespective of age and the nature and origin of their disabilities, the effective exercise of the right to independence, social integration and participation in the life of the community, the Parties undertake, in particular:

to take the necessary measures to provide persons with disabilities with guidance, education and vocational training in the framework of general schemes wherever possible or, where this is not possible, through specialised bodies, public or private;

to promote their access to employment through all measures tending to encourage employers to hire and keep in employment persons with disabilities in the ordinary working environment and to adjust the working conditions to the needs of the disabled or, where this is not possible by reason of the disability, by arranging for or creating sheltered employment according to the level of disability. In certain cases, such measures may require recourse to specialised placement and support services;

to promote their full social integration and participation in the life of the community in particular through measures, including technical aids, aiming to overcome barriers to communication and mobility and enabling access to transport, housing, cultural activities and leisure.

[649]Article 26 of the Charter of the Fundamental Rights of the European Union states that: The Union recognises and respects the right of persons with disabilities to benefit from measures designed to ensure their independence, social and occupational integration and participation in the life of the community.

Finally, the intersection between age and sexual orientation has not yet been fully developed by any European instrument. Notwithstanding, the prohibition of discrimination on age and sexual orientation is contemplated in Directive 2000/78/EC with respect to employment and occupation.[650] The CJEU has decided many cases on discrimination on the basis of age, but the case law concerning discrimination on the basis of sexual orientation is much less developed.[651] Most of the cases on this ground relate to cases where same sex-partners were excluded from work-related benefits enjoyed by heterosexual couples.[652] Even though the Directive does not refer to "multiple discrimination," it is possible that two or more grounds for discrimination, such as age and sexual orientation, could be tackled together in one case.[653] In addition to the above-mentioned Directive, Article 21 of the *Charter of Fundamental Rights* prohibits discrimination, inter alia, on the basis of age and sexual orientation.[654] On the other hand, although the *European Convention* does not specifically contemplate the grounds of age and sexual orientation in Article 14 or Article 1 of Protocol 12 to that treaty, discrimination on both of those grounds may be found on the basis of the phrase "other status."[655] The ECHR has not had an opportunity yet to decide a case where the intersection of age and sexual orientation was at stake. In *Schlumpf v. Switzerland*, however, the complaint involved the denial of the health insurance company to cover the costs of a sex-change operation because the applicant had failed to comply with a two-year waiting period.[656] The ECHR found that the denial involved a violation of the applicant's right to privacy under Article 8 because, due to her age, delaying the operation would have an influence in her gender identity.[657] Even if the claimant alleged a violation of Article 14, the ECHR decided that it did not have to review such allegation given its conclusions regarding the violation of Article 8.[658] Consequently, although the facts involved an older

[650]Council Directive 2000/78/EC of 27 November 2000 establishing a general framework for equal treatment in employment and occupation, see *supra* note 47.

[651]European Commission, *Report from the Commission to the European Parliament and the Council, Joint Report on the application of Council Directive 2000/43/EC of 29 June 2000 implementing the principle of equal treatment between persons irrespective of racial or ethnic origin* ("Racial Equality Directive") and of Council Directive 2000/78/EC of 27 November 2000 establishing a general framework for equal treatment in employment and occupation ("Employment Equality Directive"), at 7–8.

[652]*Id.*, p. 8. See in this regard, *inter alia*, C-267/06, *Tadao Maruko v. Versorgungsanstalt der deutschen Bühnen VddB*, April 1, 2008.

[653]*Id.*, at 9.

[654]For the full text of Article 21 of the *Charter of Fundamental Rights of the European Union* see *supra* note 130.

[655]For a full discussion on this issue, see section on the right of older persons to equal protection.

[656]ECtHR, *Schlumpf v. Switzerland*, *supra* note 301.

[657]Id., 110–15.

[658]Id., 117–118.

person who claimed discrimination on the basis of her sexual orientation this issue was not resolved on the merits. Lastly, *Recommendation CM/Rec(2010)5 of the Committee of Ministers on measures to combat discrimination on grounds of sexual orientation or gender identity* encourages states to ensure that domestic laws prohibiting discrimination also protect against discrimination on multiple reasons, in particular sexual orientation or gender identity in connection to other grounds, which could include age.[659]

F. Conclusion

The extensive and broad social and legal policies of the European region show an awareness of the aging of the population, and intentions to improve the lives of older individuals. While there may be vast differences in Europe, there are also conceptual common threads throughout the European states. Europeans share a deeply-embedded collective culture of belief in human rights and social protection. This identity and norm of equality may be the precise blind spot where older persons' rights are located. Older persons make up a large and growing portion of the population, but governments treat them as beneficiaries, not rights holders. The human rights frameworks in Europe can be changed with concerted litigation efforts to promote the violations older people face; not just access to pensions, which are certainly important, but access to justice when admitted to nursing homes against their will, or experiencing financial or physical violence in their homes. Rights violations occur everywhere, regardless of economic status; the real question is whether the mechanisms for justice are in place to provide justice and redress.

[659]See, *Recommendation CM/Rec(2010)5 of the Committee of Ministers to member states on measures to combat discrimination on grounds of sexual orientation or gender identity*, adopted on March 31, 2010, Appendix, Principle XII, 46.

The Americas and the Human Rights of Older Persons: Historical Overview and New Developments

A. Introduction

Persons 60 years of age or older living in Latin America and the Caribbean Region will increase to about 25 percent of the population by 2050, while it will increase to 27 % in North America.[1] This growth is occurring most rapidly among the "oldest old," those aged 80 or above.[2] By 2050, the region will have more people over the age of 60 than under the age of 15.[3]

Compared with other regions of developing countries, Latin America and the Caribbean is a social outlier in some ways. Literacy levels among older people are the highest of any developing regions, with the lowest illiteracy gender gap between men and women.[4] The region has the steepest projected dependency ratio

---

[1]United Nations Department of Economic and Social Affairs, Population Division, Population Facts, Population ageing and sustainable development, August 2014, at 1, available at http://www.un.org/en/development/desa/population/publications/pdf/popfacts/PopFacts_2014-4.pdf (last visited Dec. 30, 2014).

[2]*Id.*, at 2.

[3]United Nations Department of Economic and Social Affairs Population Division, World Population Ageing: 1950–2050, at 16, available at http://www.un.org/esa/population/publications/worldageing19502050/ (last visited Dec. 30, 2014). Also, see overall, United Nations Department of Economic and Social Affairs Population Division, World Population Aging 2013, at 18, available at http://www.un.org/en/development/desa/population/publications/pdf/ageing/WorldPopulationAgeing2013.pdf (last visited Dec. 30, 2014).

[4]United Nations Department of Economic and Social Affairs, Population Division, World Population Ageing, 2009, at 39–40, available at http://www.un.org/en/development/desa/population/publications/pdf/ageing/WorldPopulationAgeingReport2009.pdf (last visited Dec. 30, 2014).

C. Martin et al., *Human Rights of Older People*, Ius Gentium: Comparative Perspectives on Law and Justice 45, DOI 10.1007/978-94-017-7185-6_4

increases of any region in the world.[5] Despite this projection, the actual "proportions of older persons living alone tend to be higher in Latin America and the Caribbean than in the other major areas of the developing world."[6] This projected shift in demographics will likely increase demand for special care and residential arrangements.[7] Latin America and the Caribbean are also unique in that the proportion of older persons is slightly higher in urban areas than in rural areas.[8] In most parts of the developing world, those proportions are reversed.

As the population of older persons in Latin America and all over the world steadily increases, so does the need for international and regional legal instruments securing their human rights. Older persons do not enjoy the protections other groups do. Disabled persons, migrant workers, minorities and refugees each are ensured the protection of international instruments tailored to address their specific needs. The rights of older persons are at risk of violation in a variety of institutional, community and family settings. They are exposed to the risks of discrimination and marginalization in violation of their rights. In the public domain, older persons may be denied equality in opportunities available to them and in resources allocated to them. Privately, they may face violence, extreme poverty, and disability combined. The first and most essential step to empowering older persons should include measures to protect their human rights.[9] Although various international standards to protect human rights are already in place, special protections are required to protect the rights of older persons, a growing population with an increasing presence.

This chapter outlines the human rights standards that form the current framework of protection for older persons in the Americas. It concludes with thoughts for governments considering enacting domestic and regional legislation to protect the rights of older persons. Although there are general human rights standards for individuals in place, and many Inter-American treaties already refer to rights that are particularly important for older persons, a gap exists between these legal instruments and the enforceability of individual rights.

---

[5] *Id.*, at 22. From 2009 to 2050, the ratio of persons aged 65 or over to those of working age is projected to grow from 10 to 31 per 100 in Latin America. Thus, it is expected that in 2050 there will 3 persons of working age for every older person in the region. See, in this regard, United Nations Department of Economic and Social Affairs Population Division, World Population Aging 2013, *supra* note 3, at 25.

[6] United Nations Department of Economic and Social Affairs, Population Division, World Population Ageing, 2009, *supra* note 4, at 34.

[7] Latin American and Caribbean Demographic Centre—Population Division of ECLAC, *Demographic Change and its Influence on Development in Latin America and the Caribbean*, at 26, available at http://www.eclac.org/publicaciones/xml/5/33225/2008-271-SES.32-CELADE-INGLES.pdf (last visited Dec. 30, 2014).

[8] United Nations Department of Economic and Social Affairs, Population Division, World Population Ageing, 2009, *supra* note 4, at 36.

[9] Expert Meeting on Population, Inequalities and Human Rights, *Intersection of Population and Human Rights: Rights of Older Persons in the International Context*, at 6–8, available at http://www.eclac.org/celade/noticias/paginas/6/27116/KellyP.pdf (last visited Dec. 30, 2014).

Latin America and the Caribbean have produced some of the most robust arguments for a special convention on the human rights of older persons in international arenas such as the United Nations Open-Ended Working Group on Strengthening the Human Rights of Older Persons, the Inter-American Working Group for the Human Rights of Older Persons, and through vibrant civil society support.[10] Additionally, under the auspices of the Economic Commission for Latin American and the Caribbean (ECLAC), three inter-governmental regional conferences on ageing have been organized in 2003, 2007 and 2012, with the purpose of following up on the Madrid International Plan of Action on Ageing.[11] The region is a leader, and has laid the groundwork for change to improve the lives of older persons in this part of the world.

B. Regional Human Rights Frameworks

The Inter-American Conference established the Organization of American States (OAS) to strengthen the peace and security of the continent, promote democracy, settle disputes, and to devote the largest amount of resources to economic and social development of member states.[12] The US and Canada are OAS state members. The *Charter of the Organization of American States*, (*OAS Charter*) was initially signed in 1948.[13] The OAS Charter makes very few express references to human rights, but proclaims the fundamental rights of the individual without distinction as to race, nationality, creed or sex.[14] Also, it provides that states have the right to develop cultural, political and economic life freely and that in doing so they will respect the rights of the individual and the principles of universal morality.[15]

---

[10]See, in general, Israel Doron and Kate Mewhinney, *The Rights of Older Persons—A Collection of International Documents*, Jerusalem & Montreal: IFA & ESHEL, 2008. See, also, Working Group on Protection of the Human Rights of Older Persons, *Activities of the Working Group during the period 2013–2014*, Report by the Chair of the Working Group presented to the Committee on Juridical and Political Affairs during the session held May 15, 2014, OEA/Ser.G CAJP/GT/DHPM-139/14, May 14, 2014.

[11]See in this regards, Sandra Huenchuan, *Perspectivas globales sobre la protección de los derechos humanos de las personas mayores*, 2007–2013, CEPAL, 2013, at 39–43.

[12]See, in general, Article 2 of the Charter of the OAS listing the purposes of this regional organization. Charter of the OAS, as amended by the Protocol of Buenos Aires in 1967, the Protocol of Cartagena de Indias in 1985, the Protocol of Washington in 1992, and the Protocol of Managua in 1993, available at http://www.oas.org/dil/treaties_A-41_Charter_of_the_Organization_of_American_States.pdf (last visited Dec. 30, 2014).

[13]*Id.*

[14]*Id.*, Article 3(l).

[15]*Id.*, Article 17.

The OAS' human rights system is governed by two instruments: the *American Declaration on the Rights and Duties of Man* (*American Declaration*)[16] and the *American Convention on Human Rights* (*American Convention*).[17] Even though the *American Declaration* was adopted as a non-binding instrument, it is accepted today that this instrument contains and defines the fundamental human rights referred to in the OAS Charter, a binding treaty.[18] Consequently, its provisions carry significant legal weight.[19]

The *American Declaration* recognizes civil and political rights as well as specific economic, social, and cultural rights that are relevant to further the protection of older persons in the Inter-American Human Rights System (Inter-American System). In regards to economic, social and cultural rights, the *American Declaration* embodies, inter alia: the right to the protection of the family,[20] the right to the preservation of health and well-being,[21] the right to education,[22] the right to work and to a fair remuneration,[23] and the right to social security.[24] Furthermore, the *American Declaration* contains duties toward parents[25] and duties with respect to social security and welfare.[26]

The *American Convention* is the core human rights treaty in the Inter-American System. It addresses the protection of civil and political rights, with the exception of Article 26, which provides for the state duty to adopt measures towards the progressive implementation of economic, social and cultural rights. The scope of this provision has been contested in Inter-American human rights case law.[27]

The member states of the OAS have adopted additional human rights treaties that are relevant for the protection of the rights of older persons, including: the *Additional Protocol to the American Convention on Human Rights in the Area of*

---

[16]*American Declaration on the Rights and Duties of Man*, approved by the Ninth International Conference of American States, Bogota, Colombia, OAS Res. XXX.

[17]*American Convention on Human Rights*, O.A.S. Treaty Series No. 36, 1144 U.N.T.S. 123, entered into force on July 18, 1978.

[18]I/A Court H.R., *Interpretation of the American Declaration of the Rights and Duties of Man within the Framework of Article 64 of the American Convention on Human Rights*, Advisory Opinion OC-10/89 of July 14, 1989, Series A No. 10, 43, 45.

[19]*Id.*

[20]Article VI of the *American Declaration.*

[21]Article XI of the *American Declaration.*

[22]Article XII of the *American Declaration.*

[23]Article XIV of the *American Declaration.*

[24]Article XVI of the *American Declaration.*

[25]Article XXX of the *American Declaration.*

[26]Article XXXV of the *American Declaration.*

[27]See, for example, I/A Court H.R., *Case of Acevedo Buendía et al. ("Discharged and Retired Employees of the Office of the Comptroller") v. Peru*, Preliminary Objection, Merits, Reparations and Costs, Judgment of July 1, 2009, Series C No. 198.

Economic, Social, and Cultural Rights, (Protocol of San Salvador),[28] the *Inter-American Convention to Prevent and Punish Torture (Inter-American Torture Convention)*,[29] *the Inter-American Convention for the Prevention, Punishment and Eradication of the Violence Against Women, (Convention of Belém do Pará)*,[30] *the Inter-American Convention on the Elimination of all Forms of Discrimination Against Persons with Disabilities (Disabilities Convention)*,[31] *Inter-American Convention Against All Forms of Discrimination and Intolerance (Discrimination and Intolerance Convention)*,[32] and the *Inter-American Convention Against Racism, Racial Discrimination, and Related Forms of Intolerance (Racial Discrimination Convention)*.[33]

The OAS General Assembly has expressed increasing concern about the lack of studies and reports on mechanisms dealing with the problems of older persons, specifically problems arising from the infringement of their rights.[34] In 2009, it requested that the OAS Permanent Council convene a special meeting of national representatives and experts from the academic sector and civil society, as well as from international organizations, for the purpose of sharing information and best practices and also examining the feasibility of preparing an inter-American convention on the rights of older persons to evaluate and adopt measures for the protection of older people's rights.[35]

In 2010, the OAS Permanent Council held a special meeting on the human rights of older persons.[36] In 2011, a Working Group on Protecting the Human Rights of Older Persons was established.[37] Its working plan included research on the human rights of older persons as well as the drafting of an *Inter-American*

---

[28]*Additional Protocol to the American Convention on Human Rights in the Area of Economic, Social, and Cultural Rights, "Protocol of San Salvador"*, OAS Treaty Series No. 69, entered into force on November 16, 1999.

[29]*Inter-American Convention to Prevent and Punish Torture*, OAS Treaty Series No. 67, entered into force on February 28, 1987.

[30]*Inter-American Convention for the Prevention, Punishment and Eradication of the Violence against Women, "Convention of Belém do Pará,"* A-61, entered into force on March 5, 1995.

[31]*Inter-American Convention on the Elimination of all Forms of Discrimination against Persons with Disabilities*, A-65, entered into force on September 14, 2001.

[32]*Inter-American Convention against All Forms of Discrimination and Intolerance*, A 69, opened for ratification on June 6, 2013.

[33]*Inter-American Convention against Racism, Racial Discrimination, and Related Forms of Intolerance*, A-69, opened for ratification on June 6, 2013.

[34]See *AG/RES. 2455 (XXXIX-O/09)*, Human Rights and Older Persons, adopted on June 4, 2009 and *AG/RES. 2562 (XL-O/10)*, Human Rights and Older Persons, adopted June 8, 2010.

[35]*Id.*

[36]See, also, Working Group on Protection of the Human Rights of Older Persons, Activities of the Working Group during the period 2013–2014, Report by the Chair of the Working Group presented to the Committee on Juridical and Political Affairs during the session held May 15, 2014, *supra* note 10.

[37]*Id.*

Convention on the Human Rights of Older Persons (*Convention on Older Persons*).[38] The Working Group has been working on a draft of the *Convention on Older Persons*, which continues under discussion by the OAS state members as of today.[39] The last version of the draft recognizes "the need to address matters of old age and ageing from a human rights perspective that recognizes the valuable current contributions of older persons to the common good, to cultural identity, to the diversity of their communities, to human, social, and economic development, and to the eradication of poverty."[40] Also, it acknowledges that the adoption of a comprehensive treaty on the rights of older persons will significantly contribute to the promotion and protection of the rights of this group and to foster an active and fulfilling ageing process.[41] Moreover, the draft includes and gives priority to the topic of ageing in all areas of public policy, and provides for allocation and management of the human, material, and financial resources to ensure appropriate monitoring and assessment of the special measures implemented.[42]

The draft states that the purpose of the *Convention on Older Persons* is to promote, protect, and assure full enjoyment, in equality of conditions, of all the human rights and fundamental liberties of all older persons, without distinction of any kind, together with respect for their inherent dignity and their full inclusion, integration, and participation in society.[43] Where the exercise of those rights are not already protected by domestic laws, (a point of contention), states undertake to adopt all the necessary measures, including constitutional, legislative or other measures, to give them effect at the national jurisdiction.[44]

The general principles of the draft *Convention on Older Persons* include, among others, the full exercise of all the human rights of older people; equality and nondiscrimination; dignity, independence and autonomy of older persons; physical, economic, social security; gender equality; differentiated treatment for the effective enjoyment of the rights of older persons; and effective judicial protection.[45] Furthermore, Article 4 of the draft *Convention on Older Persons* provides for a set of state obligations comprising the duty to adopt affirmative measures and

---

[38] *Id.*

[39] The last Draft of the Convention of Older Persons was submitted for discussion by the Chair of the Working Group on Protection of the Human Rights of Older Persons on November 2014. See, Note from the Chair of the Working Group submitting proposed new language for the Draft Inter-American Convention on the Human Rights of Older Persons, OEA/Ser.G CAJP/GT/DHPM/INF.34/14, November 17, 2014. Also, for more information on the history of the discussions see, in general, the website of the Working Group on Protection of the Human Rights of Older Persons, http://www.oas.org/consejo/CAJP/personas%20mayores.asp (last visited Dec. 30, 2014).

[40] See, *Draft Inter-American Convention on the Human Rights of Older Persons*, *supra*, Preamble.

[41] *Id.*

[42] *Id.*

[43] *Id.*, Article 1.

[44] *Id.*

[45] *Id.*, Article 3.

necessary reasonable adjustments to implement the rights provided by this treaty. Moreover, states will adopt necessary legislative, administrative, budgetary, and judicial measures to give effect and disseminate the rights of older persons enshrined in this instrument. Also, states will pursue measures to progressively achieve the protection of the economic, social and cultural rights of older persons, and upon ratification will respect those obligations arising out of the treaty that are of immediate application. Additionally, states will promote the development of specialized institutions at the domestic level dealing with issues regarding the rights of older persons. These will have a duty to gather adequate statistical and research data to best implement public policies addressing problems faced by older persons.[46]

For those states that do not define the notion of "age," "older person" shall be considered as meaning any person aged 60 years or over.[47] Moreover, the draft Convention on Older Persons, directly defines age discrimination as any distinction, exclusion, or restriction based on age that has the effect or purpose of hindering, hampering, or annulling the recognition, enjoyment, or exercise of a person's human rights and fundamental freedoms.[48]

1. Monitoring: Supervision and Judicial Review

The OAS human rights system has two judicial review mechanisms: the Inter-American Commission on Human Rights (*IACHR*) and the Inter American Court of Human Rights (*IACtHR*). Under the *Statute of the Inter-American Commission on Human Rights*,[49] the IACHR is empowered to monitor the human rights situation of the OAS member states and elaborate and publish reports assessing whether those states respect the rights protected by the *American Declaration*, including economic, social, and cultural rights, and by the *American Convention* if the State has ratified this treaty.[50] The IACHR also conducts in situ visits to OAS member states, leading to the adoption of country reports regarding the countries visited.[51] With respect to older persons, the IACHR has denounced violations of their rights in several country reports. In 2001, in the *Third Report on the Situation of Human Rights in Paraguay* the IACHR criticized the handling of the Paraguayan Social Security Crisis.[52] In 1993, the *Second Report on the Situation*

[46] *Id.*, Article 4.

[47] *Id.*, Article 2.

[48] *Id.*, Article 5.

[49] See *Statute of the Inter-American Commission on Human Rights*, OAS G.A. Res. 447, 9th Sess. (October 31, 1979), available at http://www.oas.org/en/iachr/mandate/Basics/statuteiachr.asp (last visited Dec. 30, 2014).

[50] *Id.*, Article 18.

[51] *Id.*

[52] IACHR, *Third Report on the Situation of Human Rights in Paraguay*, OEA/Ser.L/V/II.110, Doc. 52, 9 March 2001, 42–43.

of Human Rights in Colombia noted that civilians, particularly older persons, were not protected from internal violence or violations of their right to life.[53] Also in 1993, the *Report on the Situation of Human Rights in Guatemala* criticized police for attacking older protesters.[54]

With respect to individual complaints, the IACHR has traditionally reviewed the adjudication of individual cases only with respect to the violations of certain rights of the *American Declaration*, which excluded economic, social and cultural rights.[55] On the other hand, once a state ratifies the *American Convention*, the IACHR reviews individual cases in light of the provisions and procedures established in that treaty, unless that the particular right is only protected by the *American Declaration* or the alleged violation occurred before the concerned state became bound by the treaty. In *Cirio v. Uruguay*, for example, the petitioner alleged a violation of the right to social security protected under Article XVI of the *American Declaration*.[56] The IACHR, in its analysis of admissibility, concluded that "once the American Convention entered into force... the Convention and not the Declaration became the source of legal norms for application by the Commission insofar as the petition alleges violations of substantially identical rights set forth in both instruments and those claimed violations do not involve a continuing situation."[57] However, it clarified that Article 29 of the *American Convention* "upholds the case law of the Inter-American Commission to the effect that the IACHR is competent to apply both *the American Declaration* and *the American Convention* in the same case."[58] This was of particular relevance in the *Cirio* case, where the right to social security is not protected by the *American Convention*.

[53]IACHR, *Second Report on the Situation of Human Rights in Colombia*, OEA/Ser.L/V/II.84, Doc. 39 rev., Oct. 14, 1993, Chapter VII, Section B, (chastising the Colombian government for its inability to protect the personal liberties of its citizens).

[54]IACHR, *Fourth Report on the Situation of Human Rights in Guatemala*, OEA/Ser.L/V/II.83, Doc. 16 rev., June 1, 1993, Chapter IX, Section II (discussing how the Guatemalan system has been ineffective in guaranteeing rights right to freedom of thought, expression, and assembly).

[55]Id., Article 20 (emphasizing particular attention to the following rights: life, liberty, and personal security (Article I); equality before the law (Article II); religious freedom (Article III); freedom of investigation, opinion, expression, and dissemination (Article IV); fair trial (Article XVIII); protection from arbitrary arrest (Article XXV); and due process of law (Article XXVI)).

[56]See IACHR, *Tomas Eduardo Cirio v Uruguay*, Case 11.500, Report No. 119/01, Admissibility, October 16, 2001, Annual Report 2001 (declaring the case admissible with respect to Articles of the *American Declaration* and of the *American Convention*). The case involved a retired Uruguayan military officer who was chastised for his comments opposing the government's position. The government then stripped him of his pension, health insurance, and ability to obtain credit.

[57]*Id.*, 36.

[58]*Id.*

In addition, the IACHR has varying powers for additional human rights treaties of the OAS. It entertains petitions regarding the alleged violation of provisions of the *Inter-American Torture Convention,*[59] Article 7 of the *Belém do Pará Convention,*[60] the *Discrimination and Intolerance Convention,*[61] and the *Racial Discrimination Convention.*[62] With respect to the *Protocol of San Salvador*, the IACHR can only review individual petitions regarding two provisions of that treaty, namely: Articles 8(a) and 13 protecting trade union rights and the right to education, respectively.[63]

The *Disabilities Convention* does not grant any power to the IACHR to monitor compliance with the obligations arising out of that treaty. This instrument sets up a separate Committee for the Elimination of All Forms of Discrimination against Persons with Disabilities to monitor compliance with the obligations established in the treaty.[64]

The IACtHR was created by the *American Convention* as the judicial body of the Inter-American System. It can only supervise compliance of the *American Convention* through the individual complaint mechanism for states that are parties to that treaty and have accepted the IACtHR's contentious jurisdiction to decide cases.[65] The IACtHR does not have the power to monitor compliance with the *American Declaration*, though it can use this instrument to interpret the provisions of the *American Convention.*[66] The IACtHR can also exercise its advisory jurisdiction upon request from state members to the OAS and specialized OAS agencies, particularly the IACHR.[67] The IACtHR may exercise its advisory powers to interpret the *American Convention* or other treaties regarding the protection of human rights in the Americas or to determine the compatibility of domestic laws with the *American Convention*, upon request from a State.[68] Like the IACHR, the IACtHR has the power to find violations of the provisions of the *Inter-American Torture*

[59]Even if the *Inter-American Torture Convention* does not expressly provide jurisdiction to the IACHR to review individual petitions, the IACtHR has interpreted Article 8 of that treaty to include jurisdiction of both the IACHR and the IACtHR to deal with breaches of that instrument. See, *inter alia*, I/A Court H.R., *Case of the "Street Children" (Villagrán-Morales et al.) v. Guatemala*, Merits, Judgment of November 19, 1999, Series C No. 63, 247–249.

[60]See Article 12 of the *Belém do Pará Convention.*

[61]See Article 15 of the *Discrimination and Intolerance Convention.*

[62]See Article 15 of the *Racial Discrimination Convention.*

[63]See Article 19(6) of the *Protocol of San Salvador.*

[64]See Article VI of the *Disabilities Convention.*

[65]See Article 62 of the *American Convention.*

[66]I/A Court HR, *Case of Bueno Alves v Argentina*, Merits, Reparations and Costs, Judgment of May 11, 2007, Series C no 164, 56–60.

[67]See Article 64 of the *American Convention.*

[68]*Id.*

Convention,[69] Article 7 of the *Belém do Pará Convention*,[70] the *Discrimination and Intolerance Convention*,[71] the *Racial Discrimination Convention*,[72] and Articles 8(a) and 13 of the *Protocol of San Salvador*.[73]

The current draft *Convention on Older Persons* establishes that any person or group of persons, or nongovernmental entity legally recognized in one or more member states of the OAS may present to the IACHR complaints alleging the violation by a state party of any of the provisions of that treaty.[74] Moreover, a state party may declare at the time of ratification, accession or at any time thereafter that it recognizes the jurisdiction of the IACHR to deal with inter-state petitions.[75] Also, a state party to the treaty may declare at the time of ratification, accession or at any time thereafter that it recognizes the jurisdiction of the IACtHR to deal with individual complaints alleging violations to this instrument.[76] Furthermore, state parties may seek the advisory services of the IACHR to ensure the effective implementation and application of the *Convention on Older Persons*.[77]

In addition to the individual complaint mechanism, the draft *Convention on Older Persons* provides for the creation of a Committee of Experts and a Conference of State Parties to promote implementation of this instrument.[78] The Committee of Experts will be in charge of the technical implementation of the treaty through a mechanism that will assess reports submitted by state parties.[79] The Conference of State Parties will monitor progress of implementation, monitor the activities of the Committee of Experts and foster debate among state parties by sharing successful experiments, best practices and technical cooperation.[80]

---

[69]For a discussion on the jurisdiction of the Court regarding the *Inter-American Torture Convention*, see *supra* note 59.

[70]Even if Article 12 does not expressly provide that the IACtHR has jurisdiction to deal with breaches of Article 7, the IACtHR has concluded through interpretation that Article 12 ensures its power to review breaches to the *Belém do Pará Convention*. See, in this regard, I/A Court H.R., *Case of González et al. ("Cotton Field") v. Mexico*, Preliminary Objection, Merits, Reparations and Costs, Judgment of November 16, 2009, Series C No. 205, 38–77.

[71]See Article 15 of the *Discrimination and Intolerance Convention*.

[72]See Article 15 of the *Racial Discrimination Convention*.

[73]See Article 19(6) of the *Protocol of San Salvador*.

[74]*Draft Inter-American Convention on the Human Rights of Older Persons*, *supra* note 39, Article 35.

[75]*Id.*

[76]*Id.*

[77]*Id.*

[78]*Id.*, Article 34.

[79]*Id.*

[80]*Id.*

2. Sub-Regional Frameworks

A number of other, sub-regional frameworks exist. Parties to the *Andean Charter for the Promotion and Protection of Human Rights* (*Andean Charter*)[81] explicitly commit to fulfill and enforce the rights and obligations designed to promote and protect the human rights of older adults in Article 46. Under the *Andean Charter*, states aim to improve the protection of older persons against all forms of discrimination and violence; to facilitate access to public and private entities and services; to promote the participation of older persons in decision-making on public issues concerning them; and to effectively protect the rights of older persons to social security.[82]

The Caribbean Community (CARICOM)[83] focuses largely on trade and development. The CARICOM *Caribbean Charter on Health and Ageing* is intended to be a coordinated, systematic approach for ensuring the health and full integration and participation of older persons in Caribbean societies and economies.[84]

Other groups also have a voice. For instance, the Common Southern Market, or MERCOSUR's statement to the Commission for Social Development of the United Nations Economic and Social Council, in February 2010 called for an International Convention on the Rights of Older Persons during the course of its review of relevant United Nations plans and programs pertaining to the situation of social groups. It noted that:

> There is no legally binding instrument to standardize and protect the rights of older persons. Persuaded that we need to put an end to this situation, the Heads of State of MERCOSUR, in the recent thirty-sixth meeting of the Common Market Council, committed themselves…with the goal to standardize their rights and to establish mechanisms and organs to ensure them, since this is a vulnerable sector of the population who are subjected to discriminatory practices and treatment.[85]

---

[81]*Andean Charter for the Promotion and Protection of Human Rights*, 26 July 2002, available at http://www.refworld.org/docid/3de4f94a4.html (last visited Jan. 15, 2015).

[82]*Id.*, Articles 46–47.

[83]Comprised of: Antigua and Barbuda, The Bahamas, Barbados, Belize, Dominica, Grenada, Guyana, Haiti, Jamaica, Montserrat, Saint Lucia, St. Kitts and Nevis, St. Vincent and the Grenadines, Suriname, and Trinidad and Tobago.

[84]See the *Caribbean Charter on Health and Ageing*, available at http://www.caricom.org/jsp/pressreleases/pres105_99.jsp (last visited Jan. 19, 2015).

[85]Permanent Mission of Argentina to the United Nations, "Temas Sociales 2010," quoted in ECLAC, Huenchuan & Rodríguez-Piñero, *Ageing and the protection of human rights: current situation and outlook*, (2011), at 65, http://social.un.org/ageing-working-group/documents/ECLAC_Ageing%20and%20the%20protection%20of%20human%20rights_current%20situation%20and%20outlook_Project%20document.pdf (last visited Jan. 19, 2015).

3. *Other Regional Processes Regarding the Protection of the Rights of Older Persons*

As a follow up to the Madrid International Plan of Action on Ageing adopted during the Second World Assembly on Ageing in 2002, ECLAC organized three intergovernmental regional conferences on Ageing to deal with the situation of older persons and the protection of their rights.[86]

In 2003, the first regional conference took place and the participating states adopted the *Regional Strategy for the Implementation in Latin America and the Caribbean of the Madrid International Plan of Action on Ageing*.[87] One of the most important goals of the Strategy is to promote the human rights of older persons and towards that end recommends states to adopt specific legislation to protect those rights in light of existing international human rights principles.[88] In 2007, ECLAC organized the second inter-governmental conference which adopted the *Brasilia Declaration*.[89] In this Declaration the states of the region undertook to request the United Nations Human Rights Council to appoint a special rapporteur to promote and protect the rights of older persons.[90] Additionally, states undertook to sponsor the adoption of an international convention on the rights of older persons at the universal level.[91] Finally, in 2012 during the third intergovernmental conference the states of the region adopted the *San José Charter on the Rights of Older Persons in Latin American and the Caribbean (San José Charter)*.[92] Even if this Charter is not binding, it follows the approach of a legal instrument setting out the most relevant rights and principles for the protection of older persons. Moreover, it lists a set of measures that the states of the region should follow to enhance protection of this vulnerable group, including: the adoption of public policies and the improvement of institutions in charge of supporting the rights of older persons; the provision of adequate public funding to ensure that responsible

---

[86]Sandra Huenchuan, *Perspectivas globales sobre la protección de los derechos humanos de las personas mayores*, *supra* note 11, at 39.

[87]The full text of the *Regional Strategy for the Implementation in Latin America and the Caribbean of the Madrid International Plan of Action on Action* may be reviewed at http://undesadspd.org/LinkClick.aspx?fileticket=RHZ2ZEsjCWY%3D&tabid=330 (last visited Jan. 19, 2015).

[88]Sandra Huenchuan, *Perspectivas globales sobre la protección de los derechos humanos de las personas mayores*, *supra* note 11, at 40.

[89]The full text of the *Brasilia Declaration* may be consulted at http://www.un.org/esa/socdev/ageing/documents/regional_review/Declaracion_Brasilia.pdf (last visited Jan. 19, 2015).

[90]Sandra Huenchuan, *Perspectivas globales sobre la protección de los derechos humanos de las personas mayores*, *supra* note 11, at 41.

[91]*Id.*

[92]The full text of the *San José Charter on the Rights of Older Persons in Latin American and the Caribbean* may be consulted at http://www.un.org/esa/socdev/ageing/documents/Review_and_Appraisal/CR_Carta_ENG.pdf (last visited Jan. 19, 2015).

institutions can operate effectively; the enhancement of the technical capacity of those institutions; the participation of older persons in processes of consultation and institutional accountability; and the adoption of an information system that gathers necessary data to monitor the implementation of public policies to protect the rights of older persons.[93]

As a result of these regional processes, a number of Constitutions in Latin America and the Caribbean expressly recognize the right of older persons to special protection, including Brazil, Costa Rica, Colombia, Bolivia, Dominican Republic, Honduras, Nicaragua, Panama, Paraguay and Venezuela.[94] Likewise, at least 14 states of the region have adopted legislation providing for a legal framework to protect the rights of older persons or regulate the integral attention required from the state to ensure the protection of their rights.[95]

Older persons in the region are not just waiting for legal and political change to reach them. They have organized themselves. In 2001, a regional meeting of organizations defending the rights of older persons approved the *Declaration of Lima*.[96] One of the Declaration's goals is to promote the adoption of an International Convention of Older People's Rights.[97] Also, the organizations represented in the regional meeting declared that individual members would demand that the national governments of their countries and international organizations allocate necessary and sufficient resources for the implementation of programs for older people.[98]

4. Historical Human Rights Development and the Protocol of San Salvador

Given the political situation of the Western Hemisphere during the 60–80s, it was no surprise that Inter-American human rights treaties and the work of the IACHR focused more on the civil and political rights of its citizens than on their economic,

[93]*San José Charter on the Rights of Older Persons in Latin American and the Caribbean*, *supra*, at 11.

[94]Sandra Huenchuan, *Perspectivas globales sobre la protección de los derechos humanos de las personas mayores*, *supra* note 11, at 43–44.

[95]*Id.*, at 44–46.

[96]For a text of the *Declaration of Lima* in Spanish, see http://www.gerontologia.org/noticia.php?id=240 (last visited Jan. 31, 2015).

[97]Luz Barreto, *A Report on the Status of Older People's Rights in Latin America*, available at http://www.globalaging.org/elderrights/world/2009/rightslatinamerica.pdf (last visited Jan. 31, 2015).

[98]*The Rights of Older Persons: Collection of International Documents*, Israel Doron & Kate Mewhinney, editors (2007), Latin American and Caribbean Meeting of Leaders of Organizations of Older People, Declaration of Lima (2001).

social, and cultural rights.[99] Moreover, at the time, there was an assumption that representative democracies respectful of civil and political rights should naturally translate into substantive improvements in the quality of life of the great majority, if not all, of the population.[100]

Later, however, the IACHR acknowledged that the experience in Latin American and the Caribbean showed that "there is no automatic and necessary relationship between the effectiveness of civil and political rights and the satisfaction of the basic needs of large sectors of the population,"[101] but the benefits to individuals can be tangible if a holistic approach to human rights that includes the protection of economic, social and cultural rights is implemented. It was in this regard that the OAS General Assembly instructed the OAS General Secretariat to first prepare a preliminary draft additional protocol in the area of social, economic, and cultural rights in 1982.[102] Six years later, the *Protocol of San Salvador* was adopted in November of 1988.[103] It entered into force in 1999 and as of today 16 OAS State members have ratified this treaty.[104]

The drafting history of the protocol shows a general understanding and acceptance that older persons require special protection. In its comments regarding the draft, the IACHR noted that certain minority populations deserve special attention to fully enjoy their economic, social, and cultural rights. Specifically, it noted that "it is important to include certain relevant groups in the Additional Protocol that have not yet been the object of international protection in the area of human rights such as the elderly."[105] Adequate enjoyment of rights, the IACHR noted, may require protections above-and-beyond those required for the general population.[106] Throughout the drafting process, the IACHR reiterated its concern for older persons as a group of persons that require special protection, as well as its belief that governments should pass specific measures for groups such as older

---

[99]IACHR, Annual Report 1983–1984, OEA/Ser. P/L/II.63, doc. 10 (1984), Chapter IV, Section II, 2, available at http://www.cidh.org/annualrep/83.84.eng/toc.htm (last visited Jan. 31, 2015).

[100]*Id.*

[101]*Id.*

[102]OAS General Assembly, AG/RES. 619 (XII-O/82), *Preliminary Draft Additional Protocol to the American Convention on Human Rights (Pact of San Jose)*, adopted on November 20, 1982, at 54, available at http://scm.oas.org/pdfs/agres/ag03796E01.PDF (last visited Jan. 15, 2015). See also AG/RES. 657 (XIII-O/83), *Preliminary Draft Additional Protocol to the American Convention on Human Rights (Pact of San Jose)*, adopted on November 18, 1983, at 59, available at http://scm.oas.org/pdfs/agres/ag03797E01.PDF (last visited Jan. 15, 2015) (requesting the OAS Secretary General to submit the OAS State members and the IACHR the Draft Protocol and seek their comments and observations).

[103]*Additional Protocol to the American Convention on Human Rights in the Area of Economic, Social, and Cultural Rights, "Protocol of San Salvador"*, *supra* note 28.

[104]*Id.*

[105]IACHR, Annual Report 1983–1984, *supra* note 99, 7.

[106]*Id.*

persons that go beyond those basic provisions necessary for the general majority.[107] In the *Report of the Permanent Council on the Draft Additional Protocol*,[108] comments about this protection were reaffirmed by some countries as well.[109]

In the final *Protocol of San Salvador*, Article 17 addresses the rights of older people through progressive implementation. States must take steps to realize this right, but it is not immediately enforceable upon ratification. The final version of *Article 17* states that:

> Everyone has the right to special protection in old age. With this in view the States Parties agree to take progressively the necessary steps to make this right a reality and, particularly, to:
>
> a. Provide suitable facilities, as well as food and specialized medical care, for elderly individuals who lack them and are unable to provide them for themselves;
> b. Undertake work programs specifically designed to give the elderly the opportunity to engage in a productive activity suited to their abilities and consistent with their vocations or desires;
> c. Foster the establishment of social organizations aimed at improving the quality of life for the elderly.

Article 17 expressly establishes older persons as a vulnerable group requiring special protection. Consequently, state parties must adopt special measures to ensure older persons full and equal enjoyment of human rights, as was recommended by the IACHR. The word "particularly" suggests that the rights contained therein are not exhaustive, but a minimum set of requirements. Other protections remain available for states to adopt at their option.

Part c is a particularly interesting provision. The state duty to foster the establishment of organizations designed to improve older persons' lives might suggest that states can carry out their obligations either directly or in cooperation with private institutions. In many cases, it may encourage reliance on non-state actors to fulfill the intention of the provision, though state parties will bear the final responsibility to ensure compliance with the obligations arising out of the treaty. Aside from this reliance, this provision represents a specific recognition of the

---

[107] *Id.* Also, please see IACHR, Annual Report 1985–1986, Chapter V, Section II, OEA/Ser.L/V/II.68 Doc. 8 rev. 1, September 16, 1986, available at http://www.cidh.org/annualrep/85.86eng/toc.htm (last visited Jan. 15, 2015).

[108] Report of the Permanent Council on the Draft Additional Protocol to the American Convention on Human Rights, OEA/Ser.P, AG/doc.2189/87, October 30, 1987.

[109] Observations on the Preliminary Draft Additional Protocol to the American Convention on Human Rights, Permanent Mission of Peru to the Organization of American States, OEA/Ser.G, CP/CAJP-622/85 add. 3, 9 July 1986; Observations on the Preliminary Draft of the Additional Protocol to the American Convention on Human Rights, Permanent Mission of Mexico to the Organization of American States, OEA/Ser. G, CP/CAJP-622/85 add. 6, 18 August 1986; Observations of the Government of the Argentine Republic on the Additional Protocol to the American Convention on Human Rights, OEA/Ser. G, CP/CAJP-622/85 add. 10, 1 October 1987.

importance of quality of life and social engagement for a population that is too-often isolated and marginalized. A similar provision exists for the protection of the rights of persons with disabilities.[110]

Even if, as stated above, the *Protocol of San Salvador* came into force in 1999, actual monitoring of compliance with its provisions, including Article 17, has not yet occurred in practice.[111] Article 19 of the *Protocol of San Salvador* envisions state submissions of periodic reports identifying the progressive measures adopted to ensure respect of the rights protected in this treaty. This aspect of the monitoring of the treaty obligations has not been implemented, though the procedural and substantive aspects of the supervisory mechanism have been established.[112] In 2005, the General Assembly of the OAS adopted *Standards for the Preparation of Periodic Reports Pursuant to Article 19 of the Protocol of San Salvador.*[113] Those standards outlined what the contents of the periodic reports under the *Protocol of San Salvador* should be, stating that in all cases, information provided on the specific rights enumerated should take vulnerable groups, expressly older persons, into account.[114] It further explained that the intent was to give Articles 15, 16, 17, and 18 a crosscutting effect.[115] Also, the OAS General Assembly requested the IACHR to submit a proposal to the Permanent Council regarding the progress indicators according to which states will provide information on each protected right.[116]

The IACHR prepared the *Guidelines for Preparation of Progress Indicators in the Area of Economic, Social and Cultural Rights* which also established the need to use crosscuting indicators, such as the principle of equality, to assess how historically disadvantaged or other vulnerable groups, including older persons, have enjoyed respect of the economic, social and cultural rights protected by the

[110]Article 18(d) of the *Protocol of San Salvador*.

[111]According to the Annual Report of the Working Group created to analyze the periodic reports under the Protocol of San Salvador, states were expected to submit their first reports before June 2014. However, there is no information available on the submission of those reports, so it is possible to assume that the monitoring of such reports has not started yet. See. Grupo de Trabajo encargado de analizar los Informes Nacionales Periódicos de los Estados Partes del Protocolo de San Salvador, Informe de Actividades, Período Enero-Diciembre, 2013, 28 Febrero 2014, OEA/Ser.L./XXV.2.1 GT/PSS/doc.10/13, 10 Marzo 2014, at 4, available at http://scm.oas.org/pdfs/2014/CP32419S.pdf (last visited Jan. 15, 2015).

[112]See in general Laura Pautassi, *Monitoring Access to Information from the Perspective of Human Rights Indicators*, Sur - International Journal on Human Rights, http://www.surjournal.org/eng/conteudos/getArtigo18.php?artigo=18,artigo_04.htm (last visited Jan. 20, 2015).

[113]AG/RES. 2074 (XXXV-O/05), *Standards for the Preparation of Periodic Reports pursuant to Article 19 of the Protocol of San Salvador*, adopted June 7, 2005, at 39, available at http://www.oas.org/en/sla/docs/AG02863E06.pdf (last visited Jan. 15, 2015). Article 15 protects the rights of the family, Article 16 protects the rights of the children, Article 17 protects the rights of the elderly, and Article 18 protects the rights of persons with disabilities.

[114]Id., Appendix, see in general at 43–46.

[115]Id. Appendix, at 45.

[116]Id., text of AG/RES. 2074 (XXXV-O/05), at 39–40.

Protocol of San Salvador.[117] On the basis of the principle of equality, states would be able to identify the most vulnerable groups, and within that category determine the sectors that need priority assistance, before formulating its social plans and policies to ensure access to basic economic, social and cultural rights.[118] Moreover, it would provide parameters to produce appropriate statistical information[119] and ensure an effective allocation of resources in state budgets to address the needs of those who are excluded because they are part of a vulnerable group.[120]

Moreover, the OAS General Assembly approved the creation of a working group to examine the national reports.[121] In addition, it instructed the working group to prepare progress indicators to assess state implementation of each right protected by the *Protocol of San Salvador*.[122] The Working Group was established and become operational in March 2010 and produced two sets of indicators to assess compliance with obligations arising out of the *Protocol of San Salvador*.[123] The first set of indicators relates to the protection of the rights to social security, health and education. The second set of indicators refers to the protection of the rights to work, trade union rights and the right to a healthy environment. In both cases, the Working Group followed the *Standards for the Preparation of Periodic Reports Pursuant to Article 19 of the Protocol of San Salvador* and the IACHR's *Guidelines for Preparation of Progress Indicators in the Area of Economic, Social and Cultural Rights* and adopted crosscuting indicators, including the principle of equality and non-discrimination, to ensure that vulnerable groups, including older persons, receive the appropriate differentiated attention and resources when states adopt policies to implement the rights protected in the *Protocol of San Salvador*.[124] The two sets of indicators were adopted by the OAS General

---

[117]IACHR, *Guidelines for Preparation of Progress Indicators in the Area of Economic, Social and Cultural Rights*, OEA/Ser.L/V/II.132 Doc. 14 19 July 2008, 45–47, available at http://www.cidh.org/pdf%20files/Guidelines%20final.pdf (last visited Jan. 20, 2015).

[118]*Id.*, 55.

[119]*Id.*, 58.

[120]*Id.*, 64.

[121]AG/RES. 2262 (XXXVII-O/07), *Protocol Of San Salvador: Composition and Functioning of the Working Group to Examine the Periodic Reports of the States Parties*, adopted on June 5, 2007, OEA/Ser.P, AG/doc.4771/07 rev. 1, September 14, 2007, at 22.

[122]AG/RES. 2582 (XL-O/10), *Protocol Of San Salvador: Composition and Functioning of the Working Group to Examine the Periodic Reports of the States Parties*, adopted on June 8, 2010, OEA/Ser.P AG/doc.5124/10 rev. 1, February 3, 2011, at 314, 315.

[123]Grupo de Trabajo encargado de analizar los Informes Nacionales Periódicos de los Estados Partes del Protocolo de San Salvador, *Informe de Actividades, Período Enero-Diciembre*, 2013, *supra* note 111, at 2–6.

[124]*Indicadores the Progreso para Medición de Derechos Contemplados en el Protocolo de San Salvador*, Propuesta elevada por el Grupo de Trabajo para el análisis de los informes nacionales previstos en el Protocolo de San Salvador en cumplimiento del mandato previsto en la Resolución AG/RES 2582 (XL-0-10), OEA/Ser.L/XXXV.2.1., GT/PSSI/doc.2/11, 11 marzo 2011, 43–51.

Assembly[125] and are in the process of being applied by the Working Group when the state parties to the *Protocol of San Salvador* submit their reports.[126]

On the other hand, it has been argued that Article 17 can be complemented by other provisions of the *Protocol of San Salvador*.[127] The right to social security,[128] which expressly recognizes that states should ensure those benefits to older persons; the right to health,[129] the right to food,[130] and the right to education[131] all harmonize with Article 17.

C. Human Rights of Older Persons

The organs of the Inter-American System have acknowledged the indivisibility and interdependence of human rights.[132] The *American Declaration* is an early example of such indivisibility and interdependence since it protects civil and political rights as well as economic, social and cultural rights. As explained above, this instrument, even not binding at the time of its adoption, has been considered by the IACtHR as an authoritative interpretation of human rights clauses of the OAS Charter.[133] Moreover, some arguments have been made regarding the status of customary international law of some of its provisions.[134] Although the *American*

---

[125]The first set of indicators was approved by AG/RES. 2713 (XLII-O/12), *Adoption of Progress Indicators for Measuring Rights under the Protocol of San Salvador*, adopted on June 4, 2012, OEA/Ser.P/XLII-O.2, February 1, 2013, at 56, available at http://www.oas.org/en/sla/docs/AG05796E04.pdf (last visited Jan. 20, 2015). The second group of indicators was approved in the OAS General Assembly in 2014. See, AG/doc.5422, http://scm.oas.org/ag/documentos/ (last visited Jan. 20, 2015).

[126]As stated above, the state reports regarding implementation of the first groups of rights were due before June 2014. However, there is no yet information available on whether these reports have been received and the Working Group has started the process of reviewing them. See *supra* note 111.

[127]See Diego Rodriguez-Pinzon & Claudia Martin, *The International Human Rights Status of Elderly Persons*, 18 Am. U Int'l L. Rev. 915, 969 (2002).

[128]Article 9 of the *Protocol of San Salvador*.

[129]Articles 10 of the *Protocol of San Salvador*.

[130]Article 11 of the *Protocol of San Salvador*.

[131]Article 13 of the *Protocol of San Salvador*.

[132]See, *inter alia*, IACHR, *The Work, Education and Resources of Women: The Road to Equality in Guaranteeing Economic, Social and Cultural Rights*, OEA/Ser.L/V/II.143, Doc.59, November 3, 2011, 29; I/A Court H.R., *Case of Acevedo Buendía et al. ("Discharged and Retired Employees of the Comptroller") v. Peru*, Preliminary Objection, Merits, Reparations and Costs, *supra* note 27, 101.

[133]I/A Court H.R., *Interpretation of the American Declaration of the Rights and Duties of Man within the Framework of Article 64 of the American Convention on Human Rights*. *supra* note 18.

[134]*Id.*

Declaration embodies both set of rights, the IACHR has mostly supervised compliance with civil and political rights when exercising its quasi-judicial powers under this instrument.[135] However, when exercising its political powers, the IACHR has been more willing to monitor compliance with economic, social and cultural rights in country or thematic reports.[136] On the other hand, the IACtHR has concluded that it lacks jurisdiction to apply the *American Declaration*.[137] In consequence, in spite of the recognition that human rights are indivisible and interdependent, in practice the protection of economic, social and cultural right has not been a priority at the Inter-American System. The current draft on the *Convention on Older Persons* is the first Inter-American instrument in which both set of rights will be integrated into a unique treaty. If adopted, the *Convention on Older Persons* will constitute a significant step forward towards the protection of civil and political rights and economic, social and cultural rights, particularly if the draft maintains its current approach allowing the IACHR and the IACtHR to monitor compliance with the rights protected therein through its individual petition system.

For methodological reasons and given the lack of an instrument that protects the rights of older persons based on a holistic approach, in the following sections we will maintain the traditional division of civil and political rights and economic, social and cultural rights to best reflect the existing norms, case law and policies regarding the rights of older persons in the Americas.

D. Civil and Political Rights of Older Persons

Within the scope of civil and political rights, the following rights protected by the *American Declaration* may have particular relevance for the protection of the rights of older persons: Article I (right to life, liberty and personal security); Article II (right to equality before law); Article IV(right to protection of honor, personal reputation, and private and family life); Article VI (right to a family and to protection thereof); Article IX (right to inviolability of the home); Article X (right to the inviolability and transmission of correspondence); Article XVII (right

[135]Mostly due to the language of Article 20 of the Statute of the IACHR, which provides in the pertinent part: In relation to those member states of the Organization that are not parties to the American Convention on Human Rights, the Commission shall have the following powers, in addition to those designated in Article 18: a. to pay particular attention to the observance of the human rights referred to in Articles I, II, III, IV, XVIII, XXV, and XXVI of the American Declaration of the Rights and Duties of Man. IACHR, Statute of the Inter-American Commission on Human Rights, available at http://www.oas.org/en/iachr/mandate/Basics/statuteiachr.asp (last visited Jan. 30, 2015).

[136]See, *inter alia*, IACHR, *Third Report on the Situation of Human Rights in Paraguay*, OEA/Ser.L/V/II.110, Doc.52, March 9, 2001; *Human Rights of Migrant and Other Persons in the Context of Human Mobility in Mexico*, OEA/Ser.L/V/II, Doc.48/13, December 30, 2013.

[137]I/A Court HR, *Case of Bueno Alves v Argentina*, Merits, Reparations and Costs, *supra* note 66.

to recognition of juridical personality and civil rights); Article XVIII (right to a fair trial); Article XX (right to vote and to participate in government); Article XXI (right of assembly); Article XXII (right of association); Article XXV (right of protection from arbitrary arrest); and Article XXVI (right to due process of law). Also, the following rights under the American Convention are pertinent to ensure protection of older persons: Article 4 (right to life); Article 5 (right to humane treatment); Article 7 (right to personal liberty); Article 8 (right to a fair trial); Article 11 (right to privacy); Article 13 (freedom of thought and expression); Article 15 (right of assembly); Article 16 (freedom of association); Article 17 (rights of the family); Article 21 (right to property); Article 22 (freedom of movement and residence); Article 23 (right to participate in government); Article 24 (right to equal protection); Article 25 (right to judicial protection); and Article 1(1) (obligation to respect rights).

Additionally, the draft *Convention on Older Persons* ensures the right to equality and non-discrimination for reasons of age; right to life and dignity in old age; right to independence and autonomy; right to participation and community integration, right to safety and a life free of violence of any kind; right not to be subjected to torture or to cruel, inhuman or degrading treatment or punishment; right to give prior and informed consent on health matters; rights of older persons receiving long-term care; right to personal liberty; right to freedom of expression, opinion and access to information; right to nationality and freedom of movement; right to privacy and intimacy; right to property; political rights; freedom of association and assembly; equal recognition as a person under the law; and access to justice.

1. The Right to Recognition Before the Law

Every individual has the right to recognition as a person before the law and all persons must be treated equally before the law.[138] Both the *American Declaration* and the *American Convention* require States to recognize every individual everywhere as a person having rights and obligations, and to protect the enjoyment of their basic civil rights.[139] The IACtHR has stated that "[t]he right to the recognition of juridical personality implies the capacity to be the holder of rights (capacity of exercise) and obligations; the violation of this recognition presumes an absolute disavowal of the possibility of being a holder of such rights and

[138]Articles 3 and 24 of the *American Convention*; Peggy Kelly, *UN Program on Ageing, Intersection of Population and Human Rights: Rights of Older Persons in the International Context* (Oct. 15, 2006), available at http://www.eclac.org/celade/noticias/paginas/6/27116/KellyP.pdf (last visited Jan. 15, 2015).

[139]Specifically, Article XVII of the *American Declaration* states: Every person has the right to be recognized everywhere as a person having rights and obligations, and to enjoy the basic civil rights. On the other hand, Article 3 of the *American Convention* provides: Every person has the right to recognition as a person before the law.

obligations."[140] Moreover, this right involves a determination on whether an individual, as primary holder of the right, can exercise such right in practice; "therefore, the failure to recognize or acknowledge such capability places the person in a vulnerable position in relation to the State or third parties."[141] States have a duty to provide all the means and legal conditions to ensure individuals their right to legal personality,[142] in particular those who belong to vulnerable groups, in light of the principle of equality and non-discrimination.[143] Thus, legal capacity is essential to exercise the right to legal personality under the provisions of both the *American Declaration* and the *American Convention*.

Ensuring the right to legal capacity is essential for the protection of older persons because older persons increasingly grapple with situations in which their right to make their own decisions is affected.[144] Consequently, the IACHR has concluded that the "exercise of one's right to legal capacity should be central to the discussion of the [Convention on Older Persons] to enable older persons to fully exercise their rights, under conditions of equality with other persons.[145] In that context, the IACHR has underlined the importance of preventing and avoiding situations of involuntary institutionalization.[146]

In this regard, the draft *Convention on Older Persons* reaffirms the right of older persons to recognition before the law and to enjoy legal capacity "on equal basis with others in all aspects of life."[147] The draft follows a similar approach to the one provided by Article 12 of the *Convention on the Rights of People with Disability* (CRPD)[148]; in other words, it proposes a system of supported decision making which emphasizes the autonomy and independence of the individual.

---

[140] I/A Court H.R., *Case of Bámaca Velásquez v. Guatemala*, Merits, Judgment of November 25, 2000, Series C No. 70, 179; *Case of the Girls Yean and Bosico v. Dominican Republican*, Preliminary Objections, Merits, Reparations and Costs, Judgment of September 8, 2005, Series C No. 130, 176.

[141] I/A Court H.R., *Case of Anzualdo-Castro v. Peru*, Preliminary Objection, Merits, Reparations and Costs, Judgment of September 22, 2009, Series C No. 202, 88.

[142] *Id.*

[143] *Id.*, 89. Also, see *Case of Sawhoyamaxa Indigenous Community v. Paraguay*, Merits, Reparations and Costs, Judgment of March 29, 2006, Series C No. 146, 189.

[144] Working Group on Protection of the Human Rights of Older Persons, "*Views of the Inter-American Commission on Human Rights of the Organization of American States (OAS)" regarding the Draft Inter-American Convention on the Human Rights of Older Persons* (Presented at the meeting held on November 8, 2013, in response to a request made by the Chair of the CAJP) OEA/Ser.G/CAJP/GT/DHPM 114/13, November 18, 2013, at 6, available at http://www.oas.org/consejo/CAJP/personas%20mayores.asp#Docs (last visited Jan. 20, 2015).

[145] *Id.*, at 7–8.

[146] *Id.*, at 7.

[147] *Draft Inter-American Convention on the Human Rights of Older Persons*, *supra* note 39, Article 30.

[148] *Convention on the Rights of Persons with Disabilities*, 13 December 2006, United Nations, Treaty Series, vol. 2515, p. 3, entered into force 3 May 2008.

In that light, states must provide older persons with access to the necessary support required to allow them to exercise their right to legal capacity.[149] Moreover, when providing that support, states must ensure, through adequate safeguards, that older persons enjoy their legal capacity without being subject of abuse.[150] In particular, the safeguards must ensure that older persons enjoy free will to decide on the person to support them, that the measures adopted are proportional and tailored to the needs of the beneficiary, are applied for the shortest period required and are subject to review by a competent, independent, and impartial authority.[151] Ultimately, the draft *Convention on Older Persons* requires states to ensure that older persons enjoy their right "to own or inherit property, to control their own financial affairs, and to have equal access to bank loans, mortgages, and other forms of financial credit, and [ensure] that older persons are not arbitrarily deprived of their property."[152]

Another important issue regarding the exercise of legal capacity relates to the right of older persons to provide free and informed consent with respect to medical care and their placement in residential, institutional, or psychiatric care. The notion of prior and informed consent is not currently defined in the draft *Convention on Older Persons*, but a full article is devoted to regulate this issue. Article 11 of the draft instrument protects the inalienable right of older persons to express prior and informed consent and to modify or revoke it in respect of "any medical decision, treatment, procedure or research in the area of health affecting, or that might affect, their integrity, dignity or well-being."[153] To guarantee this right, states must ensure appropriate and effective mechanisms to prevent abuse and strengthen the capacity of older persons to understand their options and the risks they face when choosing one alternative over another.[154] Specially, states must ensure older persons access to information in a non-discriminatory, timely, clear and culturally sensitive manner so that they can exercise their options in an informed manner.[155] The notion of informed consent has been extensively discussed by the IACHR in the context of women's access to reproductive health.[156]

With respect to institutionalization, the draft *Convention on Older Persons* provides that state parties should establish mechanisms to ensure that the decision to be placed in a long-term care facility be the result of the free and express will of

[149] *Id.*

[150] *Id.*

[151] *Id.*

[152] *Id.*

[153] *Draft Inter-American Convention on the Human Rights of Older Persons*, *supra* note 39, Article 11.

[154] *Id.*

[155] *Id.*

[156] IACHR, *Access to Information on Reproductive Health from a Human Rights Perspective*, OEA/Ser.L/V/II, Doc.61, November 22, 2011, 43–74.

the older person concerned.[157] Likewise, older persons should enjoy the right to end their placement at any time and upon their own decision.[158] Institutionalization should not be justified only on the basis of the age of the individual.[159]

Even if the text of the draft *Convention on Older Persons* has been the object of extensive discussions and amendments throughout the current process of negotiation, the scope of the right to legal capacity of older persons has essentially remained the same.[160] This agreement shows that there is consensus among states regarding the core nature of the right to legal capacity for the protection of older persons, as well as the need to apply to older persons the paradigm of supported decision making embodied by Article 12 of the CRPD.

2. The Right to Equality and Non-discrimination

The right to equality and non-discrimination is protected by Article II of the *American Declaration*[161] and Articles 1(1) and 24 of the *American Convention*.[162] In the case of the *American Convention*, while Article 1(1) refers to the state's duty to respect and ensure the rights enshrined in that treaty without discrimination, Article 24 protects against de facto and *the jure* discrimination in domestic laws passed by the state as well as in their application to specific situations. Thus, "if a State discriminates in the respect for or guarantee of a right contained in the Convention, it will be failing to comply with its obligation under in Article 1(1) and the substantive right in question. If, on the contrary, the discrimination refers to unequal protection by domestic laws, the fact must be analyzed in light of Article

---

[157] *Draft Inter-American Convention on the Human Rights of Older Persons*, supra note 39, Article 12.

[158] *Id.*

[159] *Id.*

[160] See, Working Group on Protection of the Human Rights of Older Persons, *Articles Considered during the Meetings Held on April 28, 29, and 30, 2014*, OEA/Ser.G/CAJP/GT/DHPM-37/12 add. 29, corr. 1, May 14, 2014, at 45.

[161] Article II of the *American Declaration* states: All persons are equal before the law and have the rights and duties established in this Declaration, without distinction as to race, sex, language, creed or any other factor.

[162] Article 1(1) of the *American Convention* provides: 1. The State Parties to this Convention undertake to respect the rights and freedoms recognized herein and to ensure to all persons subject to their jurisdiction the free and full exercise of those rights and freedoms, without any discrimination for reasons of race, color, sex, language, religion, political or other opinion, national or social origin, economic status, birth, or any other social condition. Article 24 of the *American Convention* states: All persons are equal before the law. Consequently, they are entitled, without discrimination, to equal protection of the law.

24 of the American Convention."[163] Both the IACHR and the IACtHR have consistently stated that the right to equality and nondiscrimination is a core aspect of the Inter-American human rights System.[164] Moreover, the principle of equality and non-discrimination has been characterized by the IACtHR as a *jus cogens* norm.[165]

Under Article II of the *American Declaration* discrimination is prohibited on the basis of race, sex, language, creed or any other factor; whereas, under Article 1(1) of the *American Convention* discrimination is prohibited based on the following grounds: race, color, sex, language, religion, political or other opinion, national or social origin, economic status, birth, or any other social condition. Article 24 protects equality before the law and the principle of nondiscrimination in general terms, without listing any specific grounds. The IACtHR has acknowledged that the list of grounds on which discrimination is prohibited is not exhaustive, but mostly illustrative.[166] Moreover, the phrases "any other factor" in Article II of the *American Declaration* and "any other social condition" in Article 1(1) of the *American Convention* show that there was an intention to leave the list of grounds open to incorporate additional categories, not expressly established in those provisions. With respect to "any other conditions," the IACtHR has concluded that this phrase should be interpreted "in the context of the most favorable option for the human being and in light of the evolution of fundamental rights in contemporary international law."[167] In *Atala Riffo*, for example, the IACtHR recognized that Articles 1(1) and 24 of the *American Convention* prohibited discrimination on the basis of sexual orientation.[168] Consequently, even if age is not a ground expressly protected under the Inter-American instruments, the practice of the IACHR and IACtHR demonstrates that an unjustified distinction on that base would be prohibited by the *American Declaration* and the *American Convention*.

In line with other international jurisprudence, the Inter-American organs have stated that not every distinction in treatment involves discrimination.[169] According to the IACHR, there are several conceptions of the right to equality and nondiscrimination.[170] First, there is discrimination when a distinction made between persons placed in a similar situation lacks objective and reasonable

---

[163]See, *inter alia*, I/A Court H.R., *Case of Atala Riffo and daughters v. Chile*, Merits, Reparations and Costs, Judgment of February 24, 2012, Series C No. 239, 82.

[164]IACHR, *The Situation of People of African Descent in the Americas*, OEA/Ser.L/V/II.Doc. 62, December 5, 2011, 89; I/A Court H.R., Juridical Condition and Rights of the Undocumented Migrants, Advisory Opinion OC-18/03 of September 17, 2003, Series A No.18, 109.

[165]I/A Court H.R., *Juridical Condition and Rights of the Undocumented Migrants*, *supra*, 101.

[166]I/A Court H.R., *Case of Atala Riffo and daughters v. Chile*, Merits, Reparations and Costs, supra note 163, 85.

[167]*Id.*

[168]*Id.*, 91.

[169]I/A Court H.R., *Juridical Condition and Human Rights of the Child*, Advisory Opinion OC-17/02 of August 28, 2002, Series A No. 17, 46.

[170]IACHR, *The Situation of People of African Descent in the Americas*, supra note 164, 89.

justification.[171] For a distinction to be permissible it must be based upon objective and reasonable justification and pursue a legitimate aim.[172] Moreover, there are must be a reasonable relationship of proportionality between the means employed and the aim sought to be realized.[173]

According to the IACHR "distinctions based on grounds explicitly enumerated under pertinent articles of international human rights instruments are subject to a particular strict level of scrutiny whereby states must provide an especially weighty interest and compelling justification for the distinction."[174] Thus, in light of that interpretation grounds listed in the *American Declaration* and the *American Convention* would be considered "suspect grounds" requiring a stricter scrutiny of the reasons provided by the state to justify a distinction on those bases. In contrast, those not listed, such as a distinction based on age, would not appear to require such level of scrutiny. Interestingly, the IACtHR reached a different conclusion in *Attala Riffo* where the distinction under consideration was based on sexual orientation, a ground not listed in the provisions of equal protection and nondiscrimination of the *American Convention.* In particular, this Tribunal concluded that "[a]s regards the prohibition of discrimination based on sexual orientation, any restriction of a right would need to be based on rigorous and weighty reasons."[175] Also, the IACtHR has stated that today "discriminatory treatment of any person, owning to gender, race, color, language, religion or belief, political or other opinion, national, ethic or social origin, nationality, **age**, economic situation, property, civil status, birth or other status is unacceptable."[176] Consequently, even if age as a ground of discrimination has not been so far considered a "suspect ground," the interpretations of the IACH and the IACtHR show that there is room for the development of a different approach in future interpretations. This approach may be further supported if the current draft *Convention on Older Persons* is adopted in the near future.

The prohibition of discrimination on the basis of age is one of the core issues of the draft *Convention on Older Persons* and this principle is reflected in several parts of the treaty. First, the draft *Convention on Older Persons* provides in the Preamble that state parties to this treaty reaffirm the obligation to eliminate all forms of discrimination, in particular discrimination on the basis of age.[177] Also, Article 1

---

[171]I/A Court H.R., *Juridical Condition and Rights of the Undocumented Migrants, supra* note 164, 89.

[172]IACHR, Report N° 51/01, *Case 9903, Rafael Ferrer-Mazorra et al. v. United States*, Annual Report 2000, April 4, 2001, 238. Also, I/A Court H.R., *Proposed Amendments of the Naturalization Provisions of the Constitution of Costa Rica*, Advisory Opinion OC-4/84 of January 19, 1984, Series A No. 4, 57.

[173]*Id.*

[174]IACHR, *The Situation of People of African Descent in the Americas, supra* note 164, 90.

[175]I/A Court H.R., *Case of Atala Riffo and daughters v. Chile*, Merits, Reparations and Costs, supra note 163, 124.

[176]I/A Court H.R., *Juridical Condition and Rights of the Undocumented Migrants*, at supra note 164, 101.

[177]*Draft Inter-American Convention on the Human Rights of Older Persons, supra* note 39, Preamble.

states that the objective of the treaty is to promote, protect and ensure, under equal conditions, the enjoyment of all fundamental rights and freedoms of older persons.[178] Moreover, Article 2 defines age discrimination as "any distinction, exclusion, or restriction based on age, the purpose or effect of which is to annul or restrict recognition, enjoyment, or exercise, in equality of conditions, of human rights and fundamental freedoms."[179] Article 3 lists as general principles of this convention equality and nondiscrimination.[180] Finally, Article 5 protects the right to equality and prohibits discrimination on the basis of the age of older persons.[181] The notion of discrimination incorporated in the draft *Convention on Older Persons* draws from similar definitions provided by the *International Convention on the Elimination of All Forms of Racial Discrimination* (CERD)[182] and the *Convention on the Elimination of All Forms of Discrimination against Women* (CEDAW).[183] Also, it uses similar language to the definition of discrimination embodied in three other Inter-American human rights treaties: the *Discrimination and Intolerance Convention*,[184] the *Racial Discrimination Convention*[185] and the *Disabilities*

[178] *Id.*, Article 1.

[179] *Id.*, Article 2.

[180] *Id.*, Article 3(d).

[181] *Id.*, Article 5.

[182] *International Convention on the Elimination of All Forms of Racial Discrimination, United Nations*, Treaty Series, vol. 660, p. 195, adopted on March 7, 1966, entered into force on January 4, 1969. Article 1 of *CERD* provides: In this Convention, the term "racial discrimination" shall mean any distinction, exclusion, restriction or preference based on race, colour, descent, or national or ethnic origin which has the purpose or effect of nullifying or impairing the recognition, enjoyment or exercise, on an equal footing, of human rights and fundamental freedoms in the political, economic, social, cultural or any other field of public life.

[183] *Convention on the Elimination of All Forms of Discrimination against Women*, United Nations, Treaty Series, vol. 1249, p. 13, adopted on December 18, 1979, entered into force on September 3, 1981. Article 1 of *CEDAW* states: For the purposes of the present Convention, the term "discrimination against women" shall mean any distinction, exclusion or restriction made on the basis of sex which has the effect or purpose of impairing or nullifying the recognition, enjoyment or exercise by women, irrespective of their marital status, on a basis of equality of men and women, of human rights and fundamental freedoms in the political, economic, social, cultural, civil or any other field.

[184] *Inter-American Convention Against All Forms of Discrimination and Intolerance, supra* note 32, Article 1 states that "Discrimination shall mean any distinction, exclusion, restriction, or preference, in any area of public or private life, the purpose or effect of which is to nullify or curtail the equal recognition, enjoyment, or exercise of one or more human rights and fundamental freedoms enshrined in the international instruments applicable to the States Parties."

[185] *Inter-American Convention against Racism, Racial Discrimination, and Related Forms of Intolerance, supra* note 33. Article 1 states that "Racial discrimination shall mean any distinction, exclusion, restriction, or preference, in any area of public or private life, the purpose or effect of which is to nullify or curtail the equal recognition, enjoyment, or exercise of one or more human rights and fundamental freedoms enshrined in the international instruments applicable to the States Parties. Racial discrimination may be based on race, color, lineage, or national or ethnic origin."

Convention.[186] Overall, discrimination on the basis of sex and race has been considered by other international tribunals as "suspect" and required a heighten level of scrutiny.[187] Likewise, there is some agreement that disability as a ground of discrimination is increasingly being considered as a "suspect" category.[188] Moreover, the definition of discrimination as "any distinction, exclusion or restriction" may leave little room for justifying differences in treatment. Some experts in the region have recommended that under the terms of a future *Convention on Older Persons* discrimination on the basis of age should be considered as "suspect," at least in situations in which the distinction is used to make distinctions in access to benefits, perform certain activities or exercise fundamental rights.[189] In those cases such categorization would require the measures to pass a strict scrutiny test to meet the standard of reasonability. The same experts acknowledged, however, that under certain conditions, distinctions on the basis of age could be considered legitimate and reasonable, particularly in regard to access to public positions or authorization to carry out specific jobs.[190] This "mixed approach" has been followed by the Constitutional Court of Colombia when deciding cases alleging discrimination on the basis of age.[191]

In addition to the draft *Convention on Older Persons*, at the regional level, the *Discrimination and Intolerance Convention* prohibits in Article 1 discrimination, *inter alia*, on the basis of age.[192] Also, the *Andean Charter* explicitly includes in Article 10 "age" as one of the grounds protected from all forms of discrimination and any kind of intolerance or exclusion against individuals or communities.[193]

---

[186]*Inter-American Convention on the Elimination of all Forms of Discrimination Against Persons with Disabilities*, *supra* note 31, Article 2(a) provides: The term "discrimination against persons with disabilities" means any distinction, exclusion, or restriction based on a disability, record of disability, condition resulting from a previous disability, or perception of disability, whether present or past, which has the effect or objective of impairing or nullifying the recognition, enjoyment, or exercise by a person with a disability of his or her human rights and fundamental freedoms.

[187]See, *inter alia*, European Commission Justice, *The Prohibition of Discrimination under European Human Rights Law, Relevance for the EU non-discrimination directives—an update*, Part I, Sect. 1.1.2. (2011).

[188]*Id.*

[189]Human Rights Public Policy Institute, Mercosur, *The principle of equality and non-discrimination for the protection of the rights of older persons*, Contribution to the discussion of an International Convention, OEA/Ser. G/CAJP/GT/DHPM/INF.1/11, October 28, 2011, at 6.

[190]*Id.*

[191]*Id.*, at footnote 7.

[192]*Inter-American Convention Against All Forms of Discrimination and Intolerance*, supra note 32, Article 1 provides that prohibited "[d]iscrimination may be based on nationality; age; sex; sexual orientation; gender identity and expression; language; religion; cultural identity; political opinions or opinions of any kind; social origin; socioeconomic status; educational level; migrant, refugee, repatriate, stateless or internally displaced status; disability; genetic trait; mental or physical health condition, including infectious-contagious condition and debilitating psychological condition; or any other condition.

[193]*Andean Charter for the Promotion and Protection of Human Rights*, 26 July 2002, *supra* note 81, Article 10.

Another conception of the right to equality and the principle of non-discrimination involves "the obligation of ensuring conditions of true equality for groups that have historically been excluded and are at greater risk of discrimination."[194] The IACtHR has acknowledged that under the duty to ensure the exercise of rights without discrimination provided under Article 1(1) of the *American Convention*, states must adopt positive measures, including affirmative actions, to eliminate or change existing discrimination against particular groups.[195] This includes the adoption of special measures to protect against the actions or omissions of private parties that contribute to maintain or favored discriminatory practices that affect those vulnerable groups.[196] In addition, as part of this obligation to create true or substantive equality, states must assess whether in the application of laws or policies vulnerable groups are impacted more negatively in their enjoyment of human rights, even when the formulation of the law or policy is neutral or apply to everyone without distinction.[197] Disproportionate impact upon a group resulting from an otherwise neutral law or policy may still involve discrimination in violation of the *American Declaration* and the *American Convention*.[198] The organs of the Inter-American System have concluded on the need to adopt special measures of protection to ensure the right to equal protection and equality with respect to groups disproportionately affected by discrimination, including: the rights of women,[199] racial minorities,[200] children,[201] indigenous groups,[202] and migrants.[203]

With respect to persons with disabilities, the IACHR specifically stated in the *Furlan Case* that "people with disabilities are often subject to discrimination

[194]See, *inter alia*, IACHR, *The Situation of People of African Descent in the Americas*, *supra* note 164, 89.

[195]I/A Court H.R., *Juridical Condition and Rights of the Undocumented Migrants*, supra note 164, 104; *Case of Atala Riffo and daughters v. Chile*, Merits, Reparations and Costs, *supra* note 163, 80.

[196]*Id.*

[197]See, IACHR, *The Situation of People of African Descent in the Americas*, *supra* note 164, 92; *Jessica Lenahan (Gonzales) et al. v. United States*, Report No. 80/11, Case 12.626, Merits, July 21, 2011, 109.

[198]*Id.*

[199]IACHR, *Access to Maternal Health Services from a Human Rights Perspective*, OEA/Ser.L/V/II, Doc. 69, June 7, 2010, 54; *Access to Justice for Women Victims of Violence in the Americas*, OEA/Ser.L/V//II, Doc. 68, January 20, 2007, 98–99.

[200]IACHR, *The Situation of People of African Descent in the Americas*, *supra* note 164, 97, 101–102.

[201]I/A Court H.R., *Case of the Girls Yean and Bosico v. Dominican Republic*, Preliminary Objections, Merits, Reparations and Costs, Judgment of September 8, 2005, Series C No. 130, 172–174.

[202]I/A Court H.R., *Case of the Xákmok Kásek Indigenous Community v. Paraguay*, Merits, Reparations and Costs, Judgment of August 24, 2010, Series C No. 214, 273–274.

[203]IACHR, *Human Rights of Migrants and Other Persons in the Context of Human Mobility in Mexico*, OEA/Ser.L/V/II, Doc. 48/13, December 30, 2013, 347–361.

because of their condition; therefore, States must adopt the appropriate legislative, social, educational, employment or other measures necessary to prevent all discrimination associated with mental disabilities, and to promote the full integration of such persons into society. Appropriate access to justice plays a fundamental role to address these types of discrimination.[204] Even if the organs of the Inter-American System have not issued any reports or judgments on the issue of discrimination on the basis of age, the practice of the IACHR and the IACtHR regarding the treatment of structural discrimination affecting historically affected groups, in particular persons with disabilities, will be essential to inform the application of the principle of equality and nondiscrimination to older persons.

The notion of substantive equality is also contemplated in the draft *Convention on Older Persons*. In addition to prohibit direct discrimination on the basis of age, Article 5 of the draft treaty provides that States "[i]n their policies, plans, and legislation on ageing and old age….shall develop specific approaches for older persons who are vulnerable and those who are victims of multiple discrimination."[205] Moreover, Article 3 of the draft *Convention on Older Persons* states as one of the general principles that applies to the treaty the notion of "differentiated treatment for the effective enjoyment of the rights of older persons."[206] Also, Article 4 provides the duty of state parties to adopt "affirmative measures and reasonable adjustments as may be necessary to expedite or attain de facto equality for older persons, or to ensure their full social, economic, educational, political and cultural engagement."[207] Experts participating in the current discussion on the draft *Convention of Older Persons* have indicated that the situation of structural discrimination that affects older persons as a group requires that this treaty emphasizes the notion of substantive equality, in particular the positive obligations that states must implement to overcome existing inequalities.[208] First, to be able to adopt special measures to protect a disadvantaged group such as older persons, a state must compile data and produce information on the needs and problems faced by older persons to fully ensure their human rights.[209] Next, among the special measures required to ensure full equality for older persons, experts include the need to develop an effective and sustainable social security system that incorporates both contributory and non-contributory schemes.[210] Also, states must conceive a notion of the duty of care that it is tailored to address the situations of

---

[204]I/A Court H.R., *Case of Furlan and Family v. Argentina*, Preliminary Objections, Merits, Reparations and Costs, Judgment of August 31, 2012, Series C No. 246, 135.

[205]*Draft Inter-American Convention on the Human Rights of Older Persons*, supra note 39, Article 5.

[206]*Id.*, Article 3.

[207]*Id.*, Article 4.

[208]Human Rights Public Policy Institute, Mercosur, *The principle of equality and non-discrimination for the protection of the rights of older persons*, *supra* note 189, at 7–8.

[209]*Id.*, at 10–12.

[210]*Id.*, at 12–18.

vulnerability faced by older persons, especially situations of violence.[211] Furthermore, when adopting measures that are applicable to the general population, states are required to follow a differentiated approach that takes into account the impact of those measures on specific groups and consider their particularities at the time of implementation of its policies.[212] Finally, states must ensure participation of older persons in the design and debate of the policies that benefit them to ensure that their input and demands are considered.[213]

It is interesting to note that the recently adopted *Discrimination and Intolerance Convention* and the *Racial Discrimination Convention* include definitions of "indirect discrimination"[214] and "multiple discrimination"[215] which may serve as guidelines for assessing prohibited distinctions on the basis of age that affect older persons. Moreover, these two instruments embody protection of substantial equality and provide that the adoption of special measures to ensure effective equality to disadvantaged groups does not constitute discrimination.[216] These

[211]*Id.*, at 18–20.

[212]*Id.*, at 20–21.

[213]*Id.*, at 22–24.

[214]*Inter-American Convention against All Forms of Discrimination and Intolerance, supra* note 32. Article 1 of the Discrimination and Intolerance Convention provides that "indirect discrimination shall be taken to occur, in any realm of public and private life, when a seemingly neutral provision, criterion, or practice has the capacity to entail a particular disadvantage for persons belonging to a specific group, or puts them at a disadvantage, unless said provision, criterion, or practice has some reasonable and legitimate objective or justification under international human rights law." *Inter-American Convention against Racism, Racial Discrimination, and Related Forms of Intolerance, supra* note 33. Article 1 of the *Racial Discrimination Convention* states that "indirect racial discrimination shall be taken to occur, in any realm of public and private life, when a seemingly neutral provision, criterion, or practice has the capacity to entail a particular disadvantage for persons belonging to a specific group based on the reasons set forth in Article 1.1, or puts them at a disadvantage, unless said provision, criterion, or practice has some reasonable and legitimate objective or justification under international human rights law."

[215]*Inter-American Convention against All Forms of Discrimination and Intolerance, supra* note 32. Article 1 of the *Discrimination and Intolerance Convention* provides that "multiple or aggravated discrimination is any preference, distinction, exclusion, or restriction based simultaneously on two or more of the criteria set forth in Article 1.1, or others recognized in international instruments, the objective or result of which is to nullify or curtail, the equal recognition, enjoyment, or exercise of one or more human rights and fundamental freedoms enshrined in the international instruments applicable to the States Parties, in any area of public or private life." *Inter-American Convention against Racism, Racial Discrimination, and Related Forms of Intolerance, supra* note 33. Article 1 of the *Racial Discrimination Convention* provides that "multiple or aggravated discrimination is any preference, distinction, exclusion, or restriction based simultaneously on two or more of the criteria set forth in Article 1.1, or others recognized in international instruments, the objective or result of which is to nullify or curtail, the equal recognition, enjoyment, or exercise of one or more human rights and fundamental freedoms enshrined in the international instruments applicable to the States Parties, in any area of public or private life."

[216]Article 5 of the *Inter-American Convention against All Forms of Discrimination and Intolerance, supra* note 32, and Article 5 of the *Inter-American Convention against Racism, Racial Discrimination, and Related Forms of Intolerance, supra* note 33.

conventions have not come into force yet, but the practice of the organs in charge of monitoring state compliance with their obligations will be of reference to enrich the interpretation of the prohibition of discrimination of older persons, once the draft *Convention on Older Persons* is adopted.

3. The Right Persons to Life

The right to life is protected by Article I of the *American Declaration*[217] and Article 4 of the *American Convention*.[218] The IACtHR has consistently underlined the fundamental nature of the right to life and has acknowledged that enjoyment of this right is essential for the exercise of other fundamental rights protected by the Inter-American human rights instruments. Consequently, "[d]ue to the essential nature of the right to life, no restrictive approaches thereto are to be admitted."[219] Moreover, the IACHR has concluded that the right to life is a core right of the *American Declaration*, "which has undoubtedly attained the status of customary international law."[220]

The case law and practice of the Inter-American System shows that both the IACHR and IACtHR have interpreted the scope of the right to life broadly, as encompassing both negative and positive obligations.[221] In particular, the IACtHR has consistently established that Article 4 of the *American Convention* "guarantees not only the right of every human person not to be deprived of his life arbitrarily, but also, the duty of the States to adopt such measures as may be necessary to create a legal framework which allows deterring any threat to the right to life; to establish an effective system of justice which can investigate, punish, and redress any act of deprivation of life by the State agents or private individuals; and to safeguard the right to have access to the conditions which guarantee a decent life, which includes the adoption of positive measures which prevent the violation of this right."[222]

---

[217]Article I of the *American Declaration* provides that "[e]very human being has the right to life, liberty and the security of his person."

[218]Article 4(1) of the *American Convention* provides that "[e]very person has the right to have his life respected. This right shall be protected by law and, in general, from the moment of conception. No one shall be arbitrarily deprived of his life."

[219]I/A Court H.R., *Case of Ximenes Lopes v. Brazil*, Merits, Reparations and Costs, Judgment of July 4, 2006, Series C No. 149, 124.

[220]IACHR, *Jessica Lenahan (Gonzales) et al. v. United States*, *supra* note 197, 112.

[221]See, *inter alia*, IACHR, *Leydi Dayán Sánchez v. Colombia*, Report N° 43/08, Case 12.009, Merits, July 23, 2008, 51; I/A Court H.R., *Case of the Sawhoyamaxa Indigenous Community v. Paraguay*, Merits, Reparations and Costs, Judgment of March 29, 2006, Series C No. 146, 152.

[222]I/A Court H.R., *Case of Ximenes Lopes v. Brazil*, *supra* note 219, 125.

The right of older persons not to be arbitrary deprived of life has been breach in many cases dealt by the Inter-American System, particularly in the context of the violent conflicts or social upheavals that affected the region. In *Massacre of Plan Sanchez*, the IACtHR found that about 268 persons were extra-judicially executed by state agents, among them children and older persons.[223] Likewise, in the *Mozote Massacre*, this tribunal was able to demonstrate that about 1061 persons, mostly women, children and older persons, were arbitrarily deprived of their lives by state agents.[224] In all these massacres, it was believed that members of the armed forces were coming to the communities looking for the adult males because they associated them with the armed groups that were operating in the country. When they did not find the adult males because they had escaped into the mountains, the state agents murdered those you were unable to leave such as children and older persons. Interestingly, in the *Mozote Massacre* case, the IACtHR acknowledging this situation ordered El Salvador to construct a center for older persons as one of the measures of reparation awarded in this case.[225]

In the case of *Ximenes Lopes v. Brazil*, a person suffering from a mental health disability was admitted to a private health institution hired by the state to provide mental health services to patients under the coverage of Brazil's Single Health System.[226] The victim was in excellent physical condition at the time of his admission, but died a few days after as a result of mistreatment and lack of appropriate care.[227] No medical doctor was present in the institution at the time of his death.[228]

First, the IACtHR found that the private health institution was operating as a state organ because it was exercising public authority when rendering services on behalf of the state; consequently, Brazil was responsible for the actions or omissions of the staff of the institution.[229] Next, the state had a heightened responsibility to provide the victim in this case with the appropriate health care services, taking into account the particular vulnerability to which persons with mental health disabilities are exposed when placed in psychiatric institutions.[230] Also, Brazil had a duty to regulate and supervise health care institutions as a necessary measure to ensure the protection of life and the personal integrity of patients.[231]

[223] I/A Court H.R., *Case of the Plan de Sánchez Massacre v. Guatemala*, Merits, Judgment of April 29, 2004, Series C No. 105, 42.

[224] I/A Court H.R., *Case of the Massacres of El Mozote and neighboring locations v. El Salvador*, Merits, Reparations and Costs, Judgment of October 25, 2012, Series C No. 252, 153.

[225] *Id.*, 339.

[226] I/A Court H.R., *Case of Ximenes Lopes v. Brazil*, *supra* note 219, 112(4).

[227] *Id.*, 112(5).

[228] *Id.*, 112(11).

[229] *Id.*, 100.

[230] *Id.*, 128–129, 132.

[231] *Id.*, 141.

Since the conditions of the psychiatric institution and the medical care that it provided to Ximenes Lopes were utterly inadequate to afford safe and appropriate health care services to its patients, Brazil failed in its duty to respect and ensure the right to life of the victim in this case.[232] The patient was subjected to mistreatment by the staff, excessively medicated and left in poor health conditions without medical supervision.[233] Moreover, the state failed to properly regulate and supervise the services provided by the health institution to its patients in this case.[234] After the victim's death, the state cancelled the authorization of the institution to provide services to patients of the public health services and revoked its license to operate as a psychiatric center.[235] Finally, the IACtHR found that Brazil failed to carry out an effective investigation to clarify the responsibility of the staff of the institution in the death of Mr. Ximenes Lopes and provide appropriate reparations to his next of kin.[236] In consequence, there was a violation of Article 4(1) of the *American Convention*.[237]

Interestingly, the IACtHR in this case concluded *in dicta* that the obligation to regulate and supervise also extends to private institutions even if they do not provide services under the public health care system.[238] Under the duty to ensure the right to life, states must guarantee that every health institution complies with minimum standards in their facilities and provides adequate care to its patients.[239] This obligation must be assessed more strictly in cases of private health care providers dealing with patients in situation of vulnerability, such as the case of persons who suffer mental disabilities and who are admitted to psychiatric institutions.[240]

Furthermore, as part of the positive obligations to ensure the right to life, states must create minimum living conditions that are compatible with the dignity of the human person and must avoid generating situations that endanger access to a decent life.[241] In *Yakye Axa Community*, the IACtHR specifically concluded that States have "the duty to take positive, concrete measures geared toward fulfillment of the right to a decent life, especially in the case of persons who are vulnerable and at risk, whose care becomes a high priority."[242] In that respect, the tribunal found that lack of access to basic health, food and clean water are among the measures that

---

[232]*Id.*, 120, 142–143.

[233]*Id.*, 121, 46.

[234]*Id.*

[235]*Id.*, 144, 112(63)–112(64).

[236]*Id.*, 147–149, 206.

[237]*Id.*, 150.

[238]*Id.*, 89.

[239]*Id.*, 88–89.

[240]*Id.*

[241]I/A Court H.R., *Case of the Yakye Axa Indigenous Community v. Paraguay*, Merits, Reparations and Costs, Judgment of June 17, 2005, Series C No. 125, 162.

[242]*Id.*

states must afford to those who are in precarious conditions.[243] In *Yakye Axa* and *Xakmok Kasek* the victims were members of indigenous communities who were deprived of their access to their ancestral lands and as a result lived in an inhuman situation.[244] In those cases, the IACtHR concluded that the state failure to ensure minimum living conditions breached the right to a decent life of the members of those communities.[245] Moreover, the IACtHR acknowledged the special consideration required to protect the right of the older persons of the community and provided that the state must "take measures to ensure their continuing functionality and autonomy, guaranteeing their right to adequate food, access to clean water and health care. Specifically, the state must provide care for older persons with chronic diseases and in terminal stages, to help them avoid unnecessary suffering."[246]

Compliance with the state's duty to ensure in those cases is reviewed on the basis of the principle of due diligence. According to the IACtHR, states cannot be responsible for every violation of the right to life perpetrated by private actors. Consequently, "[t]o give rise to this positive obligation, it must be established that, at the time of the facts, the authorities knew or should have known of the existence of a situation of real and immediate risk to the life of an individual or group of specific individuals, and that they did not take the necessary measures within their powers that could reasonably be expected to prevent or avoid that risk."[247] In *Xákmok Kásek*, for example, the IACtHR concluded that Paraguay was aware of the inhuman conditions in which the members of the community lived and the real and immediate risk to which they were exposed, but did nothing to ensure their right to life.[248]

Even if the *Ximenes Lopes Case* refers to a young person with mental disability admitted in a mental health institution and not to an older person, by analogy, the interpretation of the IACtHR in this case is of extremely relevance for the protection of the right to life of older individuals who are placed in public institutions or private ones hired by the state to provide health care services under the public health care system. Likewise, the two cases on indigenous communities developed a scope of the right to life of members of vulnerable groups that is crucial for protecting the rights of older persons who are left in indigence or abandonment. The *Yakye Axa* case acknowledges that enhanced protection of older persons living in extreme poverty is crucial to ensure their right to a dignified life under Article 4 of the *American Convention*.

---

[243]*Id.*, 167.

[244]I/A Court H.R., *Case of the Yakye Axa Indigenous Community v. Paraguay*, *supra* note 241, 168; *Case of the Xákmok Kásek Indigenous Community. v. Paraguay*, Merits, Reparations and Costs, Judgment of August 24, 2010, Series C No. 214, 215–216.

[245]*Id.*, 176 and 217, respectively.

[246]I/A Court H.R., *Case of the Yakye Axa Indigenous Community v. Paraguay*, *supra* note 241, 175.

[247]I/A Court H.R., *Case of the Xákmok Kásek Indigenous Community. v. Paraguay*, *supra* note 244, 188.

[248]*Id.*, 192.

Moreover, with respect to criminal cases, the *American Convention* expresses in Article 4(5) that the death penalty shall not be imposed upon persons who, at the time of the offense in question, were over 70 years of age.[249] The IACHR has found that the execution of convicted individuals who committed a crime when they were younger than 18 years old constituted an arbitrary deprivation of life in violation of Inter-American instruments.[250] This conclusion was based on Article 4(5) of the *American Convention* that also prohibits the imposition of capital punishment upon persons who, at the time the crime was committed, were under 18 years of age[251] and an understanding that this prohibition has become a *jus cogens* norm that applies to states that are only bound by the *American Declaration*.[252] Thus, even no cases have been yet resolved by the Inter-American System regarding older persons, by extension, the imposition of the death penalty to persons whom at the time of the commission of the crime were 70 years of age should likewise be considered an arbitrary deprivation of life.

The draft *Convention on Older Persons* protects the right to life and dignity in old age.[253] In particular, it imposes upon states the duty to ensure all necessary measures to guarantee effective enjoyment of older persons of their right to life and to live with dignity in old age.[254] The notion of dignity in this provision appears to be in line with the interpretation of the right to life by the IACtHR in *Yakye Axa* and *Xakmok Kasek*. Thus, it seems clear that the draft *Convention on Older Persons* spouses an approach that goes beyond the negative obligation of protecting life to include a comprehensive notion that involves clear positive obligations of states to do more in the case of older persons. The practice and case law of the IACHR and IACtHR will play a central role in shaping the application of this protection should the *Convention on Older Persons* be adopted. The right to a dignified life of older persons may also include the right to decide on their life plan, to self-realization, and to live an independent and autonomous life.

Moreover, the draft *Convention on Older Persons* includes states' obligation to afford effective protection to older persons "at the end of life."[255] End of life is not defined in the draft, though it is possible to envision that this notion may involve a number of issues, including: the right to consent to medical treatments or surgery, the decision to interrupt or suspend treatment or the decision to end life.

---

[249]Article 4(5) of the *American Convention* provides: Capital punishment shall not be imposed upon persons who, at the time the crime was committed, were under 18 years of age or over 70 years of age; nor shall it be applied to pregnant women.

[250]See, *inter alia*, IACHR, *The Death Penalty in The Inter-American Human Rights System: From Restrictions to Abolition*, OEA/Ser.L/V/II., Doc. 68, December 31, 2011, at 71–76.

[251]See Article 4(5), *supra* note 249.

[252]See, *inter alia*, IACHR, *Michael Domingues v. United States*, Report No. 62/02, Case 12.285, October 22, 2002, 85.

[253]*Draft Inter-American Convention on the Human Rights of Older Persons*, supra note 39, Article 6.

[254]*Id.*

[255]*Id.*

With respect to the first issue, Article 11 of the draft establishes that health institutions or medical personnel "may not administer any medical or surgical treatment, intervention or investigation without the prior and inform consent of the older person."[256] In regards to suspension or interruption of treatment, the same provision states expressly the right of older persons to freely consent to, refuse, or voluntary suspend treatment upon being informed of the consequences of such decision.[257] As to the right to end life and procedures to carry out such decision, the draft *Convention on Older Persons* is silent.

Another important issue regarding end of life protection of older persons relates to the issue of palliative care. Article 6 of the draft *Convention on Older Persons* establishes the duty of states to ensure that private and public institutions afford older persons comprehensive care without discrimination, including palliative care.[258] The notion of palliative care is defined in the draft as "[a]ctive, comprehensive, and interdisciplinary care and treatment of patients whose illness is not responding to curative treatment or who are suffering avoidable pain, in order to improve their quality of life until the last day of their lives. Central to palliative care is control of pain, of other symptoms, and of the social, psychological, and spiritual problems of the older persons. It includes the patient, his environment, and his family. It affirms life and considers death a normal process, neither hastening nor delaying it."[259]

The right to prior, free and informed consent regarding medical treatment or surgical proceedings is also recognized by the *San José Charter*.[260] Moreover, the Charter also emphasizes the importance of providing access to palliative care to ensure a death with dignity to older persons with terminal illnesses.[261]

4. The Right to Freedom from Violence

The prevention of violence and abuse against older persons is at the core of the draft *Convention on Older Persons*. First, the Preamble reinforces the importance of adopting and enforcement laws to prevent abuse, abandonment, negligence,

---

[256] *Id.*, Article 11.

[257] Id., Article 11 specifically provides in this regard: Older persons have the right to freely consent to, refuse, or voluntarily suspend treatment, including that provided by traditional, alternative, and complementary medicine, research or medical or scientific experiments, whether physical or psychological, and to be given clear and timely information about the potential consequences and risks of that decision.

[258] *Id.*, Article 6.

[259] *Id.*, Article 2.

[260] *San José Charter on the Rights of Older Persons in Latin American and the Caribbean*, *supra* note 92, at 10.

[261] *Id.*

mistreatment, and violence against older persons and the need to create national mechanisms to enforce those laws. Also, the draft *Convention on Older Persons* provides definitions on abuse,[262] abandonment[263] and negligence[264] to facilitate adoption and implementation of legislation geared towards the prevention of their occurrence against older persons. Next, Article 4 listing the state obligations under the draft instrument provides for the adoption of measures to prevent, punish and eradicate different forms of violence and mistreatment against older persons.[265]

The most important provision, however, is Article 9 of the draft *Convention on Older Persons* because it protects the right of older persons to live free from any form of violence or mistreatment.[266] Moreover, this provision defines violence against older persons as any act or conduct perpetrated by private persons or public authorities that causes death or physical, sexual or psychological suffering to the victim.[267] The forms of violence prohibited by this provision includes, *inter alia*, forms of physical, psychological, and sexual abuse; expulsion from the community, as well as any form of abandonment or negligence perpetrated within the family or by state agents.[268] Moreover, Article 9 sets a list of state obligations to prevent, investigate, punish and eradicate violence against older persons. Also, it provides that states should afford victims of violence access to reparation as a form of redress for the human rights violations they have experienced.[269] Consideration is made of the need to establish specialized institutions, support services, as well as train state officials, persons caring for older persons and family caregivers.[270] The provision also states the duty to ensure access to effective remedies to redress the victims.[271] The definition of the right of older persons to be free from violence as well as the other obligations listed in Article 9 draw extensively from the definition of violence against women and other state duties included in

---

[262] *Draft Inter-American Convention on the Human Rights of Older Persons*, *supra* note 39, Article 2. Abuse is defined as a single or repeated act or omission to the detriment of an older person that harms their physical, mental, or moral integrity and infringes the enjoyment or exercise of their human rights and fundamental freedoms, regardless of whether or not it occurs in a relationship of trust.

[263] *Id.* Abandonment is defined as the lack of action, deliberate or not, to comprehensively care for an older person's needs, which may jeopardize his or her life or physical, psychological or moral integrity.

[264] *Id.* Negligence is defined as involuntary error or unintentional fault, including, inter alia, neglect, omission, failure to protect, and defenselessness that causes harm or suffering to an older person, in either the public or private sphere, when necessary normal precautions in light of the circumstances have not been taken.

[265] *Id.*, Article 4.

[266] *Id.*, Article 9.

[267] *Id.*

[268] *Id.*

[269] *Id.*

[270] *Id.*

[271] *Id.*

the *Belém do Pará Convention*.[272] That treaty also requires state parties to take into special account the vulnerability of women to violence for reasons of their age, when adopting measures to implement the obligations arising out of this convention.[273] The practice and case law of the IACHR and the IACtHR dealing with issues of violence against women is of utmost relevance for interpreting the right of older persons to be free from violence.[274]

The right of older persons to be free from violence is also acknowledged by the *San Jose Charter*.[275] Specifically, the Charter provides that States must adopt the necessary legislation to prevent and eradicate all forms of violence against older persons.[276] The Charter also calls for the punishment of the perpetrators of such violence.[277] To achieve such goal, states must afford effective judicial remedies to the victims and establish mechanism of prevention and supervision, particularly in the context of institutions of care where older persons may be institutionalized.[278]

Finally, the *OAS Declaration of San Pedro Sula: Toward a Culture of Non-violence* underlines the importance of adopting measures that prevent and punish violence, exploitation and discrimination among groups in vulnerable situations, such as older persons.[279] Also, the Declaration urges states to enact domestic legislation to prevent acts of violence and provide legal assistance in order to obtain redress, as well as the elimination of all forms of discrimination.[280]

5. The Right to Freedom from Cruel, Inhuman or Degrading Treatment or Punishment

Article 5 of the *American Convention* provides that everyone has a right to be free from torture or other cruel, inhuman or degrading treatment or punishment, and that "all persons deprived of their liberty shall be treated with respect for

---

[272]See Articles 1 and 2 of the *Belém do Pará Convention*, *supra* note 30.

[273]*Id.*, Article 9.

[274]See, inter alia, IACHR, *Access to Justice for Women Victims of Violence in the Americas*, OEA/Ser.L/V//II, Doc. 68, January 20, 2007; *Jessica Lenahan (Gonzales) et al. v. United States*, *supra* note 197; I/A Court H.R., *Case of González et al. ("Cotton Field") v. Mexico*, Preliminary Objection, Merits, Reparations and Costs, Judgment of November 16, 2009, Series C No. 205; *Case of Fernández Ortega et al. v. Mexico*, Preliminary Objection, Merits, Reparations, and Costs, Judgment of August 30, 2010, Series C No. 215.

[275]*San José Charter on the Rights of Older Persons in Latin American and the Caribbean*, *supra* note 92.

[276]*Id.*, at 13–14.

[277]*Id.*, at 14.

[278]*Id.*

[279]OAS, *Declaration of San Pedro Sula: Toward a Culture of Non-Violence*, AG/DEC. 60 (XXXIX-O/09), adopted on June 4, 2009, 4, available at http://www.oas.org/sap/peacefund/resolutions/Declaration_of_San_Pedro_Sula_Culture_of_Non_Violence.pdf (last visited Jan. 20, 2015).

[280]*Id.*, 7.

the inherent dignity of the human person."[281] Moreover, "punishment shall not be extended to any person other than someone convicted of a crime that has been afforded due process of law."[282] *The Inter-American Torture Convention* explicitly bolsters the *American Convention's*, and the rule of customary international laws, prohibition of torture.[283] The *Inter-American Torture Convention* also guarantees that any person alleging to have been subjected to torture shall have the right to an impartial examination, within their jurisdiction.[284] State parties, are likewise, required to have their respective authorities immediately investigate situations where there is a well-founded reason to believe that an act of torture has been committed within their jurisdiction, and to initiate criminal proceedings, when appropriate.[285] Finally, after all domestic legal procedures of the respective state have been exhausted, a case alleging torture may be submitted to the international forum whose jurisdiction has been recognized by that state.[286]

While the *American Convention* does not provide for a definition of torture, the *Inter-American Torture Convention* provides that "torture shall be understood to be any act intentionally performed whereby physical or mental pain or suffering is inflicted on a person for purposes of criminal investigation, as a means of intimidation, as personal punishment, as a preventive measure, as a penalty, or for any other purpose. Torture shall also be understood to be the use of methods upon a person intended to obliterate the personality of the victim or to diminish his physical or mental capacities, even if they do not cause physical pain or mental anguish."[287] The IACtHR has interpreted the notion of torture in Article 5 of the *American Convention* in light of the definition provided above, as well as taking into account the definition provided by other international instruments, in particular the *United Nations Convention against Torture*.[288] In this regard, it has concluded that the constitutive elements of torture include: (1) an intentional act; (2) that causes severe physical or mental suffering; and (3) is committed with a given purpose.[289] Moreover, the IACtHR has reiterated that "torture can be inflicted not only via physical violence, but also through acts that produce severe physical,

---

[281] Article 5 of the *American Convention*.

[282] *Id.*

[283] *Inter-American Convention to Prevent and Punish Torture*, *supra* note 29, Preamble.

[284] *Id.*, Article 8.

[285] *Id.*

[286] *Id.*

[287] *Id.*, Article 2.

[288] I/A Court H.R., *Case of Bueno Alves v. Argentina*, Merits, Reparations and Costs, Judgment of May 11, 2007, Series C No. 164, 78; *Case of Maritza Urrutia v. Guatemala*, Merits, Reparations and Costs, Judgment of November 27, 2003, Series C No. 103, 90–91.

[289] See, *inter alia*, I/A Court H.R., *Case of Bueno Alves v. Argentina*, *supra*, 79.

psychological or moral suffering in the victim."[290] The prohibition is torture is considered by the IACtHR as part of the international *jus cogens*.[291]

Neither the *American Convention* nor the *Inter-American Torture Convention* provides for a definition of cruel, inhuman or degrading treatment or punishment. The IACtHR has followed the practice and case law of other international human rights bodies to conclude that the distinction between torture and other forms of ill-treatment is established on a case by case basis, and in light of the intensity of the suffering underwent by the victim.[292] Moreover, the intensity of the suffering is assessed on the basis of objective and subjective factors.[293] The objective factors include the characteristics of the mistreatment under consideration, specifically the duration, the method or manner used to cause the suffering, and the physical and psychological effects that the harm may cause.[294] On the other hand, the subjective elements involved consideration of the age, gender, health condition, and any other personal circumstance.[295] Even if there is no case yet at the Inter-American System where treatment faced by an older person has been characterized as torture on the basis of the age of the victim, the IACHR and the IACtHR both have used that ground to find that suffering experienced by children due to their age must receive special consideration for purposes of characterizing a particular ill-treatment as torture.[296] The only precedent in which older age and ill-treatment were related is *Schillizzi v. Argentina*, where an ill, older man challenged an Argentine court detention sentence on the basis of his age.[297] The Commission pointed out that personal circumstances, such as age or the state of health, do not per se render a detention a violation of Article 5 of the *American Convention*.[298]

On the other hand, in light of specific language of Article 5 of the *American Convention*, conditions of detention affecting the personal integrity of a detainee may constitute torture or other forms of ill-treatment.[299] The state oversees the

[290]See, *inter alia*, I/A Court H.R., *Case of Cantoral Benavides v. Peru*, Merits. Judgment of August 18, 2000. Series C No. 69, 100; *Case of Maritza Urrutia v. Guatemala*, *supra* note 288, 93.

[291]I/A Court H.R., *Case of Bueno Alves v. Argentina*, *supra* note 288, 76.

[292]I/A Court H.R., *Case of Caesar v. Trinidad and Tobago*, Merits, Reparations and Costs, Judgment of March 11, 2005, Series C No. 123, 67; *Case of Loayza-Tamayo v. Peru*, Merits, Judgment of September 17, 1997, Series C No. 33, 57.

[293]See, *inter alia*, I/A Court H.R., *Case of Bueno Alves v. Argentina*, *supra* note 288, 83.

[294]*Id.*

[295]*Id.*

[296]See, *inter alia*, IACHR, *Jailton Neri Da Fonseca v. Brazil*, Report No. 33/04, Case 11.634, Annual Report 2004, March 11, 2004, 314; I/A Court H.R., *Case of the Gómez Paquiyauri Brothers v. Peru*, Merits, Reparations and Costs, Judgment of July 8, 2004, Series C No. 110, 101.

[297]IACHR, *Horacio Anibal Schillizzi Moreno v. Argentina*, Report N° 22/00, Case 11.732, Annual Report 2000, March 7, 2000, 35.

[298]*Id.*

[299]As stated above, Article 5 explicitly states that "[a]ll persons deprived of their liberty shall be treated with respect for the inherent dignity of the human person."

detention centers and consequently it must ensure that inmates are held in conditions that safeguard their rights and "do not exceed the unavoidable level of suffering inherent in detention."[300] Thus, "overcrowded conditions, lacking natural light and ventilation, without a bed to rest on or adequate hygiene conditions, in isolation and incommunicado or with undue restrictions to the system of visits, constitutes a violation of that person's right to humane treatment."[301] Also, denial of appropriate medical treatment may involve a breach of the right not to be subject to any forms of ill treatment.[302] Moreover, states are forbidden from invoking lack of economic resources to justify prison conditions that breach international standards and the inherent dignity of the human person.[303]

Such violations can also be implicated by the custody of the state in nursing facilities and long-term care facilities or in its position as a guarantor when the facilities are run by third parties. In *Ximenes Lopes v Brazil*, for example, the IACtHR found the responsibility of Brazil for its failure to protect the personal integrity and life of a patient who died for lack of appropriate care in a private mental health clinic.[304] When assessing whether the treatment received by the victim in this case constituted torture, the IACtHR took into consideration the fact that he was a person with mental disability.[305] Persons with mental disability are particularly vulnerable to any health treatment, even more when they are admitted to mental institutions.[306] In addition to physical mistreatment, the victim in this case was subject to restraint.[307] The IACtHR found that in order for restraint to be in compliance with the prohibition of ill treatment under the *American Convention*, "it should be used as a last resort and with the only purpose of protecting the patient, or else the medical staff or third persons, when the behavior of the patient involved is such as to pose a threat to their safety."[308] Only qualified staff may apply restraint, the participation of other patients in enforcing this measure is forbidden.[309] Moreover, given that treatment must be assessed in each case and in light of the patient's best interest and autonomy, health care personnel should utilize the least restrictive possible restraint techniques to ensure the dignity and health of the patient concerned.[310]

---

[300] See, *inter alia*, I/A Court H.R., *Case of Vera Vera et al. v. Ecuador*, Preliminary Objection, Merits, Reparations, and Costs, Judgment of May 19, 2011, Series. C no. 226, 42.

[301] I/A Court H.R., *Case of Tibi v. Ecuador*, Preliminary Objections, Merits, Reparations and Costs, Judgment of September 7, 2004, Series C No. 114, 150.

[302] I/A Court H.R, *Case of Vera Vera et al. v. Ecuador*, *supra* note 295, 42.

[303] *Id.*

[304] I/A Court H.R., *Case of Ximenes Lopes v. Brazil*, *supra* note 219, 121–122.

[305] *Id.*, 127.

[306] *Id.*, 129.

[307] *Id.*, 136.

[308] *Id.*, 134.

[309] *Id.*

[310] *Id.*, 135.

The conditions of the mental health institutions were also assessed to conclude that they were deplorable and inadequate to care for patients.[311] The IACtHR concluded that to ensure the right to personal integrity, states have a duty to regulate and supervise the operation of the institutions that provide mental health services, regardless of whether they are public or private.[312] In the case under analysis, this Tribunal found that the state had failed to regulate and supervise the conditions in which the health institution provided mental services, thereby breaching its duty to ensure the rights protected under Article 5 of the *American Convention*.[313]

In *Ximenes Lopes*, the IACtHR also found that the state had breached its duty to ensure the victim's right not to be subject to torture and other forms of ill-treatment because, after his death, Brazil failed in its obligation to carry out an effective investigation in light of Articles 1(1) and 5 of the *American Convention*. Moreover, the state did not afford the victim's next of kin a right to an effective remedy to establish the facts that lead to his death and obtain reparation, in breach of Article 25 of the *American Convention*. The duty to investigate acts of torture or other forms of ill-treatment is also clearly provided in the provisions of the *Inter-American Torture Convention*, as stated above. The IACHR and the IACtHR have consistently found violations of Articles 6 and 8 of that treaty when state parties failed to investigate and punish the perpetrators, and redress the victims of ill-treatment.[314]

The standards set in *Ximenes Lopez,* even if they refer to a person with mental disability, are of extreme relevance for the protection of the rights of older persons residing in nursing facilities or long term care facilities. Older persons residing in such accommodations generally have reduced capacity in some way, and are therefore entitled to heightened protections due to their increased vulnerabilities. The vulnerable situation of older persons in nursing or long term facilities is contemplated extensively by the draft *Convention on Older Persons* in several provisions.

First, the notion of long term or stay facility is defined in Article 2 of the draft instrument as "a public or private establishment providing comprehensive social and health care services, over a prolonged period of time, for older persons with moderate or severe dependency, who cannot receive care in their home."[315] Next, Article 12 provides that states must ensure that long-term care facilities have specialized personnel to deal with the needs of older persons and who are trained to work with older persons without causing them any harm.[316] Moreover, states must adopt the

---

[311] *Id.*, 131–132.

[312] *Id.*, 141.

[313] *Id.*, 142–146.

[314] See, *inter alia*, I/A Court H.R., *Case of Bayarri v. Argentina*, Preliminary Objection, Merits, Reparations and Costs, Judgment of October 30, 2008, Series C No. 187, 94; IACHR, *Massacre of Corumbiara v. Brazil*, Report 32/04, Case 11.556, Merits, March 11, 2004, 287.

[315] *Draft Inter-American Convention on the Human Rights of Older Persons*, *supra* note 39, Article 2.

[316] *Id.*, Article 12.

appropriate regulatory frameworks for long-term care facilities to ensure that the personal security and integrity of older persons in these institutions is protected.[317] Also, these laws must also envision that the leadership and staff of the institutions are civilly and/or criminally responsible for any acts they perpetrated in detriment of the integrity of older persons under their care.[318] Even if the draft *Convention on Older Persons* does not specifically list the state duty to regulate and supervise long-term care facilities,[319] implementation of appropriate mechanisms to oversee the functioning of these institutions is essential to prevent incidents of ill-treatment or other forms of violence. In this respect, the IACHR in the report on *The Right of Girls and Boys to a Family, Alternative Care, Ending Institutionalization in the Americas (Rights of Girls and Boys to a Family)*[320] developed detailed standards regarding the duty to regulate and supervise residential care for children,[321] as well as minimum requirements for the delivery of services in those institutions.[322] These standards and practices could provide expert guidance to states when adopting framework laws regarding the operation of long-term care facilities for older persons, in light of the provisions enshrined in the draft *Convention on Older Persons*. The *San José Charter* also recommends states to establish mechanisms of control and supervision in long-term care facilities where older persons reside, regardless of whether these are public or private institutions.[323]

In addition to consideration of older persons admitted into long-term care institutions, Article 10 of the draft *Convention on Older Persons* affords an overall protection against torture, cruel, inhuman or degrading treatment or punishment.[324] Under this provision states must take all the necessary measures to prevent older persons to be exposed to such forms of ill-treatment.[325] Also, Article 4 when listing the state general obligations contemplates the duty to prevent, punish and eradicate practices that involve any mistreatment affecting the integrity or safety of

---

[317] *Id.*

[318] *Id.*

[319] Article 9 only provides that state parties commit to: establish or strengthen mechanisms for preventing any form of violence in the family or domestic unit, facilities in which older persons receive long-term care services, and in society, with a view to providing effective protection for the rights of older persons.

[320] IACHR, *The Right of Girls and Boys to a Family, Alternative Care, Ending Institutionalization in the Americas*, OEA/Ser.L/V/II., Doc. 54/13, October 17, 2013.

[321] *Id.*, at 154–166.

[322] *Id.*, at 179–226.

[323] *San José Charter on the Rights of Older Persons in Latin American and the Caribbean*, *supra* note 92, at 10.

[324] *Draft Inter-American Convention on the Human Rights of Older Persons*, *supra* note 39, Article 10.

[325] *Id.*

older persons.[326] Lastly, Article 9 prohibiting violence enshrines the right to have access to effective complaint mechanisms for older persons to denounce violence or other forms of mistreatment.[327] These types of mechanisms may be judicial or administrative depending on the nature of the complaint.[328]

6. The Right to Liberty and Security

The *American Declaration* states that *every* human being has the right to personal liberty and security.[329] It further ensures that no person may be deprived of liberty except in cases and according to procedures established by pre-existing law.[330]

The *American Convention* bolsters the *American Declaration* by further describing due process rights surrounding the right to personal liberty and security. The IACtHR has stated that this right protects not only the physical liberty, but also the personal safety of individuals.[331] Article 7 of the *American Convention* provides that "[n]o one shall be deprived of his physical liberty except for the reasons and under conditions established beforehand by the constitution of the State Party concerned or by a law established pursuant thereto."[332] Moreover, "[n]o one shall be subject to arbitrary arrest or imprisonment."[333] Both the IACHR and IACtHR have concluded that for a deprivation of liberty to be legal, it must comply with a material aspect embodied in the duty to arrest or detain only for reasons, cases or circumstances specifically established by law.[334] At the same time, detention or arrest must be carried out under the strict conditions established beforehand by law (formal aspect).[335]

In *Karina Montenegro, et al.* petitioner argued, *inter alia*, that Ecuador violated the right to liberty of one of the victims in the case, a woman 68 years of age, by arresting and imprisoning her in violation of Ecuadorian law.[336] The Ecuadorean Code of Criminal Procedure established that persons 65 years or over charged

---

[326] *Id.*, Article 4.

[327] *Id.*, Article 9.

[328] *Id.*

[329] Article XXV of the *American Declaration*, *supra* note 16.

[330] *Id.*

[331] I/A Court H.R., *Case of Tibi v. Ecuador*, *supra* note 301, 97.

[332] Article 7(2) of the *American Convention*.

[333] Article 7(3) of the *American Convention*.

[334] See, *inter alia*, I/A Court H.R., *Case of Bulacio v. Argentina*, Merits, Reparations and Costs, Judgment of September 18, 2003, Series C No. 100, 125; IACHR, *Ana, Beatriz and Celia González Pérez v. Mexico*, Report No. 53/01, Case 11.565, April 4, 2001, 23.

[335] *Id.*

[336] IACHR, *Karina Montenegro, et al. v. Ecuador*, Report No. 48/07, Petitions 261-03, 397-03 and 1377-04, Admissibility, Annual Report 2007, July 23, 2007, 22, 42.

with any type of crime had the right to substitute preventive detention with house arrest. The IACHR found the case admissible and concluded that the alleged failure to substitute preventive custody with house arrest as ordered by the judges in the case, as well as the failure to carry out the judgments of the Constitutional Court that ordered the judges to adopt appropriate measures for the substitution to be carried out as ordered, could constitute violation of Articles 7 and 25 of the *American Convention*.[337]

Even if the requirements provided by law are met, a deprivation of liberty may still be considered incompatible with the *American Convention*, if it is considered unreasonable, unforeseeable or out of proportion.[338] Consequently, in contrast to other human rights treaties such as the *European Convention on Human Rights*, the *American Convention* does not list the grounds on which deprivation of liberty is authorized. States seem to enjoy a margin of discretion to set those grounds, though they must prove that the selected grounds are not arbitrary or unreasonable. The IACHR has stated that authorized deprivations of liberty "are not limited to those involving the investigation and punishment of crimes, but also extend to other areas in which states may administer authority. States may, for example, detain individuals for the purpose of controlling the entry and residence of aliens in their territories or for reasons relating to physical or mental health, and, during occupations governed by international humanitarian law, may intern the civilian population as a safety measure and for imperative reasons of security."[339] Most recently, the *Principles and Best Practices on the Protection of Persons Deprived of Liberty in the Americas* (*Principles on the Protection of Persons Deprived of Liberty*) adopted by the IACHR defines deprivation of liberty as "Any form of detention, imprisonment, institutionalization, or custody of a person in a public or private institution which that person is not permitted to leave at will, by order of or under de facto control of a judicial, administrative or any other authority, for reasons of humanitarian assistance, treatment, guardianship, protection, or because of crimes or legal offenses. This category of persons includes not only those deprived of their liberty because of crimes or infringements or non-compliance with the law, whether they are accused or convicted, but also those persons who are under the custody and supervision of certain institutions, such as: psychiatric hospitals and other establishments for persons with physical, mental, or sensory disabilities; institutions for children and the elderly; centers for migrants, refugees, asylum or refugee status seekers, stateless and undocumented persons; and any other similar institution the purpose of which is to deprive persons of their liberty."[340]

---

[337] *Id.*, 64.

[338] See, *inter alia*, I/A Court H.R., *Case of Bulacio v. Argentina*, *supra* note 334, 125; IACHR, *Ana, Beatriz and Celia González Pérez v. Mexico*, *supra* note 334, 23.

[339] IACHR, *Rafael Ferrer-Mazorra et al. v. United States*, Report N° 51/01, Case 9903, Annual Report 2000, April 4, 2001, 210.

[340] IACHR, *Principles and Best Practices on the Protection of Persons Deprived of Liberty in the Americas*, Resolution 1/08, General Provision, available at http://www.oas.org/en/iachr/pdl/activities/principles.asp, (last visited Jan. 30, 2015).

Historically, older persons have not exercised the right to liberty to challenge institutionalization against their will. However, given the scope of protection of Article 7 of the American Convention and the *Principles on the Protection of Persons Deprived of Liberty* described above, there is room to argue that, under certain conditions, the placement of an older person in a long-term care without his or her consent may involve an arbitrary deprivation of liberty.

Persons detained must be brought promptly before an impartial judge or other officer authorized by law to exercise judicial power, and must be given a trial within a reasonable time or released without prejudice to the continuation of the proceedings.[341] Moreover, anyone who is deprived of liberty or who believes to be threatened with deprivation of his liberty is entitled to a remedy to challenge the legality of his or her deprivation of liberty.[342] A court must decide without delay on the lawfulness of the deprivation of liberty and order release if it founds it unlawful.[343] States may not restrict or abolish the right to a remedy to challenge the legality of detention.[344] The IACHR and the IACtHR have found that *habeas corpus* is traditionally the remedy that must be afforded by states to challenge unlawful detention, though some countries have assigned it a different name.[345] In cases of continuing detention, this necessarily requires supervision at regular intervals.[346]

If placement of an older person in an institution without his or her consent is considered deprivation of liberty for purposes of Article XXV of the *American Declaration* and Article 7 of the *American Convention*, it seems that judicial review of that placement should be provided to the person concerned. Whether *habeas corpus* should be the appropriate remedy is not clear from the practice of the organs of the Inter-American System. In the *Case of Karina Montenegro et al. v. Ecuador*, the older person resorted to a *habeas corpus* to challenge her detention in violation of the domestic laws.[347] However, this was a case in which the victim was detained for the investigation and punishment of an alleged crime.[348] It will be interesting to see how the draft *Convention on Older Persons* addresses this matter, if adopted.

---

[341] Article XXV of the *American Declaration* and Article 7 of the *American Convention*.

[342] Article 7 of the *American Convention*.

[343] *Id.*

[344] Article 27(2) of the *American Convention*.

[345] IACHR, *Report on the Human Rights of Persons Deprived of Liberty in The Americas*, OEA/Ser.L/V/II, Doc. 64, December 31, 2011, 244–246; I/A Court H.R., *Case of Tibi v. Ecuador*, *supra* note 301, 128–129.

[346] IACHR, *Rafael Ferrer-Mazorra et al. v, United States*, *supra* note 339, 212.

[347] IACHR, *Karina Montenegro, et al. v. Ecuador*, *supra* note 336, 23.

[348] *Id.*

So far the draft *Convention on Older Persons* provides for the right of older persons to personal liberty and safety in Article 13.[349] This right may be exercised in any environment in which the older persons develop his or her life. Moreover, age cannot justify alone a deprivation of liberty and the grounds on which to base such deprivation must be established by law.[350] Consequently, the draft instrument seems to consider that any situation in which the physical liberty of an older person is restricted may involve an unjustified deprivation of liberty, if it is not regulated by law or is arbitrary. This interpretation would support protection of older persons who are admitted in long-care or similar institutions.

Moreover, other provisions afford complementary protection. For example, Article 12 requires state parties to establish mechanisms to ensure that the beginning and end of long-term care services are subject to the consent of the person concerned.[351] Moreover, states must adopt legal frameworks for long-term care services that protect the personal liberty and safety of older persons admitted in those institutions.[352] Even if not clearly mentioned in the draft *Convention on Older Persons*, essential issues that should be contemplated in the domestic legal framework include the remedies available to older persons to challenge the lawfulness of a restriction to personal liberty if the admission to the institution cannot be justified. Also, other complaint mechanisms should be established to ensure that similar fundamental rights such as the right to personal integrity and privacy are respected. Likewise, availability of mechanisms to challenge the quality of services provided should be ensured. At this point, the draft *Convention on Older Persons* requires states to provide effective mechanism and remedies to deal with cases of violence against older persons and ensures an overall right to access to justice.[353] However, the draft instrument is silent as to other remedies or complaint mechanisms that would benefit older persons placed in long-term or similar care institutions.[354]

This silence in the language of the existing provisions can be filled out by the practice and standards developed by the IACHR with respect to other groups that face deprivation of liberty or some form of institutionalization. First, the *Principles on the Protection of Persons Deprived of Liberty* ensure the rights of those deprived of liberty to petition before judicial, administrative, or other

[349] *Draft Inter-American Convention on the Human Rights of Older Persons*, *supra* note 39, Article 13.

[350] *Id.*

[351] *Id.*, Article 12.

[352] *Id.*

[353] *Id.*, Article 9.

[354] The draft Convention on Older Persons only provides that the regulatory framework for long-term care service operations should ensure that older persons "are apprised of...their rights and of the legal framework and protocols governing long-term care services." See *Draft Inter-American Convention on the Human Rights of Older Persons*, *supra* note 39, Article 12.

authorities.[355] It may be exercised individually or collectively[356] and comprises a range of mechanisms to seek redress for violations of rights, failure in providing adequate conditions and services, as well as supervise the relations between inmates and management or between inmates themselves.[357] Next, the Report on *The Right of Girls and Boys to a Family* develops detailed standards regarding judicial and administrative mechanisms that should be available to institutionalized children to submit complaints denouncing breaches of their rights while deprived of liberty, as well as failure in the provision of essential services.[358]

In regard to older persons deprived of liberty as a result of judicial proceedings, Article 13 of the draft *Convention on Older Persons* ensures the right to be treated without discrimination and to receive appropriate treatment in light of the existing international human rights norms.[359] Also, states should promote alternative measures to deprivation of liberty for older persons and ensure access to special and comprehensive care programs, which includes, *inter alia*, rehabilitation mechanisms for their reintegration into society.[360] Again, the standards developed by the IACHR regarding persons deprived of liberty and children will be a useful tool to interpret the scope of protection afforded by this provision. Firstly, the *Principles on the Protection of Persons Deprived of Liberty* ensure equal protection and prohibit discrimination, *inter alia*, on the basis of age.[361] At the same time, special measures designed to protect the rights of older persons deprived of liberty would not be considered discriminatory.[362] Also, the Principles clearly state that in the case of persons with disability, deprivation of liberty in mental or psychiatric institutions will be applied as a measure of last resort, and "solely when there is a serious likelihood of immediate or imminent harm to that person or to others."[363] Alternative measures to deprivation of liberty should also be considered by the State as a substitute to institutionalization.[364] Likewise, the Report on *The Right of Girls and Boys to a Family* provides that the placement of children in institutions should be considered a measure of last resort, and that before reaching that

[355]IACHR, *Principles and Best Practices on the Protection of Persons Deprived of Liberty in the Americas*, *supra* note 340, Principles VI and VII.

[356]*Id.*, Principle VII.

[357]*Id.* Also, see, IACHR, *Report on the Human Rights of Persons Deprived of Liberty in the Americas*, *supra* note 345, 251.

[358]IACHR, *The Right of Girls and Boys to a Family, Alternative Care, Ending Institutionalization in the Americas*, *supra* note 320, at 166–176.

[359]*Draft Inter-American Convention on the Human Rights of Older Persons*, supra note 39, Article 13.

[360]*Id.*

[361]IACHR, *Principles and Best Practices on the Protection of Persons Deprived of Liberty in the Americas*, *supra* note 340, Principles II.

[362]*Id.*

[363]*Id.*, Principle III (3).

[364]*Id.*, Principle III (4).

decision other alternative measures to substitute deprivation of liberty should be explored and implemented.[365]

The *San José Charter* also recommends states to adopt the necessary measures to ensure the fundamental rights of older persons deprived of liberty.[366]

Another important right that complements the right to personal liberty in case of older persons is the right to freedom of movement. Freedom of movement is protected in the *American Declaration*[367] and the *American Convention*.[368] With respect to older persons, the draft *Convention on Older Persons* ensures their right to freedom of movement, which includes the right to choose his or her place of residence.[369] Moreover, states must adopt measures designed to guarantee older persons the effective exercise of this right.[370] Like the right to personal liberty, the right to freedom of movement is essential for older persons who are placed in long-term care institutions. This is recognized by Article 12 of the draft *Convention on Older Persons* when it requires states to afford protection to the right to freedom of movement of older persons in long-term care institutions, when adopting legal frameworks for the operation of those facilities.[371] Those frameworks should ensure mechanisms allowing older persons whose right to freedom of movement has been unduly curtailed to have access to a complaint procedure.

7. The Right to Respect of Home, Family and Private Life

The *American Declaration* protects the right to honor, personal reputation and private life.[372] Likewise, the *American Convention* ensures protection of the right to privacy, including honor and dignity, and prohibits arbitrary or abusive interference with one's private life, family, home, or correspondence, or of unlawful attacks on one's honor or reputation.[373]

Article 11 of the *American Convention* guarantees a traditional notion of privacy regarding the protection of family, home and correspondence that "is characterized

---

[365] IACHR, *The Right of Girls and Boys to a Family, Alternative Care, Ending Institutionalization in the Americas*, *supra* note 320, at 129–134.

[366] *San José Charter on the Rights of Older Persons in Latin American and the Caribbean*, *supra* note 92, at 12.

[367] Article VIII of the *American Declaration*.

[368] Article 22 of the *American Convention*.

[369] *Draft Inter-American Convention on the Human Rights of Older Persons*, supra note 39, Article 15.

[370] *Id.*

[371] *Id.*, Article 12.

[372] Article V of the *American Declaration*. The *American Declaration* also protects the inviolability of the home in Article IX and the inviolability of the correspondence in Article X.

[373] Article 11 of the *American Convention*.

by being exempt and immune from abusive and arbitrary invasion by third parties or public authorities."[374] With respect to honor and reputation, states must prohibit unlawful attacks and provide legal protection against those actions. According to the IACtHR, "[t]he right to honor relates to self-esteem and self-worth, while reputation refers to the opinion that others have of a person."[375] Furthermore, privacy embodies the protection of private life considered by the IACtHR as a very broad and undefined notion encompassing, *inter alia*, the protection of sexual life and the right to establish and develop relationships with other human beings.[376] More recently, the IACtHR has enhanced the scope of private life to include the right to personal autonomy, understood as the right to personal development, to choose one's own destiny, and establish relationship with other individuals and the external world.[377] The right to privacy and private life is not an absolute right and may restricted on the basis of permissible limitations, including those that are provided by law, pursue a legitimate aim and are necessary in a democratic society.[378]

The draft *Convention on Older Persons* embodies all the different aspects of the notion of privacy and private life that are essential for the protection of the rights of older persons. First, Article 16 ensures the right to privacy and intimacy.[379] With respect to privacy, the draft instrument provides that older persons' right to private life, family, home, domestic unit, and correspondence cannot be subject to arbitrary or illegal interference.[380] Also, older persons must not be the subject of attacks to their dignity, honor and reputation.[381] The intimacy of older persons is also another aspect of privacy that must be respected, including during bathing and other acts of personal hygiene.[382] The right to privacy requires enhanced protection in the case of older persons who reside in long-term care institutions or similar facilities.[383] In that case, Article 12 of the draft *Convention on Older Persons* bolsters the general protection afforded in Article 16 by stating that domestic legal frameworks regulating the operation of long-term care

[374]See, *inter alia*, I/A Court H.R., *Case of Escher et al. v. Brazil*, Preliminary Objections, Merits, Reparations, and Costs, Judgment of July 6, 2009, Series C No. 200, 113.

[375]*Id.*, 116.

[376]I/A Court H.R., *Case of Fernández Ortega et al. v. Mexico*, Preliminary Objection, Merits, Reparations, and Costs, Judgment of August 30, 2010, Series C No. 215, 129.

[377]I/A Court H.R., *Case of Artavia Murillo et al. (*in vitro *fertilization) v. Costa Rica*, Preliminary Objections, Merits, Reparations and Costs, Judgment of November 28, 2012, Series C No. 257, 143.

[378]I/A Court H.R., *Case of Escher et al. v. Brazil*, *supra* note 574, 116.

[379]*Draft Inter-American Convention on the Human Rights of Older Persons*, supra note 39, Article 16.

[380]*Id.*

[381]*Id.*

[382]*Id.*

[383]*Id.*

facilities must ensure that older persons residing in those institutions are not the subject of arbitrary or illegal intrusions in their private life, correspondence or any other form of communication.[384] Also, states must prevent such interferences with any other sphere in which older persons engage, including aspects of their sexuality.[385]

Also, the notion of autonomy and self-determination is at the core of the draft *Convention on Older Persons*. The right of older persons to enjoy an independent and autonomous life is recognized in the Preamble of the draft instrument as well as in the general principles of the treaty.[386] Moreover Article 7 provides a very extensive and comprehensive protection of this right.[387] First, this provision defines the right of older persons to independence and autonomy as comprising the right to make decisions, determine their life plans, and lead an independent and autonomous life in line with traditions and beliefs.[388] States have a duty to provide access to the necessary mechanisms to enable older persons to exercise their right to independence and autonomy.[389] In particular, states must provide the necessary services to ensure that older persons are able to decide their place of residence and have access to personal mobility.[390]

With respect to the protection of the family, the *American Declaration* provides that "every person has a right to establish a family, the basic element of society, and to receive protection therefore."[391] Additionally, the *American Convention* notes that the family is a natural and fundamental unit of society, and therefore entitled to protection by the state.[392] Furthermore, the *Protocol of San Salvador* emphasizes that the family is the natural and fundamental element of society and ought to be protected by the state, which should "see to the improvement of its spiritual and material conditions."[393] The IACtHR has concluded that the right to private life is closely connected to the right to a family. According to Article 17 of the *American Convention*, states have an obligation to protect and strengthen the family as such, as a way to protect its members individually. In cases of children, for example, this tribunal has acknowledged that the separation of minors from their family may involve a violation of such right. The IACtHR has consistently stated that the notion of family is very broad and involves different type of relationships in which its members are related not only on the basis of marriage.

---

384 *Id.*, Article 12.

385 *Id.*

386 *Id.*, see Preamble and Article 3 on General Principles.

387 *Id.*, Article 7 ensures the right of older persons to independence and autonomy as a separate right.

388 *Id.*

389 *Id.*

390 Id. See also Article 25 on the right of older persons to accessibility and personal mobility.

391 Article VI of the *American Declaration*.

392 Article 17 of the *American Convention*.

393 Article 15 of the *Protocol of San Salvador*.

These provisions might be used as a legal argument to ensure protection for older persons in a variety of ways. The right to family could be enforced by grandparents seeking custody of grandchildren. Also, families of older persons could enforce their right to remain in family housing, challenging institutionalization. In this situation, they could claim that since placing older persons in a care institution is a measure of last resort, the state should first provide all the necessary resources to the family so that its members could take responsibility for the care of the older person. Additionally, the state parties to the *American Convention* must take appropriate steps to ensure the equality of rights and the adequate balancing of responsibilities of the spouses as to marriage, during marriage, and in the event of its dissolution.[394]

The draft *Convention on Older Persons* contemplates the role of the family as a central element for the protection of older persons. Even if during the discussions some states proposed that there was "co-responsibility" of the state and family in ensuring the human rights of older persons, eventually and, upon strong criticism, the current draft was amended to reflect the principle that only states can be attributed responsibility for the protection of those rights under international law.[395] Currently, the draft instrument provides among the general principles of the treaty that there is "[r]esponsibility of the State and the participation of the family and the community in the active, full, and productive integration of the older person into society, and in the care of, and assistance to, the person, in accordance with domestic law."[396] In that regard, states should adopt programs, policies and actions to facilitate "the strengthening of family and social ties, and their emotional relationships and different forms of family life."[397] To achieve that goal, older persons should have access to all the necessary support services to guarantee that they remain living with family and in the community, if they choose to do so.[398] Moreover, since the incidence of domestic violence against older persons by family members is widespread, states must implement appropriate training to family caregivers and other persons providing home care services, as a measure to prevent torture and other ill-treatment, including infantilization and humiliation.[399]

In line with the provisions of the draft *Convention on Older Persons*, the *San José* Charter recommends states to adopt legal frameworks, protocols and other mechanisms to ensure the dignity and privacy of older persons living in care

[394]Article 17 of the *American Convention*.

[395]Working Group on Protection of the Human Rights of Older Persons, "Views of the Inter-American Commission on Human Rights of the Organization of American States (OAS)" regarding the Draft Inter-American Convention on the Human Rights of Older Persons, *supra* note 144, at 3.

[396]*Draft Inter-American Convention on the Human Rights of Older Persons*, *supra* note 39, Article 3.

[397]*Id.*

[398]*Id.*, Article 7.

[399]*Id.*, Article 9.

institutions.[400] Moreover, it identifies the need to assist families with special services so that they can provide for older persons living with them.[401] Also, the *San José Charter* encourages states to favor initiatives that reconcile work and family obligations for those who provide care to older persons.[402] Finally, it takes into account the situation of older persons who do not have families and recommends the state to provide support services that address their basis needs.[403]

8. The Right to Information, Association and Participation

The *American Declaration* ensures the right to freedom of opinion and of the expression and dissemination of ideas, by any medium whatsoever. Likewise, the *American Convention* provides the right to freedom of thought and expression, which includes "freedom to seek, receive, and impart information and ideas of all kinds, regardless of frontiers, either orally, in writing, in print, in the form of art, or through any other medium of one's choice." The IACHR and the IACtHR have consistently interpreted this right as having two dimensions: individual and collective.[404] The individual dimension consists of the right of each person to express her own thoughts, ideas and information, and the collective or social dimension involves society's right to obtain and receive any information, to know the thoughts, ideas and information of others, and to be well-informed.[405] In *Claude Reyes v. Chile*, the IACHR held that the right to seek and receive information under Article 13 of the *American Convention* includes the right to access state-held information and the duty of the state to provide it.[406] The type of information that falls within the scope of this right includes: "information in the State's custody, management, or possession; information the State produces, or information it is obliged to produce; information in the possession of those who manage public services or public funds; information the State receives, or is required to collect in

---

[400] *San José Charter on the Rights of Older Persons in Latin American and the Caribbean*, *supra* note 92, at 11.

[401] *Id.*

[402] *Id.*, at 12.

[403] *Id.*

[404] See, *inter alia*, IACHR, *Francisco Martorell v. Chile*, Report No. 11/96, Case 11.230, Merits, Annual Report 1996, May 3, 1996, 53; I/A Court H.R., *Compulsory Membership in an Association Prescribed by Law for the Practice of Journalism (Arts. 13 and 29 American Convention on Human Rights)*, Advisory Opinion OC-5/85 of November 13, 1985, Series A No. 5, 30.

[405] See, *inter alia*, IACHR, Office of the Special Rapporteur on Freedom of Expression, *The Inter-American legal framework regarding the right to freedom of expression*, OEA Ser.L/V/II/ CIDH/RELE/INF. 2010, 13.

[406] I/A Court H.R., *Case of Claude Reyes et al. v. Chile*, Merits, Reparations and Costs, Judgment of September 19, 2006, Series C No. 151, 77.

the course of performing its duties; and information concerning one's own personal data (habeas data) or property held by those who administer private databases or registries and are legally required to provide it."[407] The IACHR has acknowledged that the right to access to information is essential to exercise the right to prior and informed consent in regard to medical care and health treatments. In the case of reproductive health issues, the IACHR identified that access to information by the patient was essential for providing consent because: they were informed about the nature of the procedure, the treatment options, and the reasonable alternatives, including the potential benefits and risks of proposed procedures; they were able to understand the information provided; and they were able to provide consent on a free and voluntary basis.[408] Even if these standards were developed in the context of reproductive health, they can be used to interpret the right to access to information by older persons when required to provide consent to medical procedures or treatments.

Finally, the right to freedom of opinion, expression and access to information is not an absolute right and may be subject to permissible limitations, as long as they are provided by law, pursue a legitimate aim and are necessary in a democratic society.[409]

The right of older persons to freedom of expression, opinion and access to information is specifically contemplated in several provisions of the draft *Convention on Older Persons*.[410] First, Article 14 ensures these rights to older persons, including the right to seek information, impart and request such information by the medium of their choice.[411] In particular, states must disseminate and ensure older persons access to information regarding their fundamental rights as provided in the treaty.[412] With respect to persons living in long-term care institutions, domestic regulatory frameworks regulating those institutions must ensure older persons' access to their own physical or digital files.[413]

On the other hand, states will promote the gathering of statistical information and research data to design and implement policies or actions towards compliance with the duties of the draft *Convention on Older Persons*.[414] In the case of violence against older persons, for example, states must "produce and disseminate

[407]IACHR, *Access to Information on Reproductive Health from a Human Rights Perspective*, OEA/Ser.L/V/II., Doc. 61, November 22, 2011, 18.

[408]*Id.*, 44.

[409]See, *inter alia*, I/A Court H.R., *Case Kimel v. Argentina*, Merits, Reparations and Costs, Judgment of May 3, 2008. Series C No. 177, 54; *Case of Palamara-Iribarne v. Chile*, Merits, Reparations and Costs, Judgment of November 22, 2005, Series C No. 135, 79.

[410]*Draft Inter-American Convention on the Human Rights of Older Persons*, *supra* note 39, Article 14.

[411]*Id.*

[412]*Id.* Also see Article 4(d).

[413]*Id.*, Article 12(c)(i).

[414]*Id.*, Article 4(h).

information for the purpose of generating diagnostic assessments of possible situations of violence and developing prevention policies."[415] Also, Article 11 regarding prior and informed consent on health matters provides for the state duty to set out mechanisms to strengthen the capacity of older persons to make decisions freely and voluntarily, including guaranteeing their right to access to adequate information. In particular, this provision states that "[t]hose mechanisms must ensure that the information provided is appropriate, clear and timely, available on a non-discriminatory basis, in an accessible and easily understood manner, in accordance with the older person's culture identity, level of education and communication needs."[416]

The rights to assembly[417] and association[418] are also protected by the *American Declaration* and the *American Convention*. Both instruments ensure the right to associate freely for political, religious, economic, social, cultural, professional, labor union, sports, or for other purposes.[419] As with the right to freedom of expression, the right to association is not an absolute right and may be restricted on the basis of permissible limitations, provided that they are established by law, pursue a legitimate goal and are necessary in a democratic society.[420] Article 17 of the *Protocol of San Salvador* provides that states must "[f]oster the establishment of social organizations aimed at improving the quality of life for the elderly."[421]

The draft *Convention on Older Persons* protects the right of older persons to association and assembly.[422] According to Article 27, states must promote the recognition and establishment of associations by older persons and provide them with state support to strengthen their activities and leadership.[423] Moreover, states must promote the participation of those associations in the drafting, implementation and oversight of public policies and legislation adopted at the domestic level to protect the rights of older persons, in light with the duties arising out of the draft *Convention on Older Persons*.[424]

The *American Declaration* notes that every person with legal capacity is entitled to participate in the government of his or her country, directly or through a

[415] *Id.*, Article 9(b).

[416] *Id.*, Article 11(a).

[417] Article XXI of the *American Declaration* and Article 15 of the *American Convention* protect the right to freedom of assembly.

[418] Article XXII of the *American Declaration* and 16 of the *American Convention* protect the right to association.

[419] *Id.*

[420] Article 16(2) of the *American Convention*.

[421] Article 17(c) of the *Protocol of San Salvador*.

[422] *Draft Inter-American Convention on the Human Rights of Older Persons*, *supra* note 39, Article 27.

[423] *Id.*

[424] *Id.* Article 4(g).

representative, and to take part in popular elections.[425] Further, under the *American Convention,* every citizen is also entitled to take part in the conduct of public affairs, directly or through freely chosen representatives, to vote and to be elected, and to have access, under general conditions of equality, to the public service of his or her country.[426] Political participation includes a wide variety of activities that individuals perform by themselves or within an organization in order to impact the election of those who will represent them or who will be responsible for conducting the public affairs of the country, or to shape state policy regarding matters of their interest.[427] The law may regulate the exercise of the rights to political participation, *inter alia,* on the basis of age and civil and mental capacity.[428] In *Castañeda Gutman,* the IACtHR stated that these limitations are authorized as "enabling conditions that the law can impose to exercise political rights;"[429] however, these measures must ensure the principle of non-discrimination and cannot be unreasonable or disproportionate. With respect to the age limitation, the IACtHR indicates that it refers to a minimum age to exercise the right to vote or occupy certain public positions.[430] The limitation on civil and mental capacity, however, is less clear and has not been interpreted in case law or other practice. It is important to note that at the universal level the CRPD ensures an ample right to political participation of persons with disabilities.[431] It would be interesting to see how the organs of the Inter-American System interpret these restrictions in future cases, particularly if they are ever faced with a situation affecting an older person. So far, the practice of the IACtHR has been to resort to the existing international corpus juris in a particular subject when interpreting a right under the *American Convention.*[432] The *Case of Furlan v. Argentina* is a case in point, since the IACtHR interpreted the rights under the *American Convention* in light of the CRPD.[433]

In addition to interpreting the scope of limitations to the right to political participation under Article 23 of the *American Convention,* in *Castañeda Gutman* the IACtHR concluded that that provision, together with Articles 1(1) and 2 of the treaty, imposes upon states a positive obligation to ensure the right to vote and be elected.[434] For that, states must organize their electoral systems, set up specific mechanisms and adopt the necessary legal frameworks to ensure that everybody

[425]Article XXI of the *American Declaration.*

[426]Article 23 of the *American Convention.*

[427]I/A Court H.R., *Case of Castañeda Gutman v. México,* Preliminary Objections, Merits, Reparations, and Costs, Judgment of August 6, 2008, Series C No. 184, 146.

[428]Article 23(2) of the *American Convention.*

[429]I/A Court H.R., *Case of Castañeda Gutman v. México,* supra note 427, 155.

[430]*Id.*

[431]Article 29 of the CRPD.

[432]I/A Court H.R., *Case of Furlan and Family v. Argentina, supra* note 204, 124.

[433]*Id.*, 139.

[434]I/A Court H.R., *Case of Castañeda Gutman v. México, supra* note 427, 156–157.

can have access to participate in the political process.[435] Issues of accommodation for certain vulnerable groups, such as older persons, may be required under this positive obligation to guarantee that they may exercise their right to political participation, and are not prevented to do so from external conditions unrelated to the exercise of this right.

With regard to political participation, the draft *Convention on Older Persons* in its Preamble acknowledges the relevance of ensuring the participation of older persons in the political process.[436] Moreover, Article 26 protects this right by stating that older persons must be ensured their right to participate in political and public life without discrimination on the basis of age.[437] States have a duty to facilitate the conditions and means to guarantee their right to vote and be elected, in particular they must ensure that electoral procedures, facilities and voting materials are accessible and easy to understand and use.[438] Also, states must ensure that older persons may cast their vote in secret and without intimidation.[439] Exceptionally, older persons may be assisted by a person of their choice if that is necessary.[440]

In addition, the draft *Convention on Older Persons* enshrines a broader notion of participation of older persons in the community. This notion revolves around the understanding that older persons must be completely integrated in the society and must have the opportunity to participate as full members of their communities. First, Article 1 of the draft instrument provides that the goal of the treaty is to promote, protect and fulfill the fundamental rights of older persons, "in order to contribute to their full inclusion, integration, and participation on society."[441] This objective is also included in Article 3, listing the general principles of the draft treaty.[442] Furthermore, Article 4 concerning state duties provides that affirmative measures to de facto ensure the full participation of older persons in the economic, educational, political, and cultural files would not be considered discriminatory.[443] Finally a full provision is dedicated to afford protection to the right of older persons to participation and community integration. Article 8 protects "the right to active, productive, full and effective participation in the family, community and society."[444] To ensure that right, States must adopt all the necessary measures to

---

[435] *Id.*, 157–159.

[436] *Draft Inter-American Convention on the Human Rights of Older Persons*, *supra* note 39, Preamble.

[437] *Id.*, Article 26.

[438] *Id.*

[439] *Id.*

[440] *Id.*

[441] *Id.*, Article 1.

[442] *Id.*, Article 3(e).

[443] *Id.*, Article 4.

[444] *Id.*, Article 8.

create or strengthen mechanisms where older persons may participate in full equality to promote intergenerational activities.[445] Older persons' active participation in their communities will contribute to the eradication of prejudices and negative stereotypes of older persons and develop mutual support among generations as an essential component of social advancement.[446] This notion of broad participation for older persons was upheld by the IACtHR in the *Yakye Axa Indigenous Community* case when it provided that the state was under the obligation to take special measures to protect the rights of older persons in the community, because "the Yakye Axa Community oral transmission of the culture to the younger generations is primarily entrusted to the elderly."[447] Transmission of the culture in that case was at the basis of the community's survival.

The *San José Charter* also envisions a broad notion of participation in the community for older persons as a matter of right. In this regard, it recommends states to ensure participation of older persons in civil society organizations so that they actively influence the design and adoption of the public policies that affect them.[448] Also, states must ensure that older persons engage and participate in the processes of consultation and accountability of the work of the public institutions overseeing implementation of those policies.[449]

9. Access to Justice and Effective Remedy

The rights of access to justice and to an effective remedy, as crosscutting rights, are fundamental to protect the rights of older persons in the sphere of civil and political rights, as well as economic, social and cultural rights. The rights to access to justice and to an effective remedy are comprehensively protected by both the *American Declaration*[450] and the *American Convention*.[451]

The core nature of the rights to access to justice and to an effective remedy is underscored by Article 27 of the *American Convention* that considers them as

[445] *Id.*

[446] *Id.*

[447] I/A Court H.R., *Case of the Yakye Axa Indigenous Community v. Paraguay*, *supra* note 241, 175.

[448] *San José Charter on the Rights of Older Persons in Latin American and the Caribbean*, *supra* note 92, at 8.

[449] *Id.*, at 17.

[450] Articles XVIII and XXVI of the *American Declaration*.

[451] Articles 8 and 25 of the *American Convention*.

non-derogable,[452] as well as by the subsequent interpretation of the IACtHR upholding such status.[453] Moreover, the other Inter-American human treaties that afford special protection to vulnerable groups against violence and discrimination enshrine the rights to access to justice and to an effective remedy as the fundamental mechanisms to ensure that states will effectively respect and protect the rights of the members of those groups.[454] In the same vein, the IACtHR has concluded in *Furlan* that the right to access to justice plays a fundamental role in addressing those forms of discrimination.[455]

The notion of access to justice involves respecting fundamental due process rights, including the right to a hearing by a competent, independent, and impartial tribunal and the right to obtain a decision with due guarantees and within a reasonable time.[456] Also, it comprises the right to be presumed innocent; the right to defense; the right not to be compelled to testify against him or herself; and the right to appeal.[457] Among these fair trial rights, one that has been essential to older persons for accessing justice has been the right to obtain a decision within a reasonable time. In *Menéndez v. Argentina*, a case involving social security benefits, the petitioners argued that the age of the claimants had to be taken into account when assessing the reasonableness of the duration of proceedings, because if they were unduly long and the judicial remedies were not effective, that situation prevented them from enjoying their rights to social security, health, and physical and mental integrity.[458] Petitioners asserted that "courts should be all the more diligent when dealing with social security, as any delay may render the proceedings

[452]Article 27 of the *American Convention* states that this provision "does not authorize any suspension of the following articles: Article 3 (Right to Juridical Personality), Article 4 (Right to Life), Article 5 (Right to Humane Treatment), Article 6 (Freedom from Slavery), Article 9 (Freedom from Ex Post Facto Laws), Article 12 (Freedom of Conscience and Religion), Article 17 (Rights of the Family), Article 18 (Right to a Name), Article 19 (Rights of the Child), Article 20 (Right to Nationality), and Article 23 (Right to Participate in Government), **or of the judicial guarantees essential for the protection of such rights**."

[453]See, *inter alia*, I/A Court H.R., *Judicial Guarantees in States of Emergency (Articles 27(2), 25 and (8) American Convention on Human Rights)*, Advisory Opinion OC-9/87 of October 6, 1987, Series A No. 9, 30.

[454]See, inter alia, *Inter-American Convention for the Prevention, Punishment and Eradication of the Violence against Women*, *supra* note 30, Article 7(f) and (g); *Inter-American Convention Against All Forms of Discrimination and Intolerance*, *supra* note 32, Article 10; and *Inter-American Convention Against Racism, Racial Discrimination, and Related Forms of Intolerance*, *supra* note 33, Article 10.

[455]I/A Court H.R., *Case of Furlan and Family v. Argentina*, *supra* note 204, 124.

[456]See, *inter alia*, Article 8 of the *American Convention*.

[457]*Id.*

[458]IACHR, *Amilcar Menéndez, Juan Manuel Caride et al. (Social Security System) v. Argentina*, Report No. 03/01, Case 11.670, Admissibility, Annual Report 2001, January 19, 2001, 25.

futile."[459] Ultimately, this case was resolved through a friendly settlement procedure and therefore it was never decided on the merits.[460]

Similarly, in *Furlan v. Argentina*, the IACtHR analyzed whether the disability of the complainant had to be considered when assessing whether the proceedings to obtain compensation for damages were unreasonably delay. First, the IACtHR stated that the determination of the reasonableness of a proceeding is based on the following considerations: (1) the complexity of the matter; (2) the procedural activity of the interested party; (3) the conduct of the judicial authorities; and (4) the adverse effect of the duration of proceedings on the judicial situation of the interested party.[461] When analyzing the fourth element, the IACtHR concluded that when the passage of time negatively affects the legal situation of the complainant, there is a duty to carry out the proceedings with more diligence to ensure that they are completed in a shorter period of time.[462] In the case under analysis, the IACtHR concluded that the fact that the petitioner was a person with disabilities, who additionally lacked personal financial resources to pay for the treatment he needed, required the civil claim against the state to be processed more diligently.[463] The delay of more than 12 years to complete the proceedings and comply with the judgment affected the victim's capacity to access the necessary health and psychological treatment to improve his quality of life.[464] Thus, there was a violation of the right to have a judgment within a reasonable time under Article 8 of the *American Convention*.[465] Even if this case referred to a person with disability, the IACtHR in its findings cited to cases decided by the ECtHR where that tribunal concluded that the age of the complainant was an element to consider when deciding whether the delay in proceedings adversely affected the situation of the petitioner.[466] Consequently, the reasoning and findings in *Furlan* are very relevant as a precedent to deal with cases regarding older persons, in particular to assess how unreasonable delay in the resolution of proceedings in which they are parties may affect their rights to access to justice and to an effective remedy.

In another case that raised somewhat similar issues, petitioners argued that the delay in the execution of pension rights judgments obtained in their favor violated their right to an effective remedy and caused physical and emotional suffering to

[459]*Id.*

[460]IACHR, *Amilcar Menendez, Juan Manuel Caride et al. v. Argentina*, Report No. 168/11, Petition 11.670, Friendly Settlement, Annual Report 2011, November 3, 2011.

[461]I/A Court H.R., *Case of Furlan and Family v. Argentina*, *supra* note 204, 152.

[462]*Id.*, 194.

[463]*Id.*, 197–202.

[464]*Id.*, 203.

[465]*Id.*, 204.

[466]*Id.*, 195.

the complainants because those benefits were their only source of income.[467] The IACHR declared admissible the alleged violation of the right to an effective remedy resulting from the delay in the execution of a judgment on pension rights, but rejected the argument alleging that the unreasonable delay involved a breach of the pensioners' right to humane treatment.[468] Furthermore, in *Acevedo Buendia*, the IACtHR reviewed the delay in the execution of judgments issued in regard to pension benefit cases and concluded that there was a breach of the right to an effective remedy.[469] Interestingly, in this case, the tribunal did not take into consideration the fact that the complainants were pensioners. However, it confirmed a principle that could be of importance in future pension right cases, when it stated that the domestic rules regulating budget issues cannot justify an excessive delay in complying with final judgments.[470]

The draft *Convention on Older Persons* includes several references to the right to access to justice and the right to an effective remedy. First, the general principles applicable to the treaty ensure the right of older persons to effective judicial protection.[471] Also, when dealing with violence against older persons, Article 9 includes the states' duty to promote appropriate and effective mechanisms for cases of violence against older persons and strengthen administrative and judicial proceedings for dealing with cases of this nature.[472] Likewise, Article 11 regulating the right to give informed consent on health matters contemplates states' duty to establish a procedure that enables older persons to express their will in advance.[473] With respect to older persons receiving long-term care, the draft instrument establishes that states must enact laws to ensure the criminal and civil liability of those overseeing long-term care services for any acts caused to the detriment of older persons under their care.[474] Next, Article 22, regulating the right to property, states that older persons are entitled to respect of fundamental due process rights in proceedings regarding restrictions to their exercise of this right.[475]

Finally, Article 31 establishes the right of older persons to a hearing by a competent, independent and impartial tribunal and to obtain a judgment within a

---

[467] IACHR, *Agustina Alonso et al. v. Argentina*, Report No. 14/14, Petition 265-00, Admissibility, Annual Report 2014, 23, 25.

[468] *Id.*, 65, 67.

[469] I/A Court H.R., *Case of Acevedo Buendía et al. ("Discharged and Retired Employees of the Comptroller") v. Peru*, Preliminary Objection, Merits, Reparations and Costs, *supra* note 27, 74–79.

[470] *Id.*, 75.

[471] *Draft Inter-American Convention on the Human Rights of Older Persons*, supra note 39, Article 3(n).

[472] *Id.*, Article 9.

[473] *Id.*, Article 11.

[474] *Id.*, Article 12.

[475] *Id.*, Article 22.

reasonable time.[476] Moreover, it ensures maximum due diligence and reasonable proceedings in the processing of cases regarding older persons.[477] Also, states undertake to ensure the adoption of procedural accommodations for older persons to guarantee access on equal basis with others.[478] Additionally, states will develop policies and programs to promote alternative judicial mechanisms for disputes involving older persons and train police and prison personnel on the protection of the rights of older persons.[479]

The *San José Charter* envisions protection of the right to access to justice and to an effective remedy in regard to the exercise of several rights of concern to older persons. First, it recommends states to provide priority attention and preferential treatment to older persons in the processing of matters that affect them and in the execution of judgments in their favor.[480] Also, states should protect the rights of older persons in public or private long-term care institutions through the establishment of appropriate mechanisms of supervision and control.[481] Furthermore, violence against older persons should be made a crime and appropriate judicial remedies should be strengthened to prevent and punish such violence.[482] Additionally, judicial remedies should also be made available to older persons to protect them against patrimonial exploitation.[483]

E. Economic, Social and Cultural Rights of Older Persons

With respect to the protection of economic, social and cultural rights that may be pertinent for the protection of older persons, the *American Declaration* enshrines the following rights: right to education (Article XII); right to the benefits of culture (Article XIII); right to work and to fair remuneration (Article XIV); right to leisure time and to the use thereof (Article XV); and the right to social security (Article XVI). Likewise, the *Protocol of San Salvador* affords the following economic, social and cultural rights applicable to older persons: right to work (Article 6); just, equitable, and satisfactory conditions of work (Article 7); right to social security (Article 9); right to health (Article 10); right to food (Article 12); right to education (Article 13); right to the benefits of culture (Article 14); right to the formation and the protection of families (Article 15); and protection of older persons

---

[476] *Id.*, Article 31.

[477] *Id.*

[478] *Id.*

[479] *Id.*

[480] *San José Charter on the Rights of Older Persons in Latin American and the Caribbean*, *supra* note 92, at 8.

[481] *Id.*, at 11.

[482] *Id.*, at 14.

[483] *Id.*

(Article 17). Moreover, Article 26 of the *American Convention* provides for a state duty to adopt measures to achieve progressively the full realization of the economic, social, and cultural rights set forth in the OAS Charter.

As stated before, the *Protocol of San Salvador* provides that only two rights may be the basis of individual complaints before the IACHR and IACtHR, namely the right to education and the right of workers to organize.[484] Beyond those rights, the IACHR has confirmed that it does not have jurisdiction *ratione materiae* to find violations of the *Protocol of San Salvador* through the system of individual petitions, but that it can resort to the Protocol to interpret other provisions of the *American Convention*, in light of Articles 26 and 29 of the American Convention."[485]

With respect to Article 26 of the *American Convention*, the most important aspect of the debate has been whether this provision is justiciable and may be used to argue breaches to economic, social and cultural rights protected either by the OAS Charter and the *American Declaration*, and/or the *Protocol of San Salvador*. So far, the IACHR has been more willing to explore this argument and has decided in some cases that Article 26 provides the basis to find breaches to those rights, though there has not been consistency in the case law regarding this approach.[486] One aspect, however, in which both the IACHR and the IACtHR seem to agree, is that Article 26 creates a duty to take progressive measures to protect economic, social and cultural rights and that consequently, there is a duty not to adopt regressive measures.[487] The notion of regression does not imply that the state cannot restrict the exercise of a right, but it requires "a joint analysis of the individual affectation in relation to the collective implication of the measure. In that sense, not every regressive measure is incompatible with Article 26 of the *American Convention*."[488] To be compatible with the *American Convention* a measure of regression must be justified by a strong reason.[489] Furthermore, the duty not to regress can be assessed by the IACHR and the IACtHR through the individual petition system.[490] So far, however, neither the IACHR nor the IACtHR have con-

[484]See Article 19(6) of the *Protocol of San Salvador*.

[485]IACHR, *Jorge Odir Miranda Cortez, et al. v. El Salvador*, Report No. 29/01, Case 12.249, Admissibility, Annual Report 2001, 36.

[486]See, *inter alia*, IACHR, *Milton García Fajardo et al. v. Nicaragua*, Report No. 100/01, Case 11.381, Merits, Annual Report 2001, 94–101.

[487]IACHR, *Jorge Odir Miranda Cortez et al. v. El Salvador*, Report No. 27/09, Case 12.249, Merits, Annual Report 2009, 139; I/A Court H.R., *Case of Acevedo Buendía et al. ("Discharged and Retired Employees of the Comptroller") v. Peru*, Preliminary Objection, Merits, Reparations and Costs, *supra* note 27, 103.

[488]IACHR, *Jorge Odir Miranda Cortez, et al. v. El Salvador*, *supra*, 104.

[489]I/A Court H.R., *Case of Acevedo Buendía et al. ("Discharged and Retired Employees of the Comptroller") v. Peru*, Preliminary Objection, Merits, Reparations and Costs, *supra* note 27, 103.

[490]IACHR, *Jorge Odir Miranda Cortez et al. v. El Salvador*, *supra* note 487, 103; I/A Court H.R., *Case of Acevedo Buendía et al. ("Discharged and Retired Employees of the Comptroller") v. Peru*, Preliminary Objection, Merits, Reparations and Costs, *supra* note 27, 102.

cluded that specific measures breached the duty of non-regression under Article 26 of the *American Convention*.[491]

In addition, the draft *Convention on Older Persons* ensures the right to social security; right to work; right to health; right to education; right to culture; right to housing; right to a healthy environment, right to accessibility and personal mobility; and right to recreation, leisure and sports. It is important to note, that Article 4 of the draft instrument listing the state obligations under this treaty provides that states "shall adopt measures as necessary in the framework of international cooperation to the full extent of their available resources and taking their degree of development into account in order to progressively achieve, in keeping with domestic law, the full realization of these economic, social and cultural rights, without prejudice to those obligations that are immediately applicable according to international law."[492] Consequently, from the language of this duty it is possible to conclude that the drafters of the treaty envision that both obligations of immediate application and progressive application are involved in regard to the protection of economic, social and cultural rights. Ultimately, if the draft *Convention on Older Persons* maintains access to the individual petition system to supervise compliance with all the rights protected in the treaty as it stands today, it will be the task of the IACHR and the IACtHR to decide on specific cases and circumstances the nature of each obligation arising out of that Convention.[493]

1. The Right to Education and Culture

The *American Declaration* includes protection of the right to education for all. "All" includes older persons. *Every* person has the right to education, which includes the right to equality of opportunity in every case, "in accordance with natural talents, merit and the desire to utilize the resources that the state or the community is in a position to provide."[494] Therefore, educational opportunities may not be denied to persons based on their age. Also, the right to education is ensured as a measure to prepare the beneficiaries "to attain a decent life, to raise his standard of living, and to be a useful member of society."[495]

---

[491]See, *inter alia*, IACHR, *Jorge Odir Miranda Cortez et al. v. El Salvador*, *supra* note 487, 141–147.

[492]*Draft Inter-American Convention on the Human Rights of Older Persons*, *supra* note 39, Article 4.

[493]*Id.*, Article 35.

[494]Article XII of the *American Declaration*.

[495]*Id.*

Although Article 13 of the *Protocol of San Salvador* does not explicitly discuss older persons, it also expresses that *everyone* has the right to education.[496] It further elaborates that education should be directed towards the full development of the human personality and human dignity and should strengthen respect for human rights, ideological pluralism, fundamental freedoms, justice and peace.[497] Education should have the goal of enabling everyone to effectively participate in society[498]; and states should encourage the education of those who have not completed primary education.[499] With many older persons having limited literacy, education can be an important aspect for increasing their ability to participate and contribute to the communities and society they live in. Furthermore, states should make available the access to technical or vocational education, as well as higher education for all through every appropriate means, including by the introduction of free education.[500]

The right of education is one of the two rights enshrined in the *Protocol of San Salvador* for which the Inter-American Human Rights System may receive individual complaints.[501]

Beyond the right to education, the *American Declaration* provides that every person enjoys the right to take part in the cultural life of their community, to enjoy the arts, and to participate in the benefits that result from intellectual progress.[502] Under *the American Declaration*, every person also has the right to leisure time, to wholesome recreation, and to the opportunity for advantageous use of his free time to his spiritual, cultural and physical benefit.[503] State parties to the *Protocol of San Salvador* further recognize the right of everyone to take part in the cultural and artistic life of the community; to enjoy the benefits of scientific and technological progress; and to benefit from the protection of moral and material interests deriving from any scientific, literary or artistic production of which he is the author.[504] Older persons are often excluded from the benefits of progress and technology, sometimes by discrimination, and sometimes by vision, health, or literacy limitations. Efforts to adapt and include older persons in progress are to be commended.

The draft *Convention on Older Persons* includes the right to education[505] and the right to culture.[506] Article 20 states that older persons have the right to access

---

[496] Article 13(1) of the *Protocol of San Salvador.*

[497] Article 13(2) of the *Protocol of San Salvador.*

[498] *Id.*

[499] Article 13(3) of the *Protocol of San Salvador.*

[500] *Id.*

[501] Article 19 of the *Protocol of San Salvador.*

[502] Article XIII of the *American Declaration.*

[503] Article XV of the *American Declaration.*

[504] Article 14(1) of the *Protocol of San Salvador.*

[505] *Draft Inter-American Convention on the Human Rights of Older Persons*, *supra* note 39, Article 20.

[506] *Id.*, Article 21.

and take part in educational programs at all levels, and to make use of their knowledge and experience to benefit younger generations.[507] To ensure the right to education of older persons, states will, *inter alia*: reduce illiteracy among older persons, especially women and other vulnerable groups; promote life-long learning programs for older persons to ensure they adapt to changing conditions; promote and facilitate education of older persons on the new information and communication technologies; and promote the development of educational materials suitable for older persons.[508]

Also, older persons under the draft *Convention on Older Persons* have the right to participate in social and cultural life, enjoy cultural, spiritual, and material artistic and literary works, as well as the world of the plastic and visual arts.[509] States shall promote special measures to ensure that older persons have access to cultural goods and services, under easy to access formats and conditions.[510] Furthermore, the draft instrument includes the right of older persons to recreation, leisure, and sports stating that state parties shall develop social, recreational, leisure, and sports services, programs, and activities that are designed for older people, in particular for those who are in long-stay facilities, in order to promote self-realization and prevent them from being isolated.[511]

The right to education of older persons is also recognized in the *San José Charter*. This instrument favors the right of older persons to education, on the basis of a lifelong learning approach.[512] First, states should promote policies to eradicate illiteracy among older persons.[513] Also, they should encourage plans to guarantee the access of older persons to the new technologies so that they can remain integrated in society.[514] Additionally, older persons should be invited to participate in educational programs where they can share their knowledge and cultural values to the new generations.[515] Finally, states should facilitate the access of older persons to recreational, cultural and sport activities sponsored by public and private institutions.[516]

One interesting aspect that it is included both in the draft *Convention on Older Persons* and the *San José Charter* is the request to states to promote the topic of ageing and age-related issues to the different levels of education.[517]

---

[507] *Id.*, Article 20.

[508] *Id.*

[509] *Id.* Article 21.

[510] *Id.*

[511] *Id.* Article 28.

[512] *San José Charter on the Rights of Older Persons in Latin American and the Caribbean*, *supra* note 92, at 15.

[513] *Id.*

[514] *Id.*

[515] *Id.*

[516] *Id.*

[517] *Id.*, and *Draft Inter-American Convention on the Human Rights of Older Persons*, *supra* note 39, Article 20.

2. The Right to Work

The *Protocol of San Salvador*, referencing the special protection of older persons, establishes that state parties should agree to progressively implement steps necessary to ensure the rights to special protection in old age, by making available work programs, specifically designed to give older persons opportunities to engage in productive activities suited to their abilities and consistent with their vocations and desires.[518]

The *Protocol of San Salvador* also explicitly expresses the right to work, which includes the opportunity to secure the means for living a dignified and decent existence by performing a freely elected or accepted lawful activity and that the right shall be enjoyed under just, equitable and satisfactory conditions.[519] Furthermore, under the *American Declaration*, every person has the right to follow his or her vocation freely, insofar as existing conditions of employment permit.[520] Every person who works has the right to receive such remuneration as will, in proportion to his or her capacity and skill, assure him a standard of living suitable for himself and for his or her family.[521]

In the same line, the 2005 OAS *Inter-American Program for the Promotion and Protection of the Human Rights of Migrants, Including Migrant Workers and their Families* began promoting human rights for migrant workers, outlining the causes of migration, the dangers of violations to their human rights, the existing conventions, and the state actions necessary to preserve all human rights, including those of "senior citizens."[522] This is important to note, as economic productivity does not stop at retirement age for most old persons; and some may still continue to work outside their home countries. Those who choose to continue to work are entitled to the same rights as other migrant workers, including the right to social security contributions accumulated during work abroad once the age of payout is reached.[523]

Additionally, the Ministers of Labor of the OAS member states noted that the *Declaration of Mexico, People and their Work at the Heart of Globalization* declares in Paragraph 25 that equality and nondiscrimination in employment extends to older persons, and special attention must be paid to decent work for

---

[518]Article 17(b) of the *Protocol of San Salvador*.

[519]Articles 6–7 of the *Protocol of San Salvador*.

[520]Article XIV of the *American Declaration*.

[521]*Id.*

[522]OAS, *Inter-American Program for the Promotion and Protection of the Human Rights of Migrants, Including Migrant Workers and their Families*, AG/RES. 2141 (XXXV-O/05), adopted on June 7, 2005, Section B, 37.

[523]The Program promotes the negotiation of bilateral on the social security benefits of migrants and their families, so that any social security contributions may be claimed in or transferred to the state in which the migrant worker lives. *Id.*, 56.

them, as a vulnerable group.[524] Such recognition is an example of good practice in legal frameworks to protect older persons; policies such as non-discrimination, job placement services, and appropriate safety standards may keep older persons in the workforce contributing to the productivity of society for longer, under better conditions.

The draft *Convention on Older Persons* includes the right to work and ensures rights in the workplace. It states that older persons should have the opportunity to work and/or to have free access to other forms of obtaining income. Age-based labor discrimination and the setting of an upper age limit for hiring person for any job or employment is prohibited, except in those cases in which the nature of the job so requires.[525] Moreover, retirement should be voluntary, except in cases in which the health of the older person is a factor.[526] State parties should promote labor policies that ensure the older worker a formal employment with full social security benefits, without loss or suspension of prior acquired retirement pensions.[527] Transition to retirement should be gradual and organizations of workers should participate in framing the necessary programs and measures.[528]

The employment of older persons after retirement should guarantee access to all benefits and a salary equal to other workers.[529] Also, labor policies for older workers should ensure access to appropriate working conditions, environment, working hours, and organization of tasks.[530] In terms of training, states should encourage the design of knowledge-certification programs to ensure the access of older workers to more inclusive labor markets.[531] It is important to note that the right to work in the draft *Convention on Older Persons* is framed with obligations of immediate application and those of progressive application. It would be interesting to see in the future if this treaty is adopted and gives the IACHR and IACtHR the power to review individual petitions how these obligations will be interpreted in practice.

The *San José Charter* also considers the right of older persons to work and to generate income.[532] First, states are encouraged to promote equal treatment and opportunity in access to employment, particularly in regard to working conditions,

---

[524]OAS, *Report of the XIV Inter-American Conference of Ministers of Labor*, AG/RES. 2205 (XXXVI-O/06), Appendix, adopted on June 6, 2006, 24, available at http://www.iin.oea.org/2006/Res_36_AG_OAS/AGRES_2205_ingles.htm (last visited Jan. 30, 2015).

[525]*Draft Inter-American Convention on the Human Rights of Older Persons*, *supra* note 39, Article 18.

[526]*Id.*

[527]*Id.*

[528]*Id.*

[529]*Id.*

[530]*Id.*

[531]*Id.*

[532]*San José Charter on the Rights of Older Persons in Latin American and the Caribbean*, *supra* note 92, at 13.

training, and labor market accessibility.[533] Also, states should adopt labor policies that promote the work of older persons and induce legal reforms and economic benefits to those who employ older persons after retirement.[534] Beyond retirement, states should inform older persons of other opportunities to contribute to society, including volunteer opportunities.[535] Finally, states should facilitate access to credits and encourage older persons to engage in the development of new activities.[536]

3. The Right to an Adequate Standard of Living and Housing

The *American Declaration* does not include express protection of the rights to housing or food. Notwithstanding, it provides for the right to own private property that meets essential needs of decent living and that maintains dignity of the individual and of the home.[537] Toward that end, the inviolability of the home is also expressly protected.[538] The right to adequate housing may include government-subsidized housing for those who cannot afford it, or those who are incapable of maintaining housing of their own. The *Protocol of San Salvador* does not include a separate provision regarding the protection of the right to housing either. However, it enshrines the right to food which ensures the right of everyone, including older people, to adequate nutrition.[539] This right could be used as the basis to construe the scope of the right to an adequate standard of living of older persons. Article 17 of the *Protocol of San Salvador* affords special protection in old age and, in that context, states must progressively provide access of older persons to suitable facilities, either a home or an institution as well as ensure their right to food.[540]

The draft *Convention on Older Persons* includes the right to property, housing and a healthy environment, and the right to accessibility and personal mobility. With respect to the right to property, the draft treaty envisions a range of rights to ensure that older persons are not deprived of their assets at old age.[541] First, the provision states that property cannot be deprived on the basis of age and that older

[533] *Id.*

[534] *Id.*

[535] *Id.*

[536] *Id.*

[537] Article XXIII of the *American Declaration*.

[538] Article IX of the *American Declaration*.

[539] Article 12 of the *Protocol of San Salvador*.

[540] Article 17(a) of the *Protocol of San Salvador*.

[541] *Draft Inter-American Convention on the Human Rights of Older Persons*, *supra* note 39, Article 22.

persons must enjoy in full their inheritance rights.[542] States must adopt all the necessary measures to guarantee that the disposal of property by older persons is done with their full consent, and prevent the abuse or illegal transfer of their property.[543] States will recognize the intellectual property rights of older persons on an equal basis with other persons.[544]

The right of older persons to housing is also extensively protected by the draft *Convention on Older Persons*.[545] First, older persons should be ensured the right to choose their place of residence and the persons with whom they want to live, without being force to live in any particular living arrangements.[546] Also, older persons should be able to stay in their homes for as long as possible and states should facilitate the enjoyment of such right by providing integrated social and home care services.[547] Furthermore, states should protect the right of older persons to adequate and appropriate housing, and to achieve that end, they need to build or progressively adapt housing units to the address the reduced mobility and physical disabilities of older persons.[548] States should also take into account the specific needs of older persons living alone by ensuring their access to rental subsidies, housing cooperatives or support for housing renovations, among others.[549] Additionally, states must prevent the forced eviction of older persons and other forms of expulsion to avoid homelessness of this particular vulnerable group at an age where they cannot provide for themselves.[550]

To complement older persons' right to independence and autonomy, the draft *Convention on Older Persons* ensures the right to accessibility and personal mobility.[551] This right requires states to progressively adopt the necessary measures to guarantee older persons access "to the physical environment, to transportation, to information and communications…and to other facilities and services open or provided to the public, both in urban and rural areas."[552] All these measures should also be geared towards the identification and elimination of obstacles and barriers to accessibility.[553]

---

[542] *Id.*

[543] *Id.*

[544] *Id.*

[545] *Id.*, Article 23.

[546] *Id.*, Article 7.

[547] *Id.*, Article 23.

[548] *Id.*

[549] *Id.*

[550] *Id.*

[551] *Id.*, Article 25.

[552] *Id.*

[553] *Id.* This provision lists a number of measures necessary to eliminate obstacles and barriers of older persons to accessibility.

Older persons have the right to a healthy environment that guarantees access to basic sanitation services, clean water and air, and that minimizes their exposure to pollution.[554] Even if the right to an appropriate standard of living, including food, clothes, and social services, was included in the initial draft of the *Convention on Older Persons*, it was eventually eliminated in the process of discussions.[555] So far, no other provision of the draft instrument seems to address protection of those other essential rights. One option to address the silence of the draft *Convention on Older Persons* on these issues involves Articles 12 and 17 of the *Protocol of San Salvador*. These provisions protecting the right to food generally and the right to food of older persons, respectively, may be used to afford protection to older persons regarding their right to proper and adequate access to food and essential social services. Alternatively, these rights may also be protected under the right of older persons to live with dignity under Article 6 of the draft *Convention on Older Persons* or in light of the interpretation provided by the IACtHR of Article 4 of the *American Convention*.[556]

The *San José Charter* provides a section on the rights of older persons to social services, including the need to create and ensure the provisions of those services to older persons in need.[557] Also, it envisions the relevance of providing home services to promote the opportunity of older persons to remain in their homes for as long as possible. Additionally, it encourages states to design measures to support families who assume care of an older person.[558] With respect to housing, the *San José Charter* recommends states to make all necessary efforts to ensure adequate housing to older persons and prioritize older persons when assigning public housing or land, particularly in situations of emergency, forced displacement or evictions.[559] Moreover, it promotes access to mobility for older persons, particularly in services and transportation.[560]

4. The Right to Social Protection and Security

Throughout Latin America and the Caribbean, there is general agreement that older persons are a group that should, through progressive implementation, receive

[554] *Id.*, Article 24.

[555] See, *inter alia*, *Preliminary Draft Inter-American Convention on the Human Rights of Older Persons*, Article 21, OEA, Ser.G/CAJP/GT/DHPM-37/12, April 30, 2012.

[556] For a further discussion on this issue, see section on the right to life of older persons in this document.

[557] *San José Charter on the Rights of Older Persons in Latin American and the Caribbean*, *supra* note 92, at 11.

[558] *Id.*

[559] *Id.*, at 14.

[560] *Id.*

some benefits of social protection.[561] Social protection can include state-funded cash transfers, food and housing programs, as well as privately negotiated pension plans between workers and employers.[562] It is one of the areas with the most litigation on behalf of older persons. Part of the reason for this might be the pool of litigants for social security. Older persons who have a right to a pension of any sort are relatively well-off, and more likely to have access to the justice system, education, and literacy to be able to claim a right to social security.

The right to social security is outlined in the *American Declaration*, and the *Protocol of San Salvador*.[563] The *American Declaration* embodies the right to social security in Article XVI. It states that every person has the right to social security which will protect from the consequences of unemployment, old age, and any disabilities arising from causes beyond a person's control that make it physically or mentally impossible for the individual to make a living.[564] The *Protocol of San Salvador* recognizes in Article 17 that "[e]veryone has the right to special protection in old age."[565] It calls for states to take the necessary steps to progressively "make this right a reality," particularly by providing food and adequate medical services; undertaking work programs specifically designed for older persons and establishing social organizations designed to improve the quality of life of older persons.[566] Article 9 of *the Protocol of San Salvador* also provides for the right to social security for older persons. It states that everyone shall have the right to social security protecting him or her from the consequences of old age and of disability which prevents a person, physically or mentally, from securing the means for a dignified and decent existence.[567] In the event of the beneficiary's death, social security benefits shall be transfer to living dependents.[568]

Both the IACHR and the IACtHR have analyzed pension cases from the perspective of Article 21 of the *American Convention* that protects the right to property. The IACtHR has defined property very broadly, encompassing "among other

---

[561] See, *inter alia*, Sandra Huenchuan, *Perspectivas globales sobre la protección de los derechos humanos de las personas mayores*, *supra* note 11, at 64–65.

[562] Progress Indicators in Respect of Rights Contemplated in the Protocol of San Salvador, Proposal Put Forward by the Working Group to Examine the National Reports Contemplated in the Protocol of San Salvador Pursuant to the Mandate Contained in Resolution Ag/Res. 2582 (XI-O/10), OEA/Ser.GCP/CAJP-2943/11, March 14, 2011, 68–70, available at http://scm.oas.org/pdfs/2011/CP26772E-2.pdf (last visited Jan. 30, 2015).

[563] Report of the Working Group to Study the Draft Additional Protocol to the American Convention on Human Rights in the Area of Economic, Social and Cultural Rights, Article 9, sec. B, 5, Inter-Am. C.H.R. OEA/Ser. G, CP/CAJP-728/88 corr. 1, 17 Oct. 1988 (including the right to social security to protect against the consequences of old age and to preserve a dignified and decent existence).

[564] Article IX of the *American Declaration*.

[565] Article 17 of the *Protocol of San Salvador*.

[566] *Id.*

[567] Article 9 of the *Protocol of San Salvador*.

[568] *Id.*

aspects, the use and enjoyment of property defined as those material objects which are susceptible of being possessed, as well as any rights which may be part of a person's assets."[569] Moreover, it also includes "vested rights, understood as rights that have been incorporated into the patrimony of the persons."[570]

In *Five Pensioners v. Peru*,[571] the IACtHR acknowledged that the right to social security is protected as a property right and is entitled to judicial protection under the *American Convention*. In that case, domestic law ensured the pensioners in the complaint access to an equalized retirement pension and, as such, the pension was considered a vested right under the Constitution of Peru.[572] Thus, the IACtHR found that the pension rights of the complainants had been incorporated into their patrimony.[573] When the state arbitrarily changed the amounts they received, the Court found that it had affected their right to property under Article 21 of the *American Convention*.[574] Moreover, since Peru did not comply with the domestic rulings affording judicial protection to the victims in the case, it also breached Article 25 of the same instrument.[575] Moreover, in *Acevedo Buendía*, the IACtHR found that when a domestic judicial order was not observed for payment of a pension, there existed "the deterioration of the right to property, enshrined in Article 21 of the [American] Convention, the violation of which should not have occurred if said rulings had been fully and promptly complied with."[576]

In the same vein, the IACHR has concluded that "the proprietary effects of a pension regime to which persons have made contributions or met the respective legal requirements" fall within the scope of the right to property under Article 21 of the *American Convention*.[577] Moreover, when the internal law provides for a specific mechanism to increase the amount of the pension, the pensions and their corresponding equalization becomes part of the property of the alleged victim.[578]

---

[569] I/A Court H.R., *Case of Acevedo Buendía et al. ("Discharged and Retired Employees of the Office of the Comptroller") v. Peru*, Preliminary Objection, Merits, Reparations and Costs, *supra* note 27, 84.

[570] *Id.*

[571] I/A Court H.R., *Case of the "Five Pensioners" v. Peru*, Merits, Reparations and Costs, Judgment of February 28, 2003, Series C No. 98.

[572] *Id.*, 102.

[573] *Id.*

[574] *Id.*, 121.

[575] *Id.*, 141.

[576] I/A Court H.R., *Case of Acevedo Buendía et al. ("Discharged and Retired Employees of the Office of the Comptroller") v. Peru*, Preliminary Objection, Merits, Reparations and Costs, *supra* note 27, 90.

[577] IACHR, *National Association of Ex-Employees of the Peruvian Social Institute et al. v. Peru*, Report No. 38/09, Case 12.670, Admissibility and Merits, Annual Report 2009, 103.

[578] *Id.*

The right to property, however, is not an absolute right and may be consequently subject to permissible restrictions. In the case of pensions, the IACHR has considered that to be permissible a limitation to the right to property it must comply with the following criteria: (i) the restriction must be imposed through a law; (ii) it must respond to the legitimate aim to raise a social interest or to preserve the general well-being in a democratic society; and (iii) the restriction must be proportional, in the sense of being reasonable to obtain this aim and, in any case, of not sacrificing the essence of the right to a pension.[579]

Also, the IACHR has interpreted that the right to pension, as part of the right to social security, falls within the scope of protection afforded by Article 26 of the *American Convention*.[580] This provision refers to the economic, social and cultural rights enshrined in the OAS Charter. Next, the *American Declaration* specifically protects the right to social security.[581] Since the *American Declaration* is considered the instrument that spells out the rights enshrined in the OAS Charter, the right to social security is thus protected by that instrument.[582] Consequently, as one of the social rights protected by the Charter, the IACHR concluded that access to social security is one of the rights for which Article 26 imposes states the duty to implement progressively.[583]

As stated before, the obligation to progressively protect economic, social and cultural rights under Article 26 implies that states cannot adopt regressive measures, unless there are reasons of weight to justify them.[584] In the case of pension rights specifically, the principle of non-regression may be reviewed by the IACHR and the IACtHR in an individual complaint to assess whether the state concerned has adopted a measure incompatible with Article 26 of the *American Convention*.[585] In *National Association of Ex-Employees of the Peruvian Social Security Institute et al.*, for example, the IACHR found that the limitation imposed to the right to a pension by a constitutional reform met the standard of proportionality and did not amount to a deprivation of the right to property for the purposes of Article 21(2) of the *American Convention*.[586] Moreover, the Commission

[579] *Id.*, 112.

[580] *Id.*, 130.

[581] *Id.*, 132.

[582] *Id.*, 131.

[583] *Id.*, 133.

[584] I/A Court H.R., *Case of Acevedo Buendía et al. ("Discharged and Retired Employees of the Comptroller") v. Peru*, Preliminary Objection, Merits, Reparations and Costs, *supra* note 27, 103.

[585] IACHR, *National Association of Ex-Employees of the Peruvian Social Institute et al. v. Peru*, *supra* note 572, 139; I/A Court H.R., *Case of Acevedo Buendía et al. ("Discharged and Retired Employees of the Comptroller") v. Peru*, Preliminary Objection, Merits, Reparations and Costs, *supra* note 27, 102.

[586] IACHR, *National Association of Ex-Employees of the Peruvian Social Institute et al. v. Peru*, *supra* note 572, 126–127.

concluded that the restriction did not constitute a regression incompatible with Article 26 of the *American Convention* because it was of sufficient weight to justify the state's need to adopt the measure in the context of the case under consideration.[587] In the same vein, the IACtHR followed a similar interpretation regarding the principle of non-regression in Article 26 of the *American Convention* in the case of *Acevedo Buendia et al. v Peru*.[588] The IACtHR held that compliance of the principle of non-regression of economic, social and cultural rights, including social security, protected in the above-mentioned provision, could be reviewed under its contentious jurisdiction.[589] However, at last the IACtHR did not assess in the particular case whether or not the restrictions adopted breached the principle of non-regression under Article 26.

On the other hand, the IACHR has heard a number of admissibility cases on social security. Petitioners have argued that denial of old-age pensions violates the rights to preservation of health, well-being, and social security protected by Articles XI and XVI of the *American Declaration*, the right to social security under Article 9 of the *Protocol of San Salvador* and the rights to property, equal protection, access to justice and to an effective remedy ensured by Articles 21, 24, 8 and 15 of the *American Convention*.

In *Saccone v. Argentina*, the petitioners contributed to a compensatory retirement plan which provided that, in case the establishment of the fund failed, contributions made would return to the workers.[590] Several years later, the company's board decided to liquidate the fund and discontinued the benefits that had been given to retirees since its inception.[591] The claim was rejected by domestic courts on the basis that it was filed after the applicable statute of limitations had expired.[592] The petitioners brought a complaint to the IACHR, arguing violations of the rights to property and equal protection ensured by Articles 21 and 24 of the *American Convention*; their right to the preservation of health and well-being; and social security, as protected by Articles XI and XVI of the *American Declaration*.[593] The IACHR found that the contributions made by the petitioners to the fund were part of their patrimony, in proportion to the amount contributed, and fell within the protection of Article 21 of the *American Convention*.[594] Ultimately, however, it found the petition inadmissible, stating that it had no competence to

---

[587] *Id.*, 141–147.

[588] I/A Court H.R., *Case of Acevedo Buendía et al. ("Discharged and Retired Employees of the Comptroller") v. Peru*, Preliminary Objection, Merits, Reparations and Costs, *supra* note 27.

[589] *Id.*, 106.

[590] IACHR, *Saccone v. Argentina*, Report No. 8/98, Case 11.671, Admissibility, Annual Report 1998, 3.

[591] *Id.*, 5.

[592] *Id.*, 6–8.

[593] *Id.*, 1, 12.

[594] *Id.*, 25–26.

determine whether the Argentine courts erred as to the interpretation and application of domestic civil law, when applying the statute of limitation.[595]

In *Menéndez v. Argentina,* the petitioners in this case were retirees who filed suit demanding an adjustment to the amount of their pensions.[596] In 1995, a reorganization of judicial proceedings around the social security system in Argentina, provided that the state comply with these judgments until the state had exhausted all budgetary resources allocated to that end in the corresponding fiscal year.[597] Once the resources were exhausted, compliance with all other awards would be suspended until new budgetary funds were allocated at the beginning of the subsequent fiscal year.[598] Since appropriate compliance with awards was conditioned on the existence of resources, the petitioners argued a violation of the right to an effective remedy.[599] They maintained that their right to property had been affected since there had been unwarranted delays in enforcing judgments recognizing this right.[600] They also argued that the retirement system violated the right to equal protection of the law since certain groups of retirees received "privileged pensions."[601] Consequently, the petitioners argued that the application of a domestic law reorganizing the social security system in Argentina violated Articles 8, 25 and 24 of the *American Convention.*[602] Moreover, the petitioners asserted a violation of Articles XI and XVI of the *American Declaration*, which protect the right to preservation of health and well-being and social security.[603] The IACHR declared the complaints admissible.[604] First, it reiterated its prior position stating that after a state ratifies the *American Convention*, the IACHR must apply this instrument with respect to rights protected by both the *American Declaration* and the *American Convention.*[605] Consequently, it found admissible the alleged breaches to Articles 8, 25, 21 and 24 of that instrument.[606] Interestingly, with respect to the rights that are only protected by the *American Declaration*, namely the rights to health and well-being and the right to social security, the IACHR found that they were also admissible on the basis of Article 29(d) of the *American Convention* which provides that "[n]o provision of this Convention shall be interpreted… as excluding or limiting the effect that the American Declaration of the

[595] *Id.*

[596] IACHR, *Amilcar Menéndez, Juan Manuel Caride et al. (Social Security System) v. Argentina, supra* note 458, 3.

[597] *Id.*, 14.

[598] *Id.*, 17.

[599] *Id.*, 24–28.

[600] *Id.*, 31.

[601] *Id.*, 30–31.

[602] *Id.*, 1.

[603] *Id.*

[604] *Id.*, 40.

[605] *Id.*, 41.

[606] *Id.*

Rights and Duties of Man and other international acts of the same nature may have."[607] Thus, the IACHR concluded that it will review the petitioners' allegations on violations of the *American Declaration*.[608] As stated before, this case was resolved through a friendly settlement agreement reached between the parties and confirmed by the IACHR, so none of the substantive issues were address in a report.[609]

In *Agustina Alonso et al. v. Argentina*, petitioners alleged that the delay in complying with final judgments regarding the adjustments of their pensions breached their right to physical and emotional integrity under Article 5 of the *American Convention* because it affected older persons whose only form of income was the pension at stake.[610] Also, they alleged that the delay affected their right to property and their right to an effective remedy protected by Articles 21 and 25 respectively.[611] The IACHR concluded that the claims regarding the alleged violations of Articles 21 and 25, together with Article 8 of the *American Convention*, were all admissible given that the delay in the compliance with the judgments issued by domestic courts *prima facie* affected the rights to property, effective remedy and fair trial.[612] The IACHR, however, rejected the claim of breach of Article 5, because it concluded that there were not sufficient elements to substantiate the complaint.[613] The case is currently in the merit proceedings without further information being provided.

The draft *Convention on Older Persons* includes the right to social security as a way to ensure older persons access to a dignified and decent life.[614] Social security benefits should include retirement benefits as well as other complementary social protection in the case of disability, widowhood, loss of old-age pension and the lack of other source of income.[615] The draft *Convention on Older Persons* envisions the right to social security as of progressive implementation and within the limits of available resources.[616] In the same vein, the draft instrument encourages states to promote policies geared toward the equalization of pensions, without discrimination, in accordance with the existing social, cultural and economic circumstances of the particular state.[617] Finally, it provides that states should through conventions or bilateral agreements seek to facilitate the recognition of benefits,

[607] *Id*., 42.

[608] *Id.*

[609] IACHR, *Amilcar Menendez, Juan Manuel Caride et al. v. Argentina*, *supra* note 458.

[610] IACHR, *Agustina Alonso et al. v. Argentina*, *supra* note 467, 23.

[611] *Id.*, 24–25.

[612] *Id.*, 65.

[613] *Id.*, 67.

[614] *Draft Inter-American Convention on the Human Rights of Older Persons*, *supra* note 39, Article 17.

[615] *Id.*

[616] *Id.*

[617] *Id.*

pension contributions and entitlements to older persons who are immigrants.[618] In spite of the centrality of the right to social security for the protection of older persons, the draft *Convention on Older Persons* does not seem to expand on the entitlements that older persons should enjoy to ensure a dignified life in old age. The reason for that could be that the drafting of this provision has proven to be controversial and long debated by the states participating in the negotiations.[619]

The *San José Charter* provides a section regarding access to social security for older persons.[620] First, it promotes the universalization of the right to social security of older persons, taking into account the particular situation of each country.[621] Next, the *San José Charter* encourages states to ensure the sustainability of the social security systems, including contributory and non-contributory pensions.[622] Finally, it underlines the need to complement social security benefits with other social services, including the provision of health care and housing.[623]

5. The Right to Health

The right to health is not the right to be healthy. Many older persons are in perfectly good health; others will experience health problems as age increases. The *American Declaration* protects the preservation of health and well-being for all, including measures relating to food, clothing, housing and medical care, to the extent permitted by public and community resources.[624] Older persons are often at higher-risk for developing both health problems and poverty. The two processes of deteriorating health and financial security are mutually aggravating. One result of attainment of the right to health is that it disjoins basic health rights from poverty through equal access to some basic forms of health care. This, in turn, fosters better individual health and economic stability.

The *Protocol of San Salvador* also addresses the right to health.[625] It ensures that everyone shall have the right to the highest attainable standard of physical,

---

[618] *Id.*

[619] For example, the preliminary draft submitted to discussion included a provision stating that "State parties should ensure that, within available resources, all older persons receive a permanent minimum income sufficient for a dignified life." See, *Preliminary Draft Inter-American Convention on the Human Rights of Older Persons*, *supra* note 555, Article 15(c). This paragraph was eliminated from the current draft.

[620] *San José Charter on the Rights of Older Persons in Latin American and the Caribbean*, *supra* note 92, at 8.

[621] *Id.*

[622] *Id.*

[623] *Id.*, at 8–9.

[624] Article XI of the *American Declaration*.

[625] Article 10 of the *Protocol of San Salvador*.

mental and social well-being.[626] In order to ensure the exercise of the right to health, the state parties agree to recognize health as a public good and, particularly, to adopt the following measures to ensure that right: primary health care, that is, essential health care made available to all individuals and families in the community; extension of the benefits of health services to all individuals subject to the state's jurisdiction; universal immunization against the principal infectious diseases; prevention and treatment of endemic, occupational and other diseases; education of the population on the prevention and treatment of health problems, and satisfaction of the health needs of the highest risk groups, including older persons, and of those whose poverty makes them the most vulnerable.[627] Compliance with the right to health is among the first group of rights that will be monitored by the OAS Working Group to Examine the National Reports contemplated in the *Protocol of San Salvador*.[628]

According to this Working Group, Article 10 enshrines obligations of immediate application, such as the access to primary care for all, and those that are of progressive implementation requiring the development of a system that provides coverage to the entire population.[629] Assessment of compliance with this right will take into account the availability and accessibility of the health care services.[630] Within the accessibility issue, it will consider: the principle of non-discrimination; physical accessibility; economic accessibility; access to information; acceptability; and quality.[631] Furthermore, monitoring of the right to health will be carried out in light of crosscuting indicators, particularly the principle of equality and non-discrimination, to ensure that vulnerable groups, including older persons, receive the appropriate differentiated attention and resources when the state adopts policies to implement health care access.[632] This approach is supported by the duty to affor special health care services to older persons protected by Article 17 of the *Protocol of San Salvador*.[633]

Like the right to social security, the IACHR has found that the right to health falls within the scope of Article 26 of the *American Convention*.[634] As such, states

[626] Article 10(1) of the *Protocol of San Salvador*.

[627] Article 10(2) of the *Protocol of San Salvador*.

[628] Progress Indicators in Respect of Rights Contemplated in the Protocol of San Salvador, Proposal Put Forward by the Working Group to Examine the National Reports Contemplated in the Protocol of San Salvador, *supra* note 562, 8.

[629] *Id.*, 72.

[630] *Id.*, 73.

[631] *Id.*

[632] *Id.*, 9.

[633] In this respect, Article 17 indicates that States will progressively provide specialized medical care for older persons who lack them and are unable to provide them for themselves.

[634] IACHR, *Jorge Odir Miranda Cortez et al. v. El Salvador*, *supra* note 487, 79.

must adopt progressive measures to implement this right and are prevented from adopting regressive policies that curtail access, unless that these measures are justified by strong reasons.[635] In *Jorge Odir Miranda Cortez et al.*, however, the IACHR went even further an assessed whether the state action directly breach the right to health of the victims who were HIV positive.[636] After assessing the circumstances of the case, it concluded that the measures adopted by El Salvador to provide medical treatment to the victims in this case were sufficiently expeditious to ensure their right to health under Article 26.[637] Also, the IACHR concluded that there was no breached of the principle of non-regression in this case.[638]

In addition, the IACHR and the IACtHR have consistently protected situations in which the health of persons was deteriorating as a result of state action or omission under Articles 4 and 5 of the *American Convention*, protecting the right to life and humane treatment respectively.[639] In this respect, both organs of the Inter-American System have issued interim measures ordering states to ensure that the failing condition of the victims did not worsen to the point of affecting their life or integrity.[640] Even if these cases referred mostly to imprisoned individuals, they are of relevance for future situations of older persons placed in long-term care or similar institutions.

The right to health is also included in the draft *Convention on Older Persons*. Older persons have the right to enjoyment of the highest attainable standard of physical, mental, and sexual health without discrimination based on age or any other condition.[641] For its fulfillment, states must establish comprehensive policies to health promotion, prevention of illness, comprehensive care and rehabilitation, and palliative care for the terminally ill.[642] They also have to assure universal, equitable access to timely, quality comprehensive primary health services; ensure access to essential medicines; as well as the creation and implementation of integrated health and social services, particularly for older persons with chronic degenerative diseases.[643] The draft instrument defines integrated health and social care services as the "institutional benefits and entitlements to address the health care and social needs of older persons with a view to guaranteeing their dignity and well-being and promoting their

[635] *Id.*, 1105–106.

[636] *Id.*, 108.

[637] *Id.*

[638] *Id.*

[639] See, *inter alia*, IACHR, *Matter Nestora Salgado García concerning México*, Resolution 2/2015, Precautionary Measure No 455/13, January 28, 2015; *Matter of Julio César Cano Molina concerning Cuba*, Resolution No. 24/2014, Precautionary Measure 307/14, September 10, 2014; I/A Court H. R., *Matter of B. regarding El Salvador*, Order of the Inter-American Court of Human Rights of May 29, 2013.

[640] IACHR, *Matter Nestora Salgado García concerning México*, *supra*, 21; I/A Court H.R., Matter of B. regarding El Salvador, *supra*, 17.

[641] *Draft Inter-American Convention on the Human Rights of Older Persons*, *supra* note 39, Article 19.

[642] *Id.*

[643] *Id.*

independence and autonomy."[644] Additionally, states undertake to adopt the necessary legal frameworks and supervision mechanisms to ensure the right to health to older persons living in long-term care facilities.[645] Furthermore, states must develop and implement a plan to foster active ageing.[646]

The *San José Charter* includes a section regarding the right to health of older persons and promotes the universal access of older persons to health care services.[647] Furthermore, it encourages states to adopt and implement a policy of preventive health regarding primary health care services and rehabilitation.[648] States should also facilitate access of older persons to medicines and other equipment necessary for rehabilitation, depending on the economic situation of each country.[649] Additionally, states must regulate access to health and implement mechanisms of supervision to monitor respect of the right to integrity of older persons living in long-term care institutions.[650]

On the other hand, the Pan-American Health Organization (PAHO) has been an active participant in the discussions regarding the right to health of older persons. PAHO has published materials and also presented reports on the right to health for older persons before the OAS Working Group on Protection of the Human Rights of Older Persons.[651] Likewise, PAHO's *Regional Strategy and Plan of Action on Nutrition in Health and Development, 2006–2015*,[652] adopted by the OAS General Assembly,[653] recommends actions designed to improve nutritional status as people age, throughout life, especially among the poor and other vulnerable groups as a way to ensure their right to health.[654] Moreover, in 2009, PAHO adopted a *Plan of Action on the Health of Older Persons, Including Active and Healthy Aging*, a strategy to facilitate access of older persons to comprehensive health services that meet their needs.[655] The Plan of Action is comprised of four strategic areas addressing the following issues: (1) adaptation of state policy to international

---

644 *Id.*

645 *Id.*

646 *Id.*

647 *San José Charter on the Rights of Older Persons in Latin American and the Caribbean*, *supra* note 92, at 10.

648 *Id.*

649 *Id.*

650 *Id.*

651 See, *inter alia*, Presentación de Javier Vázquez, Asesor Regional en Derechos Humanos, Organización Panamericana de la Salud (OPS), during the meeting of the Working Group on Protection of the Human Rights of Older Persons held on November 4, 2011, OEA.Ser.G/CAJP/GT/DHPM/INF-8/12, April 5, 2012.

652 PAHO, *Regional Strategy and Plan of Action on Nutrition in Health and Development*, 2006–2015, CD47/18, August 21, 2006.

653 OAS, AG/RES. 2332 (XXXVII-O/07), *The Importance of Nutrition for Achieving Healthier Populations and Integral Development*, adopted on June 5, 2007.

654 PAHO, *Regional Strategy and Plan of Action on Nutrition in Health and Development*, 2006–2015, *supra* note 652, at 9.

655 PAHO, *Plan of Action on the Health of Older Persons, Including Active and Healthy Aging*, CD49/8, July 10, 2009.

human rights standards on the right to health[656]; (2) develop legal frameworks and supervision mechanisms to protect the health of older persons in long-term care services[657]; (3) training of the personnel needed to meet the health needs of older persons[658] and (4) strengthen the mechanisms to produce and gather the necessary information on the basis of which public health policy of older persons will be designed.[659] The strategy and objectives enshrined in this plan of action may serve as a roadmap to guide the implementation of the right of health of older persons under existing international standards applicable to the OAS state members or to interpret the *Convention on Older Persons* if eventually adopted.

F. Rights of Older Persons as Members of Other Vulnerable Groups and Multiple Discrimination

The IACHR has acknowledged that structural vulnerability is many times compounded by other forms of discrimination associated with being identified as a member of several vulnerable groups.[660] With respect to women migrants, for example, the IACHR found that they "suffer an intersection of forms of discrimination combined with their sex and their condition as migrants, such as their age, nationality, educational and economic level, among others; dimensions that should be examined by States in the design of interventions with the goal of better protecting their human rights in the realm of justice."[661] Aware of this intersection of forms of discrimination, most recently the *Discrimination and Intolerance Convention* and the *Racial Discrimination Convention* include definitions of "multiple discrimination" as unjustified distinctions, exclusions or restrictions that are based simultaneously on two or more prohibited grounds.[662]

---

[656] *Id.*, Annex A, at 1.

[657] *Id.*, at 3.

[658] *Id.*, at 6.

[659] *Id.*, at 7.

[660] IACHR, *Human Rights of Migrants and Other Persons in the Context of Human Mobility in Mexico*, *supra* note 203, 83.

[661] *Id.*

[662] *Inter-American Convention Against All Forms of Discrimination and Intolerance*, *supra* note 32, Article 1 of the *Discrimination and Intolerance Convention* provides that "multiple or aggravated discrimination is any preference, distinction, exclusion, or restriction based simultaneously on two or more of the criteria set forth in Article 1.1, or others recognized in international instruments, the objective or result of which is to nullify or curtail, the equal recognition, enjoyment, or exercise of one or more human rights and fundamental freedoms enshrined in the international instruments applicable to the States Parties, in any area of public or private life." *Inter-American Convention Against Racism, Racial Discrimination, and Related Forms of Intolerance*, *supra* note 33. Article 1 of the *Racial Discrimination Convention* provides that "multiple or aggravated discrimination is any preference, distinction, exclusion, or restriction based simultaneously on two or more of the criteria set forth in Article 1.1, or others recognized in international instruments, the objective or result of which is to nullify or curtail, the equal recognition, enjoyment, or exercise of one or more human rights and fundamental freedoms enshrined in the international instruments applicable to the States Parties, in any area of public or private life."

Older persons may also face compounded discrimination resulting from the intersection of age and other grounds for distinction including, inter alia, gender, sex, or disability. To address this issue, the draft *Convention on Older Persons* defines multiple discrimination as "any distinction, exclusion, or restriction toward an older person, based on two or more discrimination factors."[663] The Preliminary Draft Inter-American Convention on Protection of the Human Rights of Older Persons presented for the initial discussions included separate provisions affording protection to older women, older persons with disabilities, older persons who belong to an ethnic, racial, national, linguistic, religious or rural groups, older persons of different sexual orientation and gender identities.[664] These provisions were eventually eliminated and the protection against multiple discrimination was softened and moved to the provision protecting the right to equality and non-discrimination. The current draft instrument provides that "[i]n their policies, plans and legislation of ageing and old age, States shall develop specific approaches for older persons who are vulnerable and those who are victims of multiple discrimination, including women, persons with disabilities, persons of different sexual orientations and gender identities, migrants, persons living in poverty or social exclusion, Afro-descendants, persons pertaining to indigenous peoples, the homeless, persons deprived of liberty, persons pertaining to traditional peoples, persons who belong to ethnic, national, linguistic, and religious and rural groups, among others."[665]

1. The Rights of Older Women

Older women face the cumulative effects of multiple discriminations across the life course. The *Belem Do Para Convention* explicitly calls upon states to take special account of the vulnerability of older women to violence.[666] ECLAC has noted that information on violence against older women and young girls is often excluded from data collection, and instead tends to focus on women of childbearing age.[667]

---

[663] *Draft Inter-American Convention on the Human Rights of Older Persons*, *supra* note 39, Article 2.

[664] *Preliminary Draft Inter-American Convention on the Human Rights of Older Persons*, *supra* note 555, Articles 27–30.

[665] *Draft Inter-American Convention on the Human Rights of Older Persons*, *supra* note 39, Article 5.

[666] *Inter-American Convention for the Prevention, Punishment and Eradication of the Violence against Women*, *supra* note 30, Article 9.

[667] ECLAC, *Violence against women in couples: Latin America and the Caribbean. A proposal for measuring its incidence and trends*, 2004. at 26, available at http://www.eclac.cl/publicaciones/xml/5/22695/lcl1744i.pdf (last visited Jan. 31, 2015).

Older women face unequal access to many social pension schemes.[668] Throughout their lives, older women are more likely to take on informal work than men, such as childrearing, household work or day labor. Article 9 of the *Protocol of San Salvador* provides for the right to social security for dependents in the event of the death of a beneficiary. However, contributory pensions disproportionately exclude older women because of these contributing discriminatory factors.[669] In developing countries, older women with surviving husbands rarely qualify for pensions in their own names.[670] Brazil, for example, has made great strides with the introduction of non-contributory social security for women, successfully closing the gender gap for poverty in old age.[671] Access to social protection programs is gaining increasing ground in the international agenda.

Protection of older women is at the core of the draft *Convention on Older Persons*. The Preamble underscores the need to incorporate a gender perspective in all policies and programs implemented to protect the right of older persons, to ensure the full enjoyment of those rights without discrimination.[672] Also, among the general principles of the treaty, the draft lists the principle of gender equity and equality and provides that it will apply to this instrument and guide its interpretation.[673] Moreover, the draft *Convention on Older Persons* enshrines the right of older persons to safety and a life free of violence and without discrimination on the basis of, inter alia, gender.[674] In this regard, states must "actively promote the elimination of such rituals and all related practices as are detrimental to the dignity and integrity of older women."[675] Furthermore, when developing legal frameworks to regulate the operation of long-term care institutions, states must also take into account a gender perspective.[676] With respect to social security, the draft instrument ensures that in case of death of the beneficiary, the benefits will be transferred to the dependents.[677] Also, states will include retirement benefits and other social services in the case of widowhood or inexistence of any other income.[678]

---

[668]*Aging and Financial Inclusion: An Opportunity, Helpage and the Center for Financial Inclusion at Accion*, February 2015, at 5.

[669]*See* Ministry of Labour, Employment and Youth Development in collaboration with HelpAge International (2010). *Achieving Income Security in Old Age For All Tanzanians: a study into the feasibility of a universal social pension.* Federal Ministry for Economic Cooperation and Development, at 22.

[670]Lloyd-Sherlock, *Population Ageing and International Development,* p. 79, The Policy Press, New York (2010) (noting that Kenya's private pension assets are equal to about 18 % of the country's GDP).

[671]*Id.*

[672]*Draft Inter-American Convention on the Human Rights of Older Persons*, supra note 39, Preamble.

[673]*Id.*, Article 3.

[674]*Id.*, Article 9.

[675]*Id.*

[676]*Id.*, Article 12.

[677]*Id.*, Article 17.

[678]*Id.*

Additionally, states will promote policies geared toward the equalization of pension benefits which will be applied equally to older men and women.[679] In relation to health, the draft *Convention on Older Persons* ensures the right of older women to access sexual and reproductive health without discrimination.[680] As to education, states undertake to reduce illiteracy among older persons, especially women and other vulnerable groups.[681] Finally, the draft *Convention on Older Persons* acknowledges the importance of eliminating all administrative or financial practices that discriminate older persons, in particular older women, in the exercise of their right to property.[682]

The *San José Charter* also envisions the notion of multiple discrimination and underscores the need to eradicate gender discrimination.[683] Towards that goal, it includes specific protections for older persons. In particular, the *San José Charter* underlines the need to prevent, punish and eradicate all forms of violence against older women.[684] States should promote the role that older persons, especially older women, play in the political, social, economic and cultural development of their communities.[685] Also, states should ensure equal access to older women to social security and other forms of social benefits. Finally, states should protect the inheritance rights and property rights of older women, including those who are widows.[686]

2. The Rights of Older Persons with Disabilities

Although not all older persons are disabled, many older persons experience temporary or chronic disability during their lifetimes. The *Protocol of San Salvador* expresses that "everyone affected by a diminution of his [or her], physical or mental capacities is entitled to receive special attention designed to help him achieve the greatest possible development of his [or her] personality."[687] Under Article 18 of that instrument, states agreed to undertake programs, including work programs, consistent with the possibilities of persons with disabilities, freely accepted by them or by their legal representatives; include solutions to address the needs of persons with disabilities as a priority component of their urban development plans;

---

679 *Id.*

680 *Id.*, Article 19.

681 *Id.*, Article 20.

682 *Id.*, Article 22.

683 *San José Charter on the Rights of Older Persons in Latin American and the Caribbean*, *supra* note 92, at 15.

684 *Id.*

685 *Id.*, at 16.

686 *Id.*

687 Article 18 of the *Protocol of San Salvador*.

and encourage the establishment of groups in which persons with disabilities can be supported to participate in society.[688]

The *Disabilities Convention* seeks the prevention and elimination of all forms of discrimination against persons with disabilities and to promote their full integration into society.[689] The *Andean Charter* also includes a provision for the protection of persons with disabilities that addresses protection against all discrimination and violence against persons with disabilities; prevention of causes originating disabilities, through education, safety in the workplace and public information; realization of the right to social security and health of persons with disabilities; programs that address the particular needs of persons with disabilities; and the social integration of persons with disabilities and the opportunity of full participation in their respective communities.[690]

The *Discrimination and Intolerance Convention* protects the right to equal protection against any discrimination and intolerance in public or private life, *inter alia*, on the basis of disability.[691] Under this Convention, states undertake to prevent, eliminate, prohibit and punish all acts of discrimination and intolerance based on disability.[692] Moreover, states have a positive obligation to ensure equal access and opportunity, inclusion and progress to persons with disability.[693] Also, states undertake to make an aggravating factor those acts that lead to multiple discrimination.[694] The obligations arising out of this treaty complement the existing *Disabilities Convention* as well as the OAS policies and the case law of the Inter-American human rights organs on this issue, described below.

The Program of Action for the Decade of the Americas for the Rights and Dignity of Persons with Disabilities (2006–2016) seeks progressive implementation of protections and mechanisms that protect and respect the human rights of persons with disabilities, with each country exhibiting improvement by 2016.[695] The Program's objectives include promoting social awareness of the rights of persons with disabilities; improve their access to health, education, employment, political participation, and participation in cultural, artistic, sports, and recreational activities.[696] Also, under this Program, OAS member states undertake to ensure access to social welfare programs and anti-poverty strategies with a higher

---

[688]*Id.*

[689]*Inter-American Convention on the Elimination of all Forms of Discrimination against Persons with Disabilities*, *supra* note 31, Article II.

[690]*Andean Charter for the Promotion and Protection of Human Rights*, *supra* note 81, Article 49.

[691]*Inter-American Convention against All Forms of Discrimination and Intolerance*, *supra* note 32, Articles 1 and 2.

[692]*Id.*, Article 4.

[693]*Id.*, Articles 5 and 6.

[694]*Id.* Article 11.

[695]OAS, AG/RES. 2339 (XXXVII-O/07) *Program of Action for the Decade of the Americas for the Rights and Dignity of Persons with Disabilities (2006–2016)*, Mission Statement.

[696]*Id.*, Section Objectives.

prioritization among persons with disabilities, especially women, children, and older persons. This recognizes the detrimental effect of multiple discriminations over the life course.[697]

Also, the IACtHR has consistently established that persons with disabilities are entitled to special protection, based on the special duties that states must observe to satisfy the general obligation to respect and ensure human rights under Articles 1(1) and 2 of the *American Convention*.[698] This special protection requires states to adopt positive measures to ensure that persons with disabilities are able to participate in society on an equal footing.[699] In particular, states must promote inclusive social policies and adopt affirmative measures to overcome barriers faced by persons with disabilities, including, *inter alia*, physical or architectural, communicational, attitudinal, or socioeconomic ones.[700] Moreover, because persons with disabilities are often the target of discrimination, states must adopt all the necessary legislative, social, educational, labor or other required measures to ensure that unjustified distinctions against this group is eradicated and ensure their full integration into society.[701] Additionally, in the *Furlan* case, the IACtHR found that disability may intersect with other grounds requiring special protection, in this case childhood, imposing upon the state a reinforced duty to respect and ensure human rights.[702]

Even if the protection of older persons with disabilities appears to be an important aspect of the protection of older persons, particularly those affected by mental illnesses, the draft *Convention on Older Persons* does not include a separate provision. Older people acquire disabilities in old age, and people with disabilities may grow old, however, in the current draft there are not many references to the issue, probably due to the fact that there is a *Disabilities Convention* at the regional level[703] or because overall the OAS member states have ratified the CRPD.[704] Notwithstanding, Article 5 on equality and non-discrimination for reasons of age envisions that states should develop special approaches to address the needs of older persons with disabilities.[705] Also, when dealing with the right to social security, the draft instrument provides that states shall work progressively to ensure access to

[697] *Id.*, Section on Specific Measures, Welfare and Social Assistance (8b).

[698] See, *inter alia*, I/A Court H.R., *Case of Furlan and Family v. Argentina*, *supra* note 204, 134.

[699] *Id.*

[700] *Id.*, 133–134.

[701] *Id.*, 135.

[702] *Id.*, 136.

[703] *Inter-American Convention on the Elimination of all Forms of Discrimination against Persons with Disabilities*, *supra* note 31.

[704] *Convention on the Rights of Persons with Disabilities*, *supra* note 148. For information on OAS member states that have ratified the Convention, please see http://indicators.ohchr.org/ (last visited Ja. 30, 2015).

[705] *Draft Inter-American Convention on the Human Rights of Older Persons*, *supra* note 39, Article 5.

social benefits in the case of disability.[706] Under the protection of the right to health, states undertake to develop integrated specialized social and health services for the care of older persons with degenerative diseases such as dementia or Alzheimer's disease.[707] Also, some of the protections afforded under Article 25 regarding the right to accessibility and personal mobility may be directly applicable to provide protection to older persons with disabilities.[708] In particular, this provision states that building and other facilities open to the public should provide signage in Braille, presumably to protect those older persons with visual impairments.[709]

The *San José Charter* also makes scarce references to the protection of older persons with disabilities. In particular, it provides that states should ensure preferential and differential treatment of persons with degenerative diseases in public or private, daily or long-term care institutions.[710] Also, in regard to the prevention and protection against violence, the *San José Charter* specifies that special protection should be afforded to older persons with disabilities given that, due to their increased vulnerability, they are more prone to be subjected to mistreatment.[711]

3. Older Persons and Sexual Orientation

Protection of persons of different sexual orientation or gender identities is an issue under development in the Inter-American System. The *Discrimination and Intolerance Convention* protects the right to equal protection against any discrimination and intolerance in public or private life, on the basis *inter alia*, of sexual orientation, gender identity and expression.[712] Under this Convention, states undertake to prevent, eliminate, prohibit and punish all acts of discrimination and intolerance based on these prohibited grounds.[713] Moreover, states have a positive obligation to ensure equal access and opportunity, inclusion and progress to persons of different sexual orientation or gender identity.[714] Furthermore, states must make an aggravating factor those acts that lead to multiple discrimination.[715] This treaty is the first Inter-American instrument that explicitly protects against discrimination on the basis of sexual orientation and gender identity and, in that regard, constitutes a

[706] *Id.*, Article 17.

[707] *Id.*, Article 19.

[708] *Id.*, Article 25.

[709] *Id.*

[710] *San José Charter on the Rights of Older Persons in Latin American and the Caribbean*, *supra* note 92, at 12.

[711] *Id.*, at 14.

[712] *Inter-American Convention against All Forms of Discrimination and Intolerance*, *supra* note 32, Articles 1 and 2.

[713] *Id.*, Article 4.

[714] *Id.*, Articles 5 and 6.

[715] *Id.*, Article 11.

stepping stone to increase protection to this group. Moreover, the express acknowledgement of the notion of multiple discrimination and the articulation of this intersection between prohibited grounds as an aggravating factor will be essential to protect individuals pertaining to two or more affected groups.

In addition to the treaty, the IACtHR in the *Atala Riffo* case found that discrimination on sexual orientation is prohibited under Articles 1(1) and 24 of the *American Convention*.[716] Moreover, it concluded that sexual orientation is a suspect ground for purposes of requiring States to meet the highest level of scrutiny when making distinctions for that reason.[717]

Also, discrimination on the basis of sexual orientation and gender identity has been consistently condemned by the OAS member states and states have been urged to prevent, punish, and eradicate such discrimination.[718] OAS member states have also rejected violence and human rights violations committed against individuals on the basis of their sexual orientation and gender identity and have underscored that perpetrators of such violence must be brought to justice.[719] In February, 2014, the IACHR created a Rapporteurship on the Rights of Lesbian, Gay, Bisexual, Trans and Intersex Persons within its Executive Secretariat, in order to monitor the human rights situation of LGBTI persons and promoting the visibility of violations against their human rights; develop standards of interpretation regarding the Inter-American human treaties on issues of sexual orientation and gender identity; and provide technical assistance to states on these issues.[720]

With respect to the protection of older persons and sexual orientation and gender identity, the Preliminary Draft Inter-American Convention on the Human Rights of Older Persons included a provision on this issue.[721] This provision was later eliminated and the current draft *Convention on Older* is silent on the issue. Only Article 5 on equality and non-discrimination for reasons of age envisions that states should develop special approaches to address the needs of older persons of different sexual orientation or gender identity.[722]

---

[716] I/A Court H.R., *Case of Atala Riffo and daughters v. Chile*, Merits, Reparations and Costs, *supra* note 163, 93.

[717] *Id.*, 91.

[718] OAS, AG/RES. 2653 (XLI-O/11), *Human Rights, Sexual Orientation, and Gender Identity*, adopted on June 7, 2011, 1; AG/RES. 2721 (XLII-O/12), *Human Rights, Sexual Orientation, and Gender Identity*, adopted on June 4, 2012, 1; AG/RES. 2807 (XLIII-O/13), corr. 1, *Human Rights, Sexual Orientation, and Gender Identity and Expression*, adopted on June 6, 2013, 1; AG/RES. 2863 (XLIV-O/14), *Human Rights, Sexual Orientation, and Gender Identity and Expression*, adopted on June 5, 2014, 1. It is important to note that since the term gender expression was added, many states from the Caribbean have not joined the resolution due to objections to this notion.

[719] OAS AG/RES. 2600 (XL-O/10), *Human Rights, Sexual Orientation, and Gender Identity*, adopted on June 8, 2010, 1.

[720] See the website of the Rapporteurship at http://www.oas.org/en/iachr/lgtbi/default.asp (last visited Jan. 30, 2015).

[721] *Preliminary Draft Inter-American Convention on the Human Rights of Older Persons*, *supra* note 555, Article 30.

[722] *Draft Inter-American Convention on the Human Rights of Older Persons*, *supra* note 39, Article 5.

Also, the *San José Charter* makes limited references to the protection of older persons with many other intersecting discriminations. But, when dealing with the issue of prevention and protection against violence, it provides that special protection should be afforded to older persons of different sexual orientation or gender identity given that these individuals may be exposed to an increase level of violence or mistreatment.[723]

4. The Rights of Older Refugees and Older Persons in Emergencies

A number of OAS resolutions have addressed the situation of older persons in emergencies, and in conflict situations in particular.[724] The Protection of Refugees, Returnees, and Internally Displaced Persons in the Americas is addressed in the obligations member states have undertaken in the UN 1951 Convention relating to the Status of Refugees[725] and its 1967 Protocol.[726] The OAS General Assembly has looked to the United Nations guidance on Internally Displaced Persons, which reflects certain aspects of international human rights law and international humanitarian law.[727] This guidance supports the specific needs of children, women, older persons, farm workers, and persons with disabilities.[728] The OAS General Assembly has also requested member states to consider incorporating UN guidance into their national laws in order to promote their implementation and transparency in policies for the protection of internally displaced persons.[729] Clearer guidelines for the determination of refugee status and the treatment of asylum seekers are also being urged.[730] In this regard, States have been

---

[723] *San José Charter on the Rights of Older Persons in Latin American and the Caribbean*, *supra* note 92, 14.

[724] See, *inter alia*, OAS, AG/RES. 2578 (XL-O/10*), Internally Displaced Persons*, adopted on June 8, 2010; AG/RES. 2667 (XLI-O/11), *Internally Displaced Persons*, adopted on June 7, 2011; AG/RES. 2716 (XLII-O-12), *Internally Displaced Persons*, adopted on June 4, 2012; AG/RES. 2850 (XLIV-O/14), *Internally Displaced Persons*, adopted on June 4, 2014.

[725] The *1951 Convention relating to the Status of Refugees*, 189 U.N.T.S. 137, *entered into force* 22 April 1954.

[726] The *1967 Protocol relating to the Status of Refugees*, 606 U.N.T.S. 267, *entered into force* 4 October 1967.

[727] See, *inter alia*, OAS, AG/RES. 2578 (XL-O/10*), Internally Displaced Persons, supra* note 724, 2; AG/RES. 2716 (XLII-O-12), *Internally Displaced Persons*, 2.

[728] See, *inter alia*, AG/RES. 2667 (XLI-O/11), *Internally Displaced Persons*, *supra* note 724, 2; AG/RES. 2850 (XLIV-O/14), *Internally Displaced Persons*, 2.

[729] *Id.*

[730] See, *inter alia*, OAS, AG/RES. 2047 (XXXIV-O/04) *Protection of Refugees, Returnees, and Internally Displaced Persons in the Americas*, adopted on June 8, 2004, 1.; AG/RES. 2511 (XXXIX-O/09), *Protection of Asylum Seekers and Refugees in the Americas*, adopted on June 4, 2009, 1.

specifically urged to uphold the international principles of protection of refugees, in particular the principle of non-refoulement.[731]

The draft *Convention on Older Persons* includes a specific provision affording protection to the human rights of older persons in situations of risk, such as armed conflict, humanitarian emergencies and disasters.[732] In those situations, states are recommended to adopt specific assistance measures to address the needs of older persons during emergency, disaster, or conflict preparedness, prevention, reconstruction and recovery phases.[733]

The *San José Charter* also provides for special protection of older persons in cases of emergency or natural disasters.[734] In particular, states should provide preferential assistance to older persons in cases of disaster and during reconstruction. To achieve that objective, states should establish specific guidelines regarding priority treatment of older persons, provide the necessary resources and train first responders on the treatment that older persons should receive in those circumstances.[735]

G. Conclusion

In recent years, the Americas have produced some of the most talented and insightful advocates on behalf of older people around the world. Older persons themselves, ageing in this tradition, will undoubtedly advocate for themselves more and more in the coming years. Going forward, national governments in the region have an opportunity to remain at the vanguard by enacting and perfecting human rights protections specifically tailored for older populations in their own countries as their populations grow, in particular through the adoption of the *Convention on Older Persons*. There is a huge potential to lead by example in partnership with the populations they serve. Governments throughout the Americas would do well to capitalize on the unique history and strengths that make this opportunity possible.

---

[731] See, *inter alia*, AG/RES. 2758 (XLII-O/12), *Protection of Asylum Seekers and Refugees in the Americas*, adopted on June 5, 2012, 1; AG/RES. 2839 (XLIV-O/14), *Protection of Asylum Seekers and Refugees in the Americas*, adopted on June 4, 2014, 1.

[732] *Draft Inter-American Convention on the Human Rights of Older Persons*, *supra* note 39, Article 29.

[733] *Id.*

[734] *San José Charter on the Rights of Older Persons in Latin American and the Caribbean*, *supra* note 92, at 16.

[735] *Id.*

Africa: Regional Standards on the Protection of the Rights of Older Persons

A. An Introduction to the African Region

In Africa, the pace of growth of the population of older persons is currently slower than other regions, but the number of persons aged 60 or above is still set to quadruple from 47.9 million in 2005 to 206.8 million by 2050.[1] In response to this increase, African national aging policies must adapt as the population shifts to protect current older people, and future older people.

The most immediate protection of human rights happens through domestic systems. Ideally, constitutional recognition of human rights would exist in all the countries of Africa. This survey provides only a starting point for researchers who wish to compare domestic protections of human rights protections for older persons in African countries. The situations for which social protection is guaranteed in African constitutions include old age.[2] At least 98 references to the rights of the older persons are contained in 25 constitutions,[3] with three of those references prohibiting discrimination based on age.[4] There is no specific regional treaty that articulates clear human rights standards for older persons and provides a mechanism for implementation and accountability.[5]

Socio-economic and family-structure changes are ongoing, which can increase poverty and marginalize the aging population. One example is the rapid

[1]Magdalena Sepulveda, Report of the Independent Expert on Human Rights and Extreme Poverty (2010), 9.

[2]Constitutional human rights law in Africa: current developments by Heyns, C.H. and Kaguongo, Waruguru http://hdl.handle.net/2263/5345 22 S Afr J Hum Right 673 (2006).

[3]Angola, Burkina Faso, Cape Verde, Chad, Egypt, Gabon, Ghana, Guinea, Madagascar, Mozambique, Namibia, Rwanda, Somalia, Sudan, Tanzania, Togo and Zambia. Id. at 708.

[4]Constitutional human rights law in Africa: current developments by Heyns, C.H. and Kaguongo, Waruguru http://hdl.handle.net/2263/5345 22 S Afr J Hum Right 673 (2006), at 684.

[5]Jody Kollapan: The Rights of Older People- African Perspectives (2008), at 2–3.

C. Martin et al., *Human Rights of Older People*, Ius Gentium: Comparative Perspectives on Law and Justice 45, DOI 10.1007/978-94-017-7185-6_5

urbanization taking place in many African countries. Another is the devastating HIV/AIDS pandemic, which has created a generation of AIDS orphans who are cared for most often by their grandmothers and other older persons. UNICEF estimates 40–60 per cent of orphaned and vulnerable children in East and Southern Africa are cared for by their older people, usually their grandmothers.[6] These communities have very specific policy needs, including access to health care systems and participation in society.

Much more exploration is needed to find the most appropriate and effective policy options for understanding of the social, economic, cultural, and familial contexts older persons face. *Research on Ageing in Africa: The Need to Forge Directions*[7] lays the groundwork for research, as well as efforts to increase political awareness and will to put policies into practice. It also emphasizes necessary aging research to ensure that the emerging understandings can act as a catalyst for change.[8]

The need for social policy changes is, of course, predicated on a legal framework to justify and enforce policies. Some national constitutions recognize that older persons' rights require special protection.[9]

Those necessary changes are beginning to form. The African Union (AU), composed of 54 Member States from Algeria to Zimbabwe, has an extensive human rights system described in this chapter. In 2011, it began to develop a draft protocol on the rights of older persons.[10] At the opening of the 50th Session of the African Commission on Human and Peoples Rights, the situation of older persons were recognized as being "one of the most pressing human rights issues for states and public institutions."[11] Also in 2011, the Network of African National Human Rights Institutions held its 8th Biennial conference in Cape Town, South Africa on the theme *Advancing the Human Rights of Older Persons and Persons with Disabilities – the Role of African National Human Rights Institutions.* Further bolstering its efforts, in 2012, the African Union released a common position, recommending several actions for AU member states, regional committee, and

[6]HelpAge Intl. *Psychosocial support for older carers: a summary*, 2011. 1.

[7]Isabella Aboderin, Research on Ageing in Africa: The Need to Forge Directions http://www.ageing.ox.ac.uk/research/regions/africa/afran/news/publications.

[8]Nana Apt, 30 Years of African Research on Ageing: History, Achievements and Challenges for the Future http://www.ageing.ox.ac.uk/research/regions/africa/afran/news/publications.

[9]Cape Verde's Constitution establishes that it must provide older persons with special protection. Constitution of the Republic of Cape Verde, Article 72. Outdated English translation available here: http://landwise.landesa.org/record/835.

[10]Statement of Ms. Reine Alapini Gansou, African Commission on Human and People's Rights (April 20, 2011) http://www.globalaging.org/agingwatch/convention/un/Alapini%20-PROTECTIONDPA%20AFRIQUE%5B1%5D.pdf.

[11]Statement on Behalf of the NANHRI at the Opening Session of the African Commission on Human and Peoples Rights 50th Ordinary Session, 24 October to 7 November, 2011, Banjul, The Gambia. http://www.nanhri.org/index.php?option=content&task=view&id=236.

international development partners to implement in order to establish a right-based approach for older persons.[12]

This *Common Position* works to establish a rights-based approach to issues of concern to older people. It first identifies existing regional frameworks[13] to advocate for previously necessitated recommended measures, and then goes on to establish the AU's approach to the human rights of older persons through its own specific recommendations. It makes several recommendations to AU member states, regional committees, and international development partners.

Its appreciation of development partners' actions and the challenges of rural and urban disparities are of specific relevance to the context of Africa. It makes a number of recommendations relating to the international development architecture, including the use of the post-2015 Millennium Development Goal and Rio+20 agendas as opportunities to further include older persons into development discourse on national and international scales.

For older persons' rights, the most important commitment the *Common Position* makes is one of advocacy and support for a United Nations Convention on the Human Rights of Older Persons. This recommendation is made to all African Union member states. During the fourth session of the *Open Ended Working Group Toward Strengthening the Human Rights of Older Persons* at the UN in 2013, few African Union Member States mentioned the *Common Position*.[14]

At the national level, it recommends the integration of aging issues into national development policies through evidence-based programming. It further recommends "sensitization programs" to change the overall societal awareness and attitudes around the needs and issues of older persons. This, in turn, would create a "rights-based culture of ageing". The text is silent as to how this might occur based on a public understanding of "needs and issues."

As this introduction shows, the region of Africa is unique and changing in ways specific to challenges faced by its many cultures, climates, and populations. The overarching framework for regional human rights is the African Union's *African Charter on Human and Persons' Rights*. It is discussed at length below, along with the regional policy for older persons. The second half of this chapter on the African region examines how the rights of older persons have been interpreted among African member states of the AU, in the African Peer Review Mechanism.

---

[12]3rd Session of the AU Conference of Ministers of Social Development, *Africa's Common Position on Human Rights of Older Persons in Africa* http://social.un.org/ageing-working-group/documents/africa-older-persons2012.pdf.

[13]These include the *African Union Framework and Plan of Action on Ageing*, *African Charter of Human and People's Rights*, and the *United Nations Plan of Action on Ageing* among others.

[14]AU, Africa's Common Position on Human Rights of Older Persons in Africa, 26–30 November, 2012. http://social.un.org/ageing-working-group/documents/africa-older-persons2012.pdf.

B. The African Human Rights Framework

The African Union ("AU"),[15] formerly the Organization for African Unity, seeks to "achieve greater unity and solidarity between the African countries and the peoples of Africa." It concerns itself with the sovereignty of its members, political and socio-economic integration, international cooperation, the *UN Charter* and the UDHR. It promotes peace, security, stability, democratic principles and institutions, popular participation, good governance, and human and peoples' rights in accordance with the *Banjul Charter* and other relevant human rights instruments.[16] The AU has the power to enforce political and economic sanctions among its member states.[17] Articles 60 and 61 of the *Banjul Charter* are unique among international human rights treaties in their inclusion for interpretive purposes of African practices consistent with human and peoples' rights.

The *African Charter on Human and Peoples' Rights* ("*Banjul Charter*") was adopted in 1981, and came into force in 1986. Parties to the *Banjul Charter* recognize the rights, duties and freedoms enshrined in the Charter and undertake to adopt legislative or other measures to give effect to them.[18] Under the Charter, each State Party submits a biannual report on the legislative or other measures taken, with a view to giving effect to the rights and freedoms recognized and guaranteed by the Charter.[19]

The *Banjul Charter* emphasizes regional cultural distinctiveness, as when it refers to 'taking into consideration the virtues of African states' historical tradition and the values of African civilization.'[20] The preamble states that civil and political rights cannot be dissociated from economic, social and cultural rights in their conception as well as universality and that the satisfaction of economic, social and cultural rights is a guarantee for the enjoyment of civil and political rights. It embodies civil and political as well as economic, social, and cultural rights in the same instrument, which despite the principles of indivisibility and interrelatedness, is a connection rarely seen in one human rights instrument.

The *Banjul Charter* is referenced in the *Kigali Declaration on Human Rights in Africa, 2003*,[21] which takes on broad issues relating to development and human

[15]http://www.africa-union.org/root/au/index/index.htm.

[16]The Constitutive Act of the AU, Article 3 http://www.africa-union.org/root/au/AboutAU/Constitutive_Act_en.htm.

[17]The Constitutive Act of the AU, Article 23(2) http://www.africa-union.org/root/au/AboutAU/Constitutive_Act_en.htm.

[18]African Charter on Human and Peoples' Rights, Article 1 http://www.achpr.org/english/_info/charter_en.html.

[19]African Charter on Human and Peoples' Rights, Article 62 http://www.achpr.org/english/_info/charter_en.html.

[20]International Human Rights in Context., at 504.

[21]Kigali Declaration on Human Rights in Africa Adopted at Kigali, Rwanda on 8 May 2003 http://www.africa-union.org/Structure_of_the_Commission/Political%20Affairs/x/KIGALI%20DECLARATION%20as%20adopted%20in%20Kigali.pdf.

rights. It reaffirms respect for human rights as indispensable for the maintenance of national, regional and international peace and security and that it constitutes the fundamental bedrock for sustainable development, seeking to further human rights in Africa.[22] It contains no implementation mechanisms; it merely requests the Chairperson of the AU Commission to coordinate the follow up of the implementation of its declarations and urges Members States to submit reports on implementation to the AU Commission.[23] *Article 20* further calls upon Member States to develop a Protocol on the protection of the rights of persons with disabilities and older persons.

Focal Point

The work of the AU has been turning a more careful eye to the rights of older people. In 2007, the African Commission adopted a *Resolution on the Rights of Older Persons in Africa*.[24] It established a Focal point on the Rights of the Elderly persons in Africa and appointed Yeung Kam John Yeung Sik Yuen as the Co-coordinator of the focal point. The focal point was established with a view to drafting a protocol to the African Charter on the Rights of Older Persons in Africa.[25]

The focal point was charged with researching the problems of older people, identifying progressive measures for lessons learned and best practices, recommending preparation of a Draft Declaration on the Rights and welfare of Older Persons in Africa, and drafting of an African Charter on the Rights and Welfare of Older Persons and Persons with Disabilities.[26]

The focal point planned for the creation of a new working group.[27] It consists of two members of the Commission together with a number of independent experts brought to the fore by nominations.[28] The African Commission nominated

[22]Id. at Preamble.

[23]Id., 33.

[24]41st Ordinary Session of the African Commission held in Accra, Ghana, from 16–30 May 2007. ACHPR/Res.106 (XXXXI) 07, APCHR/Res.118(XXXXII) 07: Resolution on the Establishment of a Focal Point on the Rights of the Older Persons in Africa, http://www.achpr.org/english/resolutions/resolution118_en.htm.

[25]APCHR/Res.118(XXXXII) 07: Resolution on the Establishment of a Focal Point on the Rights of the Older Persons in Africa, http://www.achpr.org/english/resolutions/resolution118_en.htm.

[26]African Commission on Human and Peoples' Rights 23rd activity report, para. 191., p. 58. http://www.achpr.org/english/activity_reports/23rd%20and%2024th%20Activity%20Reports.pdf.

[27]Report of Commissioner Yeung Sik Yuen, November 2008-May 2009 http://www.achpr.org/english/Commissioner%27s%20Activity/45th%20OS/Commissioners/Com%20Yeun.pdf.

[28]APCHR/RES 143(XXXXV)09 adopted in May 2009, also in relation to Report on the Focal Point on the Rights of Older Persons in Africa.

the members of the Working Group.[29] It has been proposed that this Working Group replace the focal point for a two year period.[30] Most recently, the Working Group's draft protocol was reviewed by the African Commission in the AU's latest 2012 session, but was deferred to the next full session in June 2013 for reconsideration.[31]

Duties

The *Banjul Charter* is distinctive from other human rights instruments, because it emphasizes duties as well as rights. Its key theme is regional cultural distinctiveness, as when it refers to 'taking into consideration the virtues of [African states'] historical tradition and the values of African civilization.[32] In modern constitutions, as in human rights treaties, provisions conferring rights on individuals far outnumber those imposing duties.[33]

Depending on interpretation, duties and ideals of solidarity may impinge in clear and serious ways on the *Banjul Charter's* definitions of rights themselves. Moreover, the *Banjul Charter* imposes individual duties not only on the state but also on different groups or communities within (or perhaps transcending) the state.[34]

Other places where duties can be found in international human rights instruments include Article 29 of the UDHR, the preamble to the ICCPR, and the preamble to the *UN Charter*.

> The *Banjul Charter* "is the first human rights treaty to include an enumeration of, to give forceful attention to, individuals' duties. In this respect, it goes well beyond obvious duties of states that are correlative (corresponding) to individual rights and duties– for example, states' duties not to torture or to provide a structure for voting in political elections. The Banjul Charter differs by defining duties that are not simply the 'other side' of individual rights, and that run from individuals to the state as well as to other groups of individuals. Hence the Banjul Charter directly raises ... issues of universalism and cultural relativism."[35]

---

[29]Community Law Centre Newsletter, April 2009: http://www.communitylawcentre.org.za/achpr/newsletters/CLC%20ACHPR%20NEWSLETTER%20APRIL%202009.pdf/.

[30]Resolution on the Transformation of the Focal Point on the Rights of Older Persons in Africa into a Working Group on the Rights of Older Persons in Africa, *and the Appointment of a Working Group on the Rights of Older Persons* http://www.achpr.org/english/resolutions/Resolution%20on%20WGOP.pdf.

[31]Final Communique of the 51st Ordinary Session of the African Commission on Human and People's Rights, May 2012, http://www.achpr.org/sessions/51st/info/communique51/.

[32]Id., at 504.

[33]International Human Rights In Context, at 501.

[34]Id, at 506.

[35]Id, at 505.

The series of explicit duties spelled out in Articles 27 through 29 of the *African Charter* could represent a rejection of the individual "who is utterly free and utterly irresponsible and opposed to society."[36] These duties, if truly assimilated into the collective consciousness, could contribute positively to social and national solidarity.[37]

Review Mechanisms

The African Commission on Human Rights ("African Commission") is the review mechanism of human rights in the African Union. The African Commission acts to ensure the protection, promotion and interpretation of human rights through the *Banjul Charter*. When the Banjul Charter was adopted in 1981 a clear decision was taken to opt for a Commission rather than a Court as the principal institutional arrangement.[38] A court has since been formed with the Constitutive Act of the AU, Article 18, on July 11, 2000.[39]

The African Commission assessed the status of older persons in its 26th Session in 1999. There the African Commission expressed its concern about the treatment of the aged as a group affected by the various conflicts throughout the region. In the 27th Session, it noted the complaints of non-governmental organizations that listed older persons among the various victims of regional strife.[40]

The *Protocol to the African Charter on Human and Peoples' Rights on the Establishment of an African Court on Human and Peoples' Rights*[41] established an African Court on Human and Peoples' Rights within the AU. The protocol was adopted June 9, 1998, and entered into force January 1, 2004.[42] In July 2004, the

[36]Id. Essay: Makau Mutua, Human Rights and the African Fingerprint (from Mutua, Human Rights: A Political and Cultural Critique (2002), at 71. from International Human Rights In Context, at 508.

[37]Id. Essay: Makau Mutua, Human Rights and the African Fingerprint (from Mutua, Human Rights: A Political and Cultural Critique (2002), at 71. from International Human Rights In Context, at 508.

[38]International Human Rights in Context, at 1081.

[39]The Constitutive Act of the AU, Article 18 http://www.africa-union.org/root/au/AboutAU/Constitutive_Act_en.htm.

[40]African Commission on Human and Peoples' Rights http://www.achpr.org/english/_info/mandate_en.html.

[41]Protocol to the African Charter on Human and Peoples Rights on the Establishment of an African Court on Human and Peoples Rights (Protocol for Establishment of an African Court), CAB/LEG/665, Adopted June 9, 1998 and entered into force January 1, 2004, at http://www.achpr.org/english/_info/court_en.html.

[42]Id.

AU merged the African Court on Human and Peoples' Rights with the African Court of Justice,[43] established as the African Court of Justice and Human Rights (the "African Court").[44]

The African Court is intended to be an organ of the African Union, and to complement the Commission on Human Rights.[45] It rounds out the Banjul Charter by providing judicial guarantees of its rights.[46] The African Court can use any existing international human rights law from the UN or other sources.[47]

Other AU Mechanisms

The *New Partnership for Africa's Development* ("NEPAD")[48] is an organization of the AU with the primary objectives of eradicating poverty; promoting growth and development, promoting Africa's inclusion in globalization, and accelerating the women's empowerment.[49] It seeks implementation through the AU, and has yet to directly address older persons or older women.[50]

The *African Peer Review Mechanism* ("APRM")[51] is an instrument voluntarily acceded to by member states of the AU as an African self–monitoring mechanism. The APRM process is designed to be open and participatory, a neutral space for sharing between private sector, civil society and government. It is not meant to exclude or punish countries; there is no conditionality or penalty attached to the mechanism.[52] Country Review Missions (CRM's) draft the *Country Review Reports*. The overarching goal of the APRM is for all participating countries to accelerate progress towards adopting and implementing the priorities and

---

[43]Protocol on the Statute of the Court of Justice and Human Rights, http://www.africa-union.org/root/au/Documents/Treaties/text/Protocol%20on%20the%20Merged%20Court%20-%20EN.pdf.

[44]Protocol on the Statute of the Court of Justice and Human Rights, http://www.africa-union.org/root/au/Documents/Treaties/text/Protocol%20on%20the%20Merged%20Court%20-%20EN.pdf.

[45]Lyons, Scott, The African Court on Human and Peoples Rights, 10 The American Society of International Law 24 (2006).

[46]See Udombana, Toward the African Court on Human and Peoples Rights: Better Late than Never, 46 Yale Hum. Rts and Development L.J. 3:45 (2000).

[47]Id.

[48]NEPAD: http://www.nepad.org/2005/files/home.php.

[49]http://www.nepad.org/AboutNepad/sector_id/7/lang/en.

[50]Framework Document, New Partnership for Africa's Development http://www.nepad.org/images/framework.pdf.

[51]*African Union Policy Framework and Plan of Action on Ageing by HelpAge International and the African Union, December, 2003* http://www.afrimap.org/english/images/treaty/APRM-BaseDocument.pdf.

[52]Id.

programs of the NEPAD, achieving mutually agreed-upon objectives, and compliance with the best practice in respect of each of the areas of governance and development.[53]

The *African Union Labour and Social Affairs Commission* has a mission to develop cooperation, to study the labor and social problems peculiar to African Countries; examine all questions related to International Labor Organization or other specialized agencies of the United Nations whose activities have a bearing on labor and social questions; and to defend African interests both at the International Labor Conference and any other international meetings dealing with social and labor matters.[54]

The *Common Position on Human Rights of Older Persons in Africa*[55] is the African Union's specified call for establishing a rights-based approach to issues of concern to older persons. In this document, the AU calls upon previous frameworks, including the *Banjul Charter* and the *UN Plan of Action on Ageing*, among others, in order to advocate for previously established measures. The document then goes on to make specific recommendations for defending and promoting the human rights of older persons.

Foremost, the *Common Position* sets forth the AU's support for holding a United Nations-based Convention on the Human Rights of Older Persons. Other recommendations include the creation of national sensitization programs which would promote societal awareness of the rights of older persons, the inclusion of the older persons in post-2015 Millennium Development Goal dialogue and discourse, and an investment in today's youth in order to improve the outlook of future populations of older persons.

C. Civil and Political Rights of Older Persons in Africa

Human rights are interrelated and indivisible. Distinctions between civil and political rights versus economic, social and cultural rights are not helpful in the interpretation or valuation of human rights. As a matter of organization, rights, as they have been treated within the AU's legal framework, are categorized and analyzed individually below.

[53]Objectives, Standards, Criteria and Indicators for the African Peer Review Mechanism, p. 1 http://www.afrimap.org/english/images/treaty/APRM-Objectives-Standards-Indicators.pdf.

[54]Concept Note, p. 1, http://www.africa-union.org/root/ua/Conferences/2007/avril/SA/16-21%20avr/meeting.htm.

[55]http://social.un.org/ageing-working-group/documents/africa-older-persons2012.pdf.

1. The Right to Freedom from Violence

States Parties to the *Protocol to the African Charter on Human and Peoples' Rights on the Rights of Women in Africa* undertake to ensure the right of elderly women to freedom from violence, including sexual abuse, discrimination based on age and the right to be treated with dignity.[56] Freedom from violence in domestic settings is of concern for older persons around the world, including in the African region.

The country report by the CRM for Mozambique[57] noted a concerted civil society initiative for a law on domestic violence. The Assembly informed the CRM that the Assembly felt that violations must cover all manifestations of domestic violence other than those solely against women e.g. against older persons, the disabled, against children, against pregnant women, hence the delay. The Bill is reportedly being redrafted to cover all social groups.[58] Efforts continue to focus on women.[59] The Ministry of Women and Social Action is mandated to handle the concerns and welfare of older people. CEDAW has cited concerns over allegations of witchcraft in Mozambique.[60]

Accusations of witchcraft often provide license for property grabbing, violence, and even murder. Such allegations are nearly always leveled at older women. Such violence tends to increase with resource scarcity[61] and relieve older women of their property, if not their lives.[62] Philip Alston, the former Special Rapporteur on extra judicial summary or arbitrary executions, noted that the murder and persecution of persons accused of witchcraft is rarely well-documented.

In Tanzania, where the practice is relatively well-documented,[63] he has reported that "as many as a thousand, mostly elderly Tanzanian women are targeted and

---

[56] Article 22 *Protocol to the African Charter on Human and Peoples' Rights on the Rights of Women in Africa* http://www.achpr.org/instruments/women-protocol/#22.

[57] Country Review Report No. 11: Mozambique, Published: 17-11-2009 http://aprm.krazyboyz.co.za/index.php?option=com_aprm_documents&Itemid=32&page=documents-category&cid=25&nid=21&id=21.

[58] Id., at 136, 458.

[59] US Department of State, Country Report on Human Rights Practices, 2012, http://www.refworld.org/topic,50ffbce4c9,50ffbce4fc,517e6e001c,0,,,MOZ.html.

[60] CEDAW, *Concluding comments of the Committee on the Elimination of Discrimination against Women:* Mozambique 11 June 2007 CEDAW/C/MOZ/CO/2 para. 42. http://www.un.org/womenwatch/daw/cedaw/cedaw25years/content/english/CONCLUDING_COMMENTS/Mozambique/Mozambique-CO-2.pdf.

[61] Edward Miguel, *Poverty and Witch Killing*, Review of Economic Studies (2005) 72, 1153–1172, at 1162.

[62] See HelpAge International, Promo Femmes Dévelopment Solidarité and Association le TOCSIN, *Joint NGO Submission for the Initial Universal Periodic Review of Burkina Faso in December 2008, available at* http://www.upr-info.org/IMG/pdf/HAI_BFA_UPR_S3_2008_HelpAgeInternational_Etal_uprsubmission.pdf.

[63] UNHCR, *Witchcraft allegations, refugee protection and human rights: a review of the evidence*, New Issues in Refugee Research, Research Paper No. 169, Jill Schnoebelen, 2009.

killed annually."[64] According to a report of the Legal and Human Rights Centre, from 2004 to 2009, there were a total of 2,585 killings of older women in 8 regions of Tanzania where such activities are common.[65] In an interview, one Tanzanian activist reported that "unlike other crimes, violence against older women is not just tolerated but accepted."[66] The social acceptability of such violence is problematic, but it is the lack of response from systems of justice after years of data collection that shows abrogation of state duties.

2. The Right to Equality and Non-discrimination

The African Commission prohibits discrimination through Articles 2 and 3 of the *Banjul Charter.* Article 2 lists specific types of prohibited grounds for discrimination. It does not name age as a specific prohibited ground, but it does include the provision for "other status."[67] Article 3 articulates that "1. Every individual shall be equal before the law. 2. Every individual shall be entitled to equal protection of the law."[68]

Articles 2 and 3 have been designated as non-derogable provisions necessary in order for anyone to enjoy all the other rights under the *Banjul Charter*.[69] Certainly the Vienna Declaration and Programme of Action of 1993, which famously declared that *"all human rights are universal, indivisible, interdependent, and interrelated*" would support this concept.

Article 18 of the African Charter on Human and Peoples' Rights includes a clause stating that "[t]he aged and disabled shall also have the right to special measures of protection in keeping with their physical or moral needs."[70] Article 22 of the *Protocol to the African Charter on Human and Peoples' Rights on the Rights of Women in Africa* ("*Maputo Protocol*") covers Special Protection of Elderly Women, stating that "[t]he States Parties undertake to: (a) provide protection to elderly women and take specific measures commensurate with their physical, economic and social needs as well as their access to employment and professional training; (b) ensure the right of elderly women to freedom from violence, including sexual abuse, discrimination based on age and the right to be treated with dignity."

[64]OHCHR, http://www.ohchr.org/EN/NEWSEVENTS/Pages/Witches21stCentury.aspx.

[65]Legal and Human Rights Centre, *Tanzania Human Rights Report*, 2009, p. 98. http://www.humanrights.or.tz/wp-content/uploads/2010/10/Tanzania-Human-Rights-Report-2009.pdf.

[66]HelpAge International, *No Country for Old Women*, http://www.helpage.org/newsroom/features/nocountry-for-old-women/.

[67]Article 2, Banjul Charter. http://www1.umn.edu/humanrts/instree/z1afchar.htm.

[68]Article 3, Banjul Charter, http://www1.umn.edu/humanrts/instree/z1afchar.htm.

[69]Sixteenth Annual Activity Report of the ACHPR 2002–2003, 241/2001—Purohit and Moore p. 68, para 49. http://www.achpr.org/english/activity_reports/activity16_en.pdf.

[70]African Union, *African (Banjul) Charter on Human and Peoples' Rights*, 1981, www.africa-union.org/official_documents/…/Banjul%20Charter.pdf.

In *159/96 Union Inter Africaine des Droits de l'Homme,* et al. *v. Angola,*[71] the African Commission concluded that the *Banjul Charter* obligates States Parties to ensure that persons living on their territory enjoy the rights guaranteed in the Charter. In this case, the victim's rights to equality before the law were violated based on their origin.[72]

This law becomes increasingly strong as courts draw on each others' opinions. The African Commission, in *Zimbabwe Lawyers for Human Rights and the Institute for Human Rights and Development/Republic Of Zimbabwe*, drew on *Brown v. Board of Education of Topeka,*[73] and that decision's pronouncement that "equal protection of the law refers to the right of all persons to have the same access to the law and courts and to be treated equally by the law and courts, both in procedures and in the substance of the law. It is akin to the right to due process of law, but in particular applies to equal treatment as an element of fundamental fairness."[74] This echoes the broad-ranging right to non-discrimination articulated in the *Maputo Protocol*, however, it has a potentially broader impact, because it applies to men as well as women, and because it affects all states party to the ACPHR, not just those states party to the *Maputo Protocol.*

3. The Right to Freedom from Cruel, Inhuman or Degrading Treatment or Punishment

Article 5 of the *Banjul Charter* provides that every individual shall have the right to the respect of dignity inherent in a human being free from torture, cruel, inhuman or degrading punishment and treatment. This was highlighted by the case against Rwanda for the conditions of detention under which children, women and older persons were held during the country's genocide. It was found that their physical and psychological integrity was violated, constituting a violation of *Article 5.*[75]

---

[71]Union Inter Africaine des Droits de l'Homme, Fédération Internationale des Ligues des Droits de l'Homme, Rencontre Africaine des Droits de l'Homme, Organisation Nationale des Droits de l'Homme au Sénégal and Association Malienne des Droits de l 'Homme vs. Angola, 11th Annual Report of the African Commission on Human and Peoples' Rights, 159/96 http://www1.umn.edu/humanrts/africa/comcases/159-96.html.

[72]Id., 18.

[73]347 U.S 483 (1954).

[74]26th Activity Report, p. 28 http://www.achpr.org/english/activity_reports/26th%20Activity%20 report.pdf.

[75]10th Annual Report 27/89, 46/91, 49/91, 99/93, Organisation Mondiale Contre La Torture and Association Internationale des Juristes Democrates, Commission Internationale des Juristes (C.I.J) Union Interafricaine des Droits de l'Homme v. Rwanda, Merits, Par 25, 26 http://www1.umn.edu/humanrts/africa/comcases/27-89.html.

4. The Right to Liberty and Security

Article 6 of the *Banjul Charter* reads: "Every individual shall have the right to liberty and security of his person. No one may be deprived of his freedom except for reasons and conditions previously laid down by law. In particular, no one may be arbitrarily arrested or detained." Freedom to live free of interference may be implicated with involuntary evictions of older persons. Country Review Reports have found that the needs of vulnerable groups should be taken into account in systems of land tenure.[76]

5. The Right to Respect of Home, Family and Private Life

The Banjul Charter protects rights and prescribes duties at the same time. It is unique in that it provides for an individual's duty to preserve the harmonious development of the family, to work for its cohesion and respect and to respect his parents at all times and is a unique attribute of the *Banjul Charter*. In that sense, there is a duty corresponding to the right protected in Article 18(4), which is articulated in Article 29 regarding the family. This provision states: "The individual shall also have the duty: 1. To preserve the harmonious development of the family and to work for the cohesion and respect of the family; to respect his parents at all times, to maintain them in case of need." Articles 27–29 provide for the specific duties of an individual towards his family, others and towards his nation and other communities including the international community. Article 18(4) of *the Banjul Charter* provides that the aged and the disabled shall also have the right to special measures of protection in keeping with their physical or moral needs.

These articles have been applied to the case of Algeria, where the CRM recommended that the state accelerate implementation of its National Strategy for the Protection of the Elderly. The CRM noted that a legal system was not yet in place to provide support to grandparents caring for grandchildren as a way to strengthen family solidarity, improve institutional aid and provide indispensable health protection.[77]

6. The Right to Information, Association and Participation

The CRM has highlighted some excellent examples of the application of Article 9 of the *Banjul Charter*, which deals with the right to information. It provides that

---

[76]Country Review Report No. 1: Ghana, Published: 06-10-2005 http://aprm.krazyboyz.co.za/index.php?option=com_aprm_documents&Itemid=32&page=documents-category&cid=25&nid=21&id=21.

[77]Country Review Report No. 5: Algeria Findings of the CRM, at 144.

every individual shall have the right to receive information and to express and disseminate his (or her) opinions to the extent the law allows.

In Uganda, historical power structures limited the right of all to receive information and express his or her opinions. It has introduced affirmative measures for groups historically excluded, including, explicitly, older people. These measures support the right of older persons to information for their participation in political and economic processes. Key challenges facing the country in this regard include addressing social exclusion and marginalization, as well as political will of leadership and the persons to transform attitudes, norms and practices that reinforce institutionalized discrimination and marginalization.[78]

In Algeria, older persons have local associations from the village to district levels. Associations mobilize older persons to participate in development; they promote interaction and intergenerational linkages; they enable the government to extend services to older persons; and they can help to elect chosen representatives to the local government councils.[79] By contrast, Benin's CRM highlighted the lack of associations that exist for the older persons in the country and recommended the government's implementation of measures which will promote the self-fulfillment of older persons.[80]

D. Economic, Social and Cultural Rights of Older Persons in Africa

All human rights are necessarily interrelated and indivisible. This is clear in the African Commission's 2005 decision against Nigeria. The African Commission found that Nigeria had contravened the rights of an indigenous people, the Ogoni, through the military government's involvement in oil production with the state oil company, violating a number of articles of the *Banjul Charter*.[81] It threatened the destruction of Ogoni sources of livelihood with permissive policies of environmental degradation. It was found that Nigeria violated Article 2, the right to non-discrimination, by its targeting and wanton violation of the rights of the Ogoni persons both individually and collectively.[82]

---

[78]Country Review Report No. 7: Uganda, Published: 21-08-2009 http://aprm.krazyboyz.co.za/index.php?option=com_aprm_documents&Itemid=32&page=documents-category&cid=25&nid=21&id=21, P. 40.

[79]Country Review Report No. 5: Algeria Findings of the CRM.

[80]Country Review Report No. 6: Benin, Published: 05-06-2008 at 126.

[81]Communication 155/96; Decision handed down at the 30th Ordinary session of the Commission held in The Gambia. For text, see University of Minnesota Human Rights Library website, online: <http://www1.umn.edu/humanrts/africa/comcases/155-96b.html> (last accessed: March 16, 2010) ["Communication 155/96"].

[82]Nwobike, Justice C., The African Commission on Human and Peoples' Rights and the Demystification of Second and Third Generation Rights under the African Charter: Social and Economic Rights Action Center (SERAC) and the Center for Economic and Social Rights (CESR) v. Nigeria, 1 Afr. J. Legal Stud. 2 129–146 (2005).

This decision further entrenches and promotes the indivisibility and interrelatedness of each right, put "beyond question the competence of a group to seek the protection and enforcement of rights, which accrued to them as a persons under the Charter,"[83] and recognized the economic, social and cultural rights as justiciable. It contained a generally purposive interpretation of the Charter, whereby it implied the right to food from a combined reading of such provisions as Article 4's right to life, Article 16's right to health, and article 22's right to economic, social and cultural development.

1. The Right to Sustainable Livelihood

Article 15 of the *Banjul Charter* outlines the right to work under satisfactory conditions. Nigeria's country report noted the need for securement of an adequate means of livelihood without discrimination.[84] While it does not explicitly prohibit discrimination based on age, this could be read to prohibit age discrimination. Older persons who are involved in formal employment, for example, have mandatory retirement ages of 60 or 65 years old in some countries.[85] Nigeria has also addressed the adequacy and appropriateness of work, and older persons' protection against any exploitation whatsoever. Moral and material neglect are also addressed,[86] which, for some older people, is certainly influenced by their access to livelihoods.

2. The Right to Education and Culture

Education and culture have been interpreted as equal access to literacy, as well as recognition of the role as keepers of cultural knowledge that older persons may hold for a society. In a discussion of Benin that is instructive well beyond its borders, the CRM recognized that "if today the youthful population is benefiting from the right to education, training and work, the elderly – known as the "illiterate" sector of the population – is disadvantaged."[87] As this observation indicates, the right to education touches the right to work.

---

[83]Nwobike at 140.

[84]Nigeria's third periodic country report: 2005–2008, September 2008, 82. http://www.achpr.org/english/state_reports/Nigeria/3_Periodic%20Rpt.pdf.

[85]Monica Ferreira & Sebastiana Kalula, Human Rights and Ageing in South Africa, International Longevity Centre, 2007. www.ilcsa.uct.ac.za/Human%20rights%20paper.doc.

[86]Nigeria's third periodic country report: 2005–2008, September 2008, 82. http://www.achpr.org/english/state_reports/Nigeria/3_Periodic%20Rpt.pdf.

[87]Country Review Report No. 6: Benin, Published: 05-06-2008 http://aprm.krazyboyz.co.za/index.php?option=com_aprm_documents&Itemid=32&page=documents-category&cid=25&nid=21&id=21.

The right to culture is articulated in the protection of Tuareg guides of North Africa and their knowledge of their territory, which has been recognized to compete even with the most sophisticated geographical information systems.[88] Their knowledge of biological diversity and rational use of natural resources (e.g. plants as traditional medicine) is a knowledge often held and handed down by older persons in a culture.

3. The Right to an Adequate Standard of Living and Housing, Social Protection and Security

Articles 14, 16 and 18(1) of the *Banjul Charter* have been interpreted by the African commission as defining a right to shelter or housing.[89] Older persons are given explicit mention in Article 18, which states in relevant part, "The aged and the disabled shall also have the right to special measures of protection in keeping with their physical or moral needs." Article 24, the right to a general satisfactory environment favorable to development, is also relevant. According to the African Commission, the care of the aged falls squarely on family and community members. This requirement has its roots in the past: it was unthinkable to abandon a parent or relative in need, says the African Commission.[90] This is unfortunately far from reality, in many instances. Recently, one court case shows, older persons have been disproportionately evicted from their homes.[91]

Many countries are struggling with the unmet needs of older persons in their communities. Algeria has developed a recent interest in a policy for promoting and protecting this older people. It developed a special program for some 3.5 million old people, as a result of criticisms and protests over the past years.[92] The CRM also recommended the improvement of the quality of infrastructures and services rendered at protection centers for older persons, notably through the creation in all the "wilayas," or multidisciplinary mobile teams in charge of providing medical care.[93]

---

[88]Country Review Report No. 6: Benin, Published: 05-06-2008 http://aprm.krazyboyz.co.za/index.php?option=com_aprm_documents&Itemid=32&page=documents-category&cid=25&nid=21&id=21.

[89]Communication 155/96; Decision handed down at the 30th Ordinary session of the Commission held in The Gambia. For text, see University of Minnesota Human Rights Library website, online: <http://www1.umn.edu/humanrts/africa/comcases/155-96b.html> (last accessed: March 16, 2010) ["Communication 155/96"] 64.

[90]Id.

[91]Justice C. Nwobike, African Commission on Human and Peoples' Rights and the Demystification of Second and Third Generation Rights Under the African Charter: Social and Economic Rights Action Center (SERAC) and the Center for Economic and Social Rights (CESR) v. Nigeria, African Journal of Legal Studies, 137, 2005.

[92]Country Review Report No. 5: Algeria Findings of the CRM.

[93]Country Review Report No. 5: Algeria Findings of the CRM, at 144.

South Africa's APR Panel noted the need for capacity to protect vulnerable groups, and use education and sensitization campaigns to increase public awareness of rights of vulnerable groups and the need for tolerance.[94] In this same panel, the right of older persons to an Old Age Pension Scheme was noted with approval.[95]

4. The Right to Health

Older persons' equal access to what provisions for health the states may provide is guaranteed in Article 16 of the *Banjul Charter*. It states that "every individual shall have the right to enjoy the best attainable state of physical and mental health." An important aspect of access to health across the life course includes the *Resolution on Access to Health and Needed Medicines in Africa.*[96] It recognizes that "access to needed medicines for treatment, prevention and palliative care is a necessary condition for leading a healthy and dignified life…" It further calls on States to promote access to medicines by refraining from measures that negatively affect access, such as: denying or limiting equal access to medicines for marginalized individuals or communities. Member states must not discriminate in their provision of access to health care.

With regard to Articles 16 and 24 on the right to health and a general satisfactory environment favorable to development, the African Commission found that Nigeria did not properly regulate oil production affecting the Ogoni people. The Ogoni persons were excluded from health facilities and basic infrastructure including health information in connection with oil exploration. The African Commission put it thusly: "Instead of warding off threats to the peaceful enjoyment by the Ogonis of their protected rights, the government has promoted an atmosphere of terror in a bid to silence the demand for observance of their rights."[97]

In the Republic of Gambia, the African Commission found violations of the right to health,[98] and urged the government of Gambia to repeal the Lunatics

---

[94]Country Review Report No. 4: South Africa, Published: 19-10-2007 p. 119 http://aprm.krazyboyz.co.za/index.php?option=com_aprm_documents&Itemid=32&page=documents-category&cid=25&nid=21&id=21.

[95]Country Review Report No. 4: South Africa, at 240.

[96]ACHPR/Res.141 (XXXXIIII)08: Resolution on Access to Health and needed Medicines in Africa http://www.achpr.org/english/resolutions/resolution141_en.htm.

[97]Para. 56 reads "…the Government has not taken such steps as would to protect the Ogoni population from harms done by the NNPC-Shell consortium. It has instead actively used its security forces to facilitate and compound the damage. Furthermore, it has not provided nor permitted studies of potential or actual environmental and health risks caused by oil operations in Ogoni Communities. Instead of warding off threats to the peaceful enjoyment by the Ogonis of their protected rights, the government has promoted an atmosphere of terror in a bid to silence the demand for observance of their rights."

[98]*Articles* 2, 3, 5, 7 *(1)(a) and (c), 13(1), 16* and *18(4)* of *the Banjul Charter* and.

Detention Act stripping those with mental illness of their rights, "and replace it with a new legislative regime for mental health in Gambia compatible with the African Charter."[99] This improved monitoring has important implications for older persons as the population increases throughout Africa. Many older persons have disabilities, Alzheimer's disease, and mental disabilities in their old age; support for as much participation as possible, with review mechanisms, will be an important part of the landscape of future protections of older persons' rights.

In addition to responding to violations, reviews have also recommended positive actions: in 2008, the APR Panel recommended that Benin speed up the implementation of the national action plan for active aging, establish a system of health care for older persons, encourage the social and medical mentoring of older persons, and establish a system of social support to facilitate the care of the older persons within their own families.[100] In 2010, Angola's new Constitution mandated the promotion and guarantee of "the measures needed to ensure the universal right to medical and health care, as well as the right to ... care in illness, disability, old age and in situations [where persons] are unable to work."[101] Such language represents the beginning of the realization of the right to health. Goods and services which are available, accessible, acceptable and of good quality[102] underpin this right.

E. The Rights of Older Persons as Members of Vulnerable Groups in Africa

This section examines some of the multiple discriminations faced by many older persons throughout Africa. Refugees, women and persons with disabilities all have specific legal frameworks protecting their rights as groups with specific needs.

Some AU member states are taking the lead on designation of older persons as a vulnerable group. The Central African Republic noted the vulnerability of older persons in its creation of a bill on the protection of older persons, and was commended by Djibouti on efforts to promote the rights of vulnerable groups.[103] For Chad, Brazil noted that there were possible discriminatory laws in legislation

[99]Sixteenth Annual Activity Report of the ACHPR 2002–2003, 241/2001—Purohit and Moore, at 74 : http://www.achpr.org/english/activity_reports/activity16_en.pdf.

[100]Country Review Report No. 6: Benin, Published: 05-06-2008 http://aprm.krazyboyz.co.za/index.php?option=com_aprm_documents&Itemid=32&page=documents-category&cid=25&nid=21&id=21.

[101]Angola Constitution, Article 77, January 21, 2010.

[102]CESCR, General Comment N° 14 (2000) on the right to health, adopted by the Committee on Economic, Social and Cultural Rights. U.N. Doc. E/C.12/2000/4, http://www.unhchr.ch/tbs/doc.nsf/%28symbol%29/E.C.12.2000.4.En.

[103]A/HRC/12/2 Central African Republic's Working Group Report.

contravening the rights of vulnerable groups, and Switzerland noted more measures are necessary to protect vulnerable groups.[104]

Ghana was compelled to take further steps to address discrimination against vulnerable groups; its LEAP Program acknowledges the needs of vulnerable groups in the lowest 20th percentile, including, expressly, older persons.[105]

1. The Rights of Older Refugees

In times of emergency, older persons are often the most likely to be left behind. As recent events such as the severe drought in East Africa have shown, older persons are particularly invisible in "young" countries. In countries looking to the future, older persons can be overlooked.[106] Those who can seek refuge have specific needs distinct from the general displaced population. Humanitarian actors have a duty to create accessible shelter and latrines, programmatic livelihood support that does not discriminate against older people, accessible and appropriate healthcare, distribution of food and other items that is accessible, registration of older persons and collection of and use of disaggregated data.[107]

The *Kigali Declaration on Human Rights in Africa* calls upon its member states to ensure the protection of older persons in situations of armed conflict.[108] The African Commission on Human and Persons' Rights has explicitly included older persons as a vulnerable group in emergencies.[109] In creating a Special Rapporteurship for refugees, asylum seekers, and internally displaced persons in Africa, the AU declared that "in spite of the adoption of the 1969 OAU Convention Governing the Specific Aspects of Refugees Problems in Africa, refugees in Africa continue to face untold suffering … the elderly being the most vulnerable among refugees…"[110]

Older persons also have an important role to play in post-conflict reconstruction efforts. During the Rwandan Genocide, the African Commission found the

---

[104] A/HRC/12/5 Chad's Working Group Report.

[105] A/HRC/8/36 Ghana's Working Group Report.

[106] Country Review Report No. 5: Algeria Findings of the CRM: The authorities appear not to have paid adequate attention to this category because of other priorities and crises, and also because this section of the population represented only a very small proportion of the total population.

[107] HelpAge International, Protecting Older People in Emergencies, a Good Practice Guide, 2012, www.helpage.org/download/4f2bcb851f1b4/.

[108] Id., 17.

[109] Resolution on the Human Rights Situation in Uganda, 20th Activity Report, 25–29 June 2006, http://www.achpr.org/english/activity_reports/20th%20Activity%20Report.pdf.

[110] The African Commission Resolutions: Resolution on the Mandate of the Special Rapporteur on Refugees, Asylum Seekers and Internally Displaced Persons in Africa http://www.achpr.org/english/_doc_target/documentation.html?../resolutions/resolution77_en.html.

detention of thousands of persons in various parts of the country by the Rwandan security forces, including older persons.[111] The APR Panel recommended to Uganda in 2009 that the rights of vulnerable groups, including older persons, should be included in the development programs, and post-conflict reconstruction efforts.[112] Unfortunately, the APR panel also found that the national machinery responsible for these issues is not only overburdened with responsibilities, but remains one of the least-resourced ministries.[113]

2. The Rights of Older Women

The 2003 Protocol to the African Charter on Human and Persons on the Rights of the Women ("Maputo Protocol")[114] requires that States Parties shall combat all forms of discrimination against women through appropriate legislative, institutional and other measures and to modify the social and cultural patterns of conduct. States parties undertake to provide special protection to older women in measures commensurate with their physical, economic and social needs as well as their access to employment and professional training.[115] They should ensure the rights of older women to freedom from violence, including sexual abuse, discrimination based on age and the right to be treated with dignity.[116]

In Article 22, older women are given special protection for measures commensurate with their physical, economic and social needs as well as their access to employment and professional training. It further ensures the right of older women to freedom from violence, including sexual abuse, discrimination based on age and the right to be treated with dignity.[117]

Sadly, there have been many situations of social and environmental distress in Africa over the past decades. In Article 24 on special protection of women in distress, the States parties undertake to: "ensure the protection of poor women and women heads of families including women from marginalized population groups and provide an environment suitable to their condition and their special physical,

[111] 10th Annual Report 27/89, 46/91, 49/91, 99/93, Organisation Mondiale Contre La Torture and Association Internationale des Juristes Democrates, Commission Internationale des Juristes (C.I.J) Union Interafricaine des Droits de l'Homme/Rwanda.

[112] http://www.africa-platform.org/sites/default/files/resources/Uganda%20Country%20 Report.pdf.

[113] Country Review Report No. 7: Uganda, Published: 21-08-2009 http://aprm.krazyboyz.co.za/ index.php?option=com_aprm_documents&Itemid=32&page=documents-category&cid=25&nid=21&id=21.

[114] Protocol to the African Charter on Human and People's on the Rights of the Women in Africa, 2003: http://www.achpr.org/english/_info/women_en.html.

[115] Article 22, Id.

[116] Article 22, Id.

[117] Article 22, Id.

economic and social needs." Older women have specific needs, and are an often-marginalized population group. Article 3 on the right to dignity and Article 4 on the right to life, integrity and security of person for women are of particular relevance in conflict and disasters.

The Country Review Report for Benin[118] found a belief in Benin that many women are witches. A recent study noted that these "witches" are nearly always comparatively older, and nearly always women.[119] A report from the UN High Commissioner for Refugees found that witchcraft allegations increase where removal of the witch would result in the occupation of property by other family members.[120] When witchcraft was again raised Benin's 2012 Universal Periodic Review, only witchcraft allegations against children were raised. Older persons, they reported, were given "forums on civil status and the identification of the specific needs of older persons, which had resulted in recommendations that would contribute to improving civil policy and everyday life for that sector of society."[121] Violence against older women as a result of allegations of witchcraft was not addressed.

With regard to property held by widows, while it is obvious that not all widows are older persons, but a great many are; therefore, the specific treatment of widow's rights within the *Maputo Protocol* is relevant to this discussion of the rights of older people. Widows have been specifically recognized as being at risk for inhuman, humiliating or degrading treatment. They are conferred the right to automatic guardianship of their children in most cases, and the right to remarry in all cases.[122] A widow is also entitled to equal inheritance rights.[123] Despite advances in inheritance laws, social practice is often very different from the law.[124] Widows of all ages face the taking of their household's land and property upon the death of their husband without regard for written law.

The 2008 Country Review Report for Benin also noted a 50 % increase in the numbers of older persons in the country, and that "the vast majorities of senior citizens are illiterate and live in very poor conditions… the social conditions of

---

[118]Country Review Report No. 6: Benin, Published: 05-06-2008 http://aprm.krazyboyz.co.za/index.php?option=com_aprm_documents&Itemid=32&page=documents-category&cid=25&nid=21&id=21.

[119]Poverty and Witch Killing by Edward Miguel in Review of Economic Studies, Vol 72, No. 4, pp. 1153–1172, October 2005 at 1153.

[120]Jill Schnoebelen, Witchcraft allegations, refugee protection and human rights: a review of the evidence, UNHCR, New Issues in Refugee Research, Research Paper No. 169, 2009.

[121]Report of the Working Group on the Universal Periodic Review, Benin, U.N. Doc. No. A/HRC/22/9 http://daccess-ods.un.org/access.nsf/Get?Open&DS=A/HRC/22/9&Lang=E.

[122]Article 20, Protocol to the African Charter on Human and People's on the Rights of the Women in Africa, 2003: http://www.achpr.org/english/_info/women_en.html.

[123]Article 21, Id.

[124]Country Review Report No. 6: Benin, Published: 05-06-2008 http://aprm.krazyboyz.co.za/index.php?option=com_aprm_documents&Itemid=32&page=documents-category&cid=25&nid=21&id=21.

older women seem more unfavorable."[125] In Benin, two out of every five older women are widows, while only one older man out of every 20 is a widower. It is suggested that this is likely due to the practice of polygamy in the country.[126]

3. The Rights of Older Persons with Disabilities

The confluence of disability and older persons has not yet been adequately examined by a legal framework, but it represents the interconnection of many different rights areas. The *Kigali Declaration*, for example, urges the development of a new Protocol to the *Banjul Charter* on the protection of the rights of persons with disabilities and older persons.[127] Attention to disability in African legal frameworks has been minimal, but picking up momentum. The Convention on the Rights of Persons with Disabilities has been ratified by twenty-six countries in Africa.[128] Most of ratifications have been added within the last 2 years.[129]

In the *Maputo Protocol,* States parties undertake to ensure the protection of women with disabilities and take specific measures commensurate with their physical, economic and social needs to facilitate their access to employment, professional and vocational training as well as their participation in decision making, as well as to ensure the right of women with disabilities to freedom from violence, including sexual abuse, discrimination based on disability and the right to be treated with dignity.[130] In thinking about multiple discriminations, older women, who live longer and are more likely to be widowed and poor, are also at increasing risk for disability at older ages. As better-disaggregated data is available, a better picture of the human rights implications of these compounded life situations will emerge.

A fine example of what can be shown with data disaggregation is in the Uganda Population and Census Report of 2002, reported by the CRM. It found that 16 % of persons in Uganda are persons with disabilities, a 50 % increase from the 11 % registered with the 1991 census. The census also found that the prevalence of

[125]Country Review Report No. 6: Benin, Published: 05-06-2008, at 128.

[126]Id.

[127]Id., 20.

[128]Algeria, Burkina Faso, Cape Verde, Ethiopia, Gabon, Guinea, Kenya, Lesotho, Malawi, Mali, Mauritius, Morocco, Namibia, Niger, Nigeria, Republic of South Africa, Rwanda, Senegal, Seychelles, Sierra Leone, Sudan, Togo, Tunisia, Uganda, Tanzania and Zambia http://treaties.un.org/pages/ViewDetails.aspx?src=TREATY&mtdsg_no=IV-15&chapter=4&lang=en.

[129]Id.

[130]Article 23, Protocol to the African Charter on Human and People's on the Rights of the Women in Africa, 2003: http://www.achpr.org/english/_info/women_en.html.

disability varied by region and increased with age, from a low of 2 % among those less than 18 years of age, to 18 % among older populations.[131] So much remains to be done to improve and strengthen the human rights of marginal and vulnerable groups.[132] Disaggregated data collection allows a more complete picture to emerge. The *Common Position's* stance that data is the first step toward better policies is correct. Better data also informs the monitoring of equal protection of older persons' human rights.

Future

When older persons' rights are respected, they can realize their full potential in their families, communities and society. Older persons with realized rights improve society in very direct ways; older persons (mostly older women) care for over half of the orphans throughout Africa.[133] Older persons' participation in local decision making can have long-term benefits for the resilience and growth of communities. They will not be accepted or heard to contribute this value if norms of marginalization and discrimination persist.

F. Conclusion

Within the African system, certain countries have been praised for their efforts to protect the human rights of older persons under the *Banjul Charter*, and many more have gone unnoticed. Older persons' rights need protection not just out of morality, or justice, but also because older persons are foundational for society. The youth bulge of today will likely be an age bulge tomorrow; an opportunity is presenting itself in this population shift to older age. The African Union has the chance to define and strengthen its body of law as one specific to the unique cultures and history of the African continent. Some signs of a normative shift, at least in international relations, are beginning to emerge. In a recent UPR review,

[131]Country Review Report No. 7: Uganda, Published: 21-08-2009, 445, P. 177 http://aprm.krazyboyz.co.za/index.php?option=com_aprm_documents&Itemid=32&page=documents-category&cid=25&nid=21&id=21.

[132]Country Review Report No. 1: Ghana.

[133]UNICEF, Africa's Orphaned and Vulnerable Generations, 2006. p. 15. http://www.unicef.org/publications/files/Africas_Orphaned_and_Vulnerable_Generations_Children_Affected_by_AIDS.pdf.

African countries both appreciated what policy changes for older persons have been put into place by other African countries, and exhorted each other to better protect their rights.[134] Recognition of the human rights of older persons would reflect the position of respect they have traditionally held.

[134]Report of the Working Group on the Universal Periodic Review, Burkina Faso, A/HRC/24/4. 8 July 2013, Para. 135.63. Senegal to Burkina Faso: "Strengthen the rights of vulnerable categories of the population, including the rights of the elderly" http://ap.ohchr.org/documents/dpage_e.aspx?si=A/HRC/24/4; Report of the Human Rights Council twenty-third session, A/HRC/23/2 27 August 2013, Para. 452., Gabon's expression of appreciation of Botswana's efforts to protect the rights of the elderly http://www.ohchr.org/EN/HRBodies/HRC/RegularSessions/Session23/Pages/23RegularSession.aspx.

Regional Overview of the Rights of Older Persons in Asia and the Pacific

A. An Introduction to Asia

Asia has the fastest growing proportion of older persons in the world. Between 2000 and 2050, the population of older persons in Asia will increase at five times the rate that it did between 1950 and 2000.[1] In South East Asia, by 2025, the percentage of older persons compared to the rest of the population will have increased to 10 percent, and by 2050, that percentage is projected to double, reaching 19 percent.

Asia is the only region without a human rights treaty.[2] While many Asian countries have constitutional guarantees of the right to equality and non-discrimination, only a handful of them have gone beyond rhetoric by applying effective enforcement procedures, particularly in cases of the rights of older persons.[3] It is also a region of diverse cultures and unique phenomena around aging. The region can be carved out under any number of lines; religious, ethnic, geographical, or otherwise. One joining factor is that filial piety exists throughout much of East and Southeast Asia; many have "similar cultural prescriptions related to the obligation to repay parents, especially when they reach older ages…"[4]

---

[1] A/HRC/14/31, Report of the independent expert on the question of human rights and extreme poverty, Magdalena Sepulveda Carmona, 31 March, 2010, 9.

[2] Kieran Fitzpatrick, Asia Pacific Forum, Presentation to the Open Ended Working Group on Strengthening the Human Rights of Older Persons, 2 August 2011.

[3] P. 5, The Rights of Older People in Asia, Niti Saxena, January 2009.

[4] J. Knodel, N. Chayovan, *Population Ageing and the Well-being of Older Persons in Thailand*, Population Studies Center Research Report 08-659, 2008. P. 60. *Citation omitted.*

C. Martin et al., *Human Rights of Older People*, Ius Gentium: Comparative Perspectives on Law and Justice 45, DOI 10.1007/978-94-017-7185-6_6

There is some evidence that these norms are changing. Research has pointed to many reasons for this change; rapid rural-to-urban migration has been explored as a cause of the erosion of filial piety among urban cultures.[5] The United Nations Economic and Social Commission for Asia and the Pacific ("ESCAP") highlights that the region is facing evolving changes in migration patterns, urbanization patterns, changes in labor dynamics, family structures and even epidemiology.[6]

Some research has shown that filial piety is significantly associated with the likelihood of coresidence between parents and their adult children.[7] Over 65 % of older persons studied lived with three or more generations in one household. A study of rural and urban depression in older persons in China found that the risk of depression was higher in rural communities than urban ones, possibly owing to loss of tradition and culture.[8] In 2006, 59 percent of older persons in China lived with their grown children or other relatives.[9] This is a changing paradigm. Without long-term care facilities to turn to, older persons remain dependent on adult children. In December 2012, the Chinese National People's Congress passed an amendment to a parental support law mandating that adult children must visit and assist older parents on a regular basis.[10] Such laws point to the particularly stark demographic shifts brought on by China's one-child policy, and a great increase in life expectancy.

The region's sometimes-severe socio-economic disparities can have grave implications for the health conditions of older persons. Aging itself has the largest impact on the health of older persons, often in the form of non-communicable (also known as chronic) disease. Older persons were included in the 2011 United Nations High Level Summit on Non-Communicable Disease Political Declaration.[11] The World Bank recently reported on positive measures being taken in countries throughout Southeast Asia to combat non-communicable diseases as

---

[5]C. Cheung, *The erosion of filial piety by modernization in Chinese cities*, Ageing & Society 29, 2009, 179–198, 193, 2009.

[6]Regional Review of the Implementation of MIPAA in Asia and the Pacific, 2007 http://www.unescap.org/esid/psis/meetings/ageingmipaa2007/RegionalReviewofMIPAA.pdf; ESCAP is made up of 53 Member States and 9 Associate Members, with a geographical scope that stretches from Turkey in the west to the Pacific island nation of Kiribati in the east, and from the Russian Federation in the north to New Zealand in the south, see http://www.unescap.org/about.

[7]Zhenmei Zhang, Danan Gu, Coresidence with Aging Parents in Contemporary Rural and Urban China: Attitudes and Behaviors, 21 (Nov. 19, 2010).

[8]R. Chen, et al, *A Meta-Analysis of Epidemiological Studies in Depression of Older People in the People's Republic of China*, 14 Int. J. Geriat. Psychiatry 821, 828, 1999.

[9]Richard Jackson, Keisuke Nakasima & Neil Howe, CHINA'S LONG MARCH TO RETIREMENT REFORM, 13 Center for Strategic and International Studies (2009).

[10]China passes legislation requiring adult children to care for elderly parents, Benjamin Minegar, December 28, 2012, http://jurist.org/paperchase/2012/12/china-passes-legislation-requiring-adult-children-to-care-for-elderly-parents.php.

[11]Political Declaration of the High-level Meeting on the prevention and control of non-communicable diseases, 2011. http://www.un.org/en/ga/ncdmeeting2011/documents.shtml.

the population continues to age.[12] The speed of aging will have a tremendous impact on the resources allocated to health care. Further, as result of such challenges to both family and health care system, the countries in the region need to develop institutions required for assisted living.[13]

National legislation focusing on older persons in Asia tends to focus only on the dimension of protecting and caring for older people. A focus on rights, or even the "soft law" of the United Nations Principles of Older Persons and the Madrid Plan of Action on Ageing for independence, participation and self-fulfillment, could drastically improve the lives of millions.[14] Evidence of this incorporation is difficult to collect.

Social change may mitigate some challenges of these arrangements and exacerbate others. Such evolutions must be taken into account when shaping policy and law to protect older persons. For example, greater urbanization and aging in rural provinces in Japan has spurred the study of active aging, public long-term public care insurance and preparations for a model plan for habitation for older persons.[15]

This chapter will briefly examine the organizations that have the most impact on the rights of older persons in Asia. It then expands upon the regional plan for implementation of the Madrid Plan of Action on Ageing, comparing it with a regional action plan for Asia completed three years before the Madrid Assembly on Ageing in 2002.

B. Regional Human Rights Frameworks

National governments in Asia have signed and ratified various UN-based international human rights treaties, but as noted above, no regional human rights framework exists in Asia. Regional institutions are significant in addressing the needs of the aging population in the region. In 2009, the Association of Southeast Asian Nations ("ASEAN") created an Asian Intergovernmental Commission on Human Rights (the "AICHR") pursuant to Article 14 of the ASEAN Charter.[16] According to the AICHR Terms of Reference (the "TOR"), the purposes of this body include

[12]World Bank, *Capitalizing on the Demographic Transition: Tackling Non-Communicable Diseases in South Asia* (2011).

[13]*Ageing in South Asia: An Overview and Emerging Issue for Policy Planning and Research* http://www.unescap.org/ESID/psis/meetings/Ageing_Change_Family/Ageing%20in%20South%20Asia%20An%20overview%20and%20emerging%20issues%20for%20policy%20planning%20and%20research%20by%20Mr%20Moneer%20Alam.pdf.

[14]UNDESA, World Population Ageing Report, 2009, p 11 (noting that by 2009, the number of persons aged 60 or over was737 million).

[15]T. Ogawa, *Japanese rural aging in transition: Evaluation of the model plan of habitation for older persons*, 4 Geriatrics and Gerontology International S63–S64, 2004.

[16]For more information, see http://www.aichr.org (last visited July 3, 2015).

promoting and protecting the human rights and fundamental freedoms of the persons of the region, as well as upholding the international human rights treaties to which ASEAN member states are party.[17] The main role of the AICHR is consultative and therefore no enforcement is mandated in the constitutive instrument.[18] Since its creation, the AICHR has been dedicated to the drafting of an ASEAN Human Rights Declaration.[19] This was approved in November 2012.[20] Moreover, pursuant to of one aspect of its mandate that includes the preparation of studies on thematic issues, the AICHR has been working on two reports. The first examines corporate responsibility and the second explores the right to peace.[21] Also, as part of its rights-promotion mandate, the AICHR has held several trainings and conferences on issues of particular relevance to the region.[22]

Even if the region has been disappointed with the limited powers granted to the AICHR and by its intergovernmental composition,[23] organizations advocating for the rights of older persons can use this newly established institution for human rights promotion. For example, civil society organizations could work with the AICHR to produce a thematic report on the right of older persons and the status of their rights in domestic constitutional law and practices. Further, AICHR could use its function training human rights trainers to add to human rights practitioners' awareness of the inequalities older persons experience in the region.

Taking advantage of the AICHR mandate to promote the implementation of human rights treaties ratified by the ASEAN Member states, civil society organizations could request the AICHR to report on the incorporation of the standards protecting the rights of older persons in domestic laws and practices. Support within the AICHR for the rights of older persons could galvanize participation of ASEAN member states in the debate regarding the adoption of an international convention on the rights of older people.

In addition to the AICHR, in 2010, a new ASEAN Commission on the Promotion and Protection of the Rights of Women and Children (the "ACWC") was inaugurated.[24] Like the AICHR, the ACWC is an intergovernmental body created with the purpose of promoting the rights of women and children in the

[17]ASEAN Intergovernmental Commission on Human Rights, TOR. The TORs were adopted in July 2009.

[18]TOR, Article 3.

[19]TOR, Article 4. Also, Press Release of the Seventh Meeting of the ASEAN Intergovernmental Commission on Human Rights (AICHR), Bali December 2, 2011.

[20]Phnom Penh Statement on the Adoption of the ASEAN Human Rights Declaration, http://aichr.org/documents/.

[21]Id.

[22]Id.

[23]See, Michelle Staggs Kelsall, *The New ASEAN Intergovernmental Commission on Human Rights: Toothless Tiger or Tentative First Step?*, Asia Pacific Issues, Analysis from the East-West Center No. 90, September 2009.

[24]Inaugurated: ASEAN Commission on the Promotion and Protection of the Rights of Women and Children, Ha Noi, 7 April 2010, at http://www.aseansec.org/ACWC.htm.

region.[25] One of the main mandates of the ACWC is to promote the implementation of international instruments protecting the rights of women and children, in particular the CEDAW Convention and the Convention on the Rights of the Child.[26] In this respect, for example, the ACWC has the power to encourage ASEAN Member States to collect and analyze disaggregated data by sex and age.[27] Also, the ACWC may promote studies and research related to the situation and well-being of women and children to foster effective implementation of the rights of women and children in the region.[28] Similar to the AICHR, civil society organizations working with the rights of older persons could lobby ACWC to devote some attention to the rights of older women in the ASEAN region as part of its mandate. These efforts could be based, for example, on the existing work done by the CEDAW Committee, particularly the General Comment No. 27 on Older Women and their Protection of Human Rights.[29] Or the intensive women's human rights workshops it has carried out with UN Women.[30]

Beyond the frameworks, at the substantive level, efforts to protect and fulfill rights obligations for older persons have been fragmented.[31] ESCAP generally focuses on areas of social development for older persons. The Association of Southeast Asian Nations ("ASEAN") has focused on the policy implications of aging populations. Among its strategic objectives, ASEAN aims to safeguard rights, provide equal opportunities, and enhance the quality of life and standards of living for older persons, who are explicitly listed among the other vulnerable groups in its publications.[32] In its *Roadmap for an ASEAN Community 2009–2015 - Social Justice and Rights*, ASEAN declared its commitment to promoting social justice and integrating the rights into its policies, including the rights of disadvantaged, vulnerable and marginalized groups.[33] This would have to include many poor, widowed, disabled, and marginalized older persons.

[25]ASEAN Commission on the Promotion and Protection of the Rights of Women and Children, Terms of Reference ("TOR"), Article 4.

[26]TOR, Article 5.

[27]Id.

[28]Id.

[29]Committee on the Elimination of Discrimination against Women, General Recommendation No 27 on Older Women and the Protection of their human rights, CEDAW/C/GC/27, 16 December 2010.

[30]Intensive Learning Workshop on Women's Human Rights (WHR) for the Regional and National Secretariats to ASEAN Human Rights' Bodies, March 2013, http://aichr.org/report/press-release-intensive-learning-workshop-on-womens-human-rights-for-the-regional-and-national-secretariats-to-asean-human-rights-bodies/.

[31]The Rights of Older People in Asia, Niti Saxena, January 2009, Background briefing paper.

[32]Id., 27.

[33]*Roadmap for an ASEAN Community 2009-20 - Social Justice and Rights* http://www.aseansec.org/publications/RoadmapASEANCommunity.pdf; made up of Indonesia, Malaysia, Philippines, Singapore, and Thailand. Brunei Darussalam joined on 8 January 1984, Vietnam on 28 July 1995, Laos and Myanmar on 23 July 1997, and Cambodia on 30 April 1999, see http://www.asean.org/asean/asean-member-states.

Older persons in general, though not their rights, have been addressed with increasing focus in the region. In 2008, an Asia/Pacific Regional Conference on Ageing brought together representatives of regional governments and international organizations, as well as civil society. Promotion of the inclusion of older persons in policies and in practices was discussed. Their conclusions further highlight the need to promote older persons and conduct impact analyses to mobilize resources.[34] A conference like this offers valuable insight into the shortcomings that attendees see in the lives of older persons from their countries, as well as insight into how these shortcomings can be addressed. The aims that came out of the conference harmonize with the right to participation, the right to social protection, the right to culture, and the right to non-discrimination.

The outcome of the 2008 Asia/Pacific Regional Conference on Ageing came to conclusions about the growing needs of older persons in the region that were built on over a decade of reports that had indirectly examined the growing aging population. For example, the outcome document of the Fourth Asia and Pacific Population Conference, the *Bali Declaration on Population and Sustainable Development, 1992* discussed that population aging is closely related to the processes of demographic and socio-economic change, that has a tremendous impact on the character and stability of private families. It further recognized that elderly persons are self-reliant and fully capable of contributing to the prosperity of their families and communities.[35]

At an ESCAP meeting in 2003, participating governments acknowledged the challenges of the changing age structure in the region. A key concern was rapid urbanization's impact on family structures, in particular as they become more nuclear, excluding extended family as younger populations search for employment and economic opportunities away from their traditional home. Along with decreasing family support through these changes, many countries with significant older populations did not have resources to establish the necessary national policies and institutional frameworks to provide additional care for older persons.[36] No frameworks have been established, but a 2010 meeting underlined the importance of accessible care, and proper training of caregivers at all levels.[37]

---

[34]www.helpage.org/Resources/Manuals/main_content/O2Mn/RC2008.pdf.

[35]www.unescap.org/esid/psis/population/5appc/doc/bali.pdf.

[36]UNESCAP, Fifth Asian and Pacific Population Conference Report and Plan of Action on Population and Poverty, 2003. http://www.unescap.org/esid/psis/population/popseries/apss159/index.asp.

[37]*Moving toward Advancing Health and Active Aging in the Asia-Pacific*, November 2010. www.unescap.org/sdd/…/Advancing-health-by-ACM-Chan&HK-Ma.pdf.

C. Regional Overview of the Rights of Older Persons

Asian cultures have traditionally highlighted the principles of respect, duty and harmony that stem from the multicultural and philosophical canvas of the region. At national levels, the majority of existing legislation for older persons is directed toward the context of family environments.[38] As has been noted previously, even such deep rooted respect in society does not protect older persons against age discrimination and other infringements, ranging from exclusion from decision-making processes, to acts of psychological abuse or physical violence.[39]

The *Shanghai Implementation Strategy consists of the Regional Implementation Strategy for the Madrid International Plan of Action on Ageing 2002,*[40] as well as the *Macao Plan of Action on Ageing for Asia and the Pacific 1999.*[41] Though "soft law," this highlights the importance of legal protection of older persons. In its section on neglect, abuse and violence against older persons, this document elaborates upon abuse in its many forms, including physical, sexual, psychological, emotional and financial abuse. Further, it recognizes the right of older persons as a consumer group to have access to goods and services amenable to their common needs, interests and preferences. The strategy suggests a number of actions in this regard for governments to follow, in particular to ensure that older persons have access to information about their rights, facilitate and regulate market development by providing safeguards to prevent the exploitation of older persons and finally combat discrimination, abuse and violence against older persons.[42] Further, the *Roadmap for an ASEAN Community 2009–2010-Social Justice and Rights* linked the welfare, quality of life and well-being of the elderly with the increased access to protection and medical care, including social security standards.[43]

The issues of aging often impact the fulfillment of two or more rights at the same time. For instance, the care of older persons, especially chronically ill older persons, is a major emerging issue affecting the right to health in the region. Older persons with chronic illnesses are often cared for in the home. The burden of care almost invariably falls on women. While improved education and employment opportunities for women have changed families for the better in many ways, these

---

[38]The Rights of Older People in Asia, Niti Saxena, supra note 31, at 3.

[39]E/CN.5/2009/5, Further Implementation of the Madrid International Plan of Action on Ageing: Strategic Implementation Framework - Report of the Secretary-General, 14.

[40]Shanghai Implementation Strategy: Regional Implementation Strategy for the Madrid International Plan of Action on Ageing (2002). www.unescap.org/esid/psis/Ageing/strategy/shanghai.pdf.

[41]See Shanghai Implementation Strategy Summary, http://www.unescap.org/esid/psis/ageing/strategy/index.asp.

[42]Id., 25-26.

[43]*Roadmap for an ASEAN Community 2009–20 - Social Justice and Rights* http://www.aseansec.org/publications/RoadmapASEANCommunity.pdf.

new families struggle with traditional care when one adult does not remain in the home to care for older family members.[44]

Policies are beginning to reflect this new reality. One ESCAP regional review found that programs for continuing education and work after retirement age are beginning to be developed.[45] The majority of countries in Southeast Asia have developed national programs and policies on aging.[46] Many are members of the Asia Pacific Forum, which consists of 15 independent national human rights institutions from the region.[47] It seeks as its mission to strengthen and support national human rights institutions and protect human rights throughout the region.[48]

As this section shows, some regional organizations have taken on tasks relating to protecting rights. This region will be an important factor in aging policies going forward, by sheer population numbers alone. At the UN, the region has been increasingly actively engaged in the Open-Ended Working Group discussions on the human rights of older persons. India and Japan both made short statements at the end of the August 2011 Open-Ended Working Group.[49] In 2012 and 2013, Bangladesh, China, Indonesia, Nepal and Thailand made statements.[50]

D. Economic and Social Commission for Western Asia

Western Asia is not separated into its own chapter in this book. The Economic and Social Commission for Western Asia ("ESCWA") prepared a regional investigative study on the situation of older persons in the Arab region in preparation for the Second World Assembly on Ageing.[51] In response, it drafted the Arab Plan of Action on Aging. Despite these efforts, creating legal protection of older persons' economic rights and visibility of those rights remains a challenge.[52] In its report for the Second Review and Appraisal of the Madrid International Plan of Action

[44]*Southeast Asia's Population in a Changing Asian Context: Policy Implications, October 2002, International Union for the Scientific Study of Population*, http://www.iussp.org/Bangkok2002/5APPC.pdf.

[45]*The Rights of Older Persons in Asia, By International Federation on Ageing and HelpAge International* (2009) Par. 8. http://www.globalaging.org/elderrights/world/2009/humanrightsasia.pdf.

[46]Id., 10.

[47]http://www.asiapacificforum.net/about.

[48]Id.

[49]http://social.un.org/ageing-working-group/govstatementsec.shtml.

[50]Records on file with authors.

[51]Arab preparatory meeting for the second world assembly on aging Beirut, 5–8 February 2002 the Arab Plan of Action on Aging to the year 2012, E/ESCWA/SD/2002/WG.1/8, 7 February 2002.

[52]Ageing in the Arab Countries Research Policy and Development, Expert Group Meeting, Proceedings of the Conference, Beirut Lebanon 18–19 November 2009, at 6, 9, 11, 13, 14, 16.

on Aging,[53] it highlighted the lack of data with which to assess the implementation of MIPAA.[54] This alone is telling, as paragraph 119 of MIPAA indicates that data collection is one of the "crucial elements of implementation."[55]

The scope and protection of the rights of older persons in Arab countries has been affected by a number of factors. The religion of Islam currently accepts the principle of equality irrespective of color, race or language. It refuses to accept this principle with respect to men and women.[56] The *Cairo Declaration on Human Rights in Islam,*[57] for example, fails to fully and unequivocally endorse the principle that all persons are equal and that discrimination based on religion or sex is impermissible.[58]

Various Islamic human rights schemes do not recognize women as fully equal human beings with the same rights and freedoms as men.[59] Therefore, while the purpose of international human rights law is to afford equal protection for human rights, some frameworks actually restrict or deny women's rights to equality and to equal treatment under the law.[60]

Discrimination in the Arab region creates gender-based neglect that is particularly acute for older women.[61] Because of the successful reduction of infant, child and maternal mortality, the population in the region is aging, and the number of widows is expected to continue to rise.[62] The demographic change in the region will continue to have a significant impact on family-related issues and on international calls for the promotion of aging issues.[63]

The human rights efforts in the Western Asia region have not directly addressed old age. This may be due in part to the traditional degree of care and support of the family. An increase in nuclear living arrangements among the young diminishes

---

[53]*Report on the second regional review meeting of the Madrid International Plan of Action on Ageing, 7 June, 2012, U.N. Doc. E/ESCWA/SDD/2011/WG.8/2/Report* http://www.escwa.un.org/information/pubaction.asp?PubID=1214.

[54]Id., at 8.

[55]MIPAA, 119.

[56]Human Rights in Arab Thought – edited by Salma K. Jayyusi- 2009, at 9, 22, 55, 389, 497, 524, 545, 553, 556, 571.

[57]Cairo Declaration on Human Rights in Islam Aug. 5, 1990, U.N. GAOR, World Conf. on Hum. Rts., 4th Sess., Agenda Item 5, U.N. Doc. A/CONF.157/PC/62/Add.18 (1993) http://www1.umn.edu/humanrts/instree/cairodeclaration.html.

[58]Ann Elizabeth Mayer, Islam and Human Rights, 2007, at 112.

[59]Steiner, Alston & Goodman, *supra* note 77, General sub- chapter-in; Human rights in the Muslim world: Sharia and the human rights of women, at 531–539.

[60]Ann Elizabeth Mayer, Islam and Human Rights, 2007, at 143, 172.

[61]ECSWA, 44; AAPA 2002.

[62]Economic and Social Commission for Western Asia, E/ESCWA/SDD/2004/WG.1/2, Aging in the Arab Countries: regional variations, policies and programmes, at 11, 13.

[63]ECSWA, Par. 18; AAPA 2002 and, Ageing in the Arab Countries- Research Policy and Development, 2009, at 46.

the extended family system and weakens care and support to the elderly,[64] without a balancing increase in states recognizing their duty to uphold older persons' basic human rights. For example, the League of Arab States, the regional organization for Arab countries has given attention to regional instruments and social work strategies on aging.[65] In its review of the *Arab Charter for Social Work*,[66] amended in 2001, it affirmed that care for older persons within families is vital. This is a social document, not a rights document, and its accompanying call for government support of families is to be encouraged. It does not, however, replace the protected relationship of individual older persons and the state under the auspices of human rights.

E. Asian Regional Action Plan on Aging

Without an enforceable regional framework on human rights, the *Shanghai Implementation Strategy: Regional Implementation Strategy for the Madrid International Plan of Action on Ageing 2002*[67] (the "Shanghai Regional Strategy"), adopted at the Asia-Pacific Seminar on Regional Follow-up to the Second World Assembly on Ageing in 2002, is of elevated importance in interpreting regional norms for older persons. The Shanghai Regional Strategy includes a specific section on older persons and development, and highlights the difficulties many countries have encountered in implementing effective aging policies due to insufficient resources. It identifies priorities and key actions to enhance the implementation of commitments made under the *Regional Implementation Strategy for the Madrid International Plan of Action on Ageing* (2002) and the earlier-agreed *Macao Plan of Action on Ageing for Asia and the Pacific* (1999) (the "Macao Plan of Action"),[68] and goes on to adopt both of those respective plans. The Macao Plan of Action elaborates upon a number of key areas of concern regarding older persons, including their social and family positions, health, housing, and transportation. It is a reflection of the best intentions of the region.

---

[64]E/ESCWA/SDD/2004/WG.1/2, Aging in the Arab Countries: Regional Variations, Polices and Programmes, at 24, 26, 27, 28.

[65]Rodrigo Tavares, Regional Security the capacity of international organizations, Global institutions, 2010, at 105–116.

[66]Arab Charter for Social Work, Arab Strategy for Social Work and Arab Charter for Comprehensive Social Development, Arab League.

[67]Shanghai Implementation Strategy: Regional Implementation Strategy for the Madrid International Plan of Action on Ageing (2002). www.unescap.org/esid/psis/Ageing/strategy/shanghai.pdf.

[68]*Macao Plan of Action* on Ageing for Asia and the Pacific (1999) (hereinafter the Macao Plan of Action) http://www.unescap.org/ageing/Macao/plan_of_action.htm.

Distinct from the Macao Plan of Action, the *Macao Declaration on Ageing for Asia and the Pacific*[69] (the "Macao Declaration") is significant in that it is a tailored expression of the intentions of the region, and that it was created before - and committed to in addition to – MIPAA regional plan for the region (the Shanghai Regional Strategy). It states that effectively addressing the issue of aging requires a basic change in social attitudes toward older persons in the Asian and Pacific region.

1. Social Position of Older Persons

The *Macao Plan of Action* notes negative images of older persons throughout the region as weak and dependent. Older persons are often at risk of social and economic marginalization in countries that experience rapid development. The Macao Plan of Action emphasizes the positive contributions older persons make to their families and societies, taking into account the *United Nations Principles for Older Persons*.[70] This represents a policy reflecting the real impact that the social construction of age can have on older persons' ability to access their rights. Where they become marginalized and discriminated against based on age, they are not just perceived as weak, and without due equal rights, society enforces it.

2. Older Persons and the Family

The status of older persons and their families is vital to the central unit of social and economic life; governments should commit to "home nursing services for older persons, facilities for respite care, and promoting programs on counseling, professional guidance and emotional support."[71] In order to ensure and enable supportive environments, in particular within the family, a key action contained in the *Shanghai Regional Strategy* focuses on multigenerational solidarity including community-based services that support older persons with or without families.[72] Additionally, with regard to social service and community, key actions call upon governments to ensure quality in formal care settings and take an active role in formulating accreditation and quality assurance systems.[73]

The *Macao Plan of Action* calls for the inclusion of older persons in disaster preparedness planning, and overall, programs should strive to ensure that

---

[69]Id.

[70]Id., 12–14.

[71]Id., 18.

[72]Id., 19, 20.

[73]Id., 21.

inter-generational relationships are interdependent.[74] Through the twentieth century, older persons have borne a particularly heavy burden in the region's armed conflicts, particularly in south Asia, that lead to a lack of food, adequate shelter, clothing and deteriorating health conditions faced by internally displaced persons (IDP's).[75] The *Shanghai Regional Strategy* recognizes the particular vulnerability of older persons in the emergencies but also highlights their potential roles as primary care givers and leaders in reconstructing damaged communities.[76] A related key action calls for older persons without families who are unaccompanied to be actively sought by governments during and after natural disasters, recognizing the particular vulnerabilities they may face.

3. Older Persons' Health and Nutrition

The cost of medical care, without a change in course, is a formidable concern for every government. As populations age, innovative approaches to health care financing, including "co-payment for medical costs between employees and employers, government-run medical funds and insurance systems,"[77] are needed. Early diagnosis and mass screening programs accessible to all ages are encouraged, and health and nutrition are identified as key areas for educational initiatives and lifestyle interventions. Further, there should be a system of coordinated care to evaluate and treat psychological and mental needs.[78]

In the area of health and long-term care, key actions outlined by the *Shanghai Regional Strategy* include addressing the unique needs of older persons living alone or in rural and remote areas. It outlines developing support systems to help families care for older members, as well as providing integrated care services designed to help individuals remain in their communities for as long as possible, promoting older persons' mental and physical well-being.[79]

4. Housing, Transportation and the Built Environment

The *Shanghai Regional Strategy* calls for actions to support "aging in place," recognizing that some older persons prefer to remain in their homes rather than move to an institution.[80] The *Macao Plan of Action* calls on governments to commit to

---

[74],Id., 15–20.

[75]The Rights of Older People in Asia, Niti Saxena, supra note 31, at 10.

[76]Id., 10.

[77]*Macao Plan of Action*, supra note 68, 25.

[78]Id., 21–25.

[79]Id., 18.

[80]Id., 22, 23.

UN Habitat's *Global Strategy for Shelter for the Year 2000,*[81] protecting and improving housing conditions. The particular challenges faced by older persons should receive adequate resources, including public housing.[82] Governments are also called upon to improve the physical environment for older persons commuting by public transport, bolster transportation infrastructure to facilitate the mobility of older persons, and tailor road safety measures to the needs of older persons.[83]

5. Older Persons and the Market

The *Macao Plan of Action* calls upon governments to facilitate and regulate a service market sensitive to the needs of elderly persons to ensure minimum standards to avoid abuse or unfair market practices. Governments should protect the elderly from harmful market practices, as provided by the *United Nations Guidelines for Consumer Protection.*[84] This would include such issues as predatory lending and financial abuse. This is much stronger than the community-oriented education approach of MIPAA.[85] Neither addresses financial abuse from a criminal standpoint.

6. Income Security, Maintenance and Employment

Social security or pension benefits can guarantee an adequate standard of living. Few older persons in Asia have savings or other resources of their own for old age.[86] The *Macao Plan of Action* encourages saving for old age in the informal economy and in rural areas. Coverage should extend to the older destitute, widows, and the disabled. It also encourages the provision of flexible retirement schemes and legal protection to enable capable older persons to remain economically active, and foster job creation and retention programs aimed at enhancing the employability of older workers. Finally, governments must constantly review the adequacy of benefits and, if possible, expand the resources devoted to social security schemes.[87] The *Shanghai Regional Strategy notes that* older persons are to be

[81] U.N. Habitat, *Global Strategy for Shelter for the Year 2000,* 20 December 1988 U.N. Doc A/RES/43/181, http://www.unhabitat.org/downloads/docs/1393_76192_other1.htm.

[82] *Macao Plan of Action*, supra note 68, 26. Par. 26, http://www.unescap.org/esid/psis/ageing/guideline.pdf.

[83] Id., 26–28.

[84] Id., 29–30.

[85] MIPAA, 107.

[86] Id., 16.

[87] *Macao Plan of Action, supra note 68,* 31–35.

included as a target group in all poverty alleviation programs, including income generation schemes and savings and credit programs, with emphasis on high-risk groups such as women.[88]

Governments in Asia are only beginning to explore increasing retirement ages, despite increasing longevity.[89] Older women face particular hardship, due to even longer life expectancy than men, decreased time in the formal workforce and fewer opportunities for advancement.[90] Additionally, older women tend to be less well-off financially than older men, due to lower education levels and interrupted or lower labor-force participation rates.[91]

The *Shanghai Regional Strategy* also notes that countries should establish viable long-term pension systems and adapt social protection to increasingly nuclear family structures and increasing life expectancies. Toward alleviating poverty in old age, it suggests that social protection, and the increasingly informal nature of work point to the need for income security that is not entirely dependent on family structures.[92] Financial resources, it posits, could be generated through taxes and other measures.[93]

In the formal employment sector, the *Shanghai Regional Strategy* suggests increased national productivity could be achieved with environments for longer employment and flexible retirement arrangements.[94] In information economies, especially, education throughout the life course is a necessary component of longer working lives. Governments should ensure that education, advocacy, and legislation to prevent discrimination, and in order to overcome the negative stereotypes of older persons as "frail, disabled and dependent."[95]

7. Social Services and the Community

Lastly, the *Macao Plan of Action* focuses on social services and the community and suggests that external support be available to help older persons in order for them to adjust to changes in their circumstances. Day care, rehabilitation,

---

[88] *Shanghai Implementation Strategy*, 9.

[89] Hand Book of Asian Aging, P. 13. www.baywood.com/intro/316-x.pdf.

[90] Id.

[91] Ch.5 Economic Status, Work, and Retirement in China Economic status of older women (p. 153).

[92] Id., Par 8, 9.

[93] http://www.unescap.org/ESID/psis/meetings/Ageing_Change_Family/Ageing%20in%20South%20Asia%20An%20overview%20and%20emerging%20issues%20for%20policy%20planning%20and%20research%20by%20Mr%20Moneer%20Alam.pdf.

[94] Id., 13, 14.

[95] Id., 12.

counseling and dietary advice are all highlighted, along with guidelines for the level of services that must be available to older persons in both the rural and urban areas.[96]

In its final section, the *Shanghai Regional Strategy* suggests establishing comprehensive frameworks at the national level and enhancing the participation of NGO's, older persons' associations and other civil groups in reviewing and implementing national policies and programs on aging,[97] while the *Macao Plan of Action* emphasizes empowerment.[98]

With regard to regional and international cooperation, the *Macao Plan of Action* calls on ESCAP to continue its focus on aging and encourage cross-country collaboration and sharing of best practices.[99]

E. The Rights Older Persons as Members of Other Vulnerable Groups: Women

Older persons in the Asian regions face the same structural barriers to equal rights and face the same discriminations as older persons in other regions. Additionally, the effects of multiple discriminations as older persons also identify with other groups are equally apparent.

With regard to the "double jeopardy" of being an older woman,[100] older women often experience shrinking roles and declining status in their families and communities[101] as a result of old age discrimination. Older women may face particular difficulty accessing their rights in patrilineal societies which require women to leave their families and move in with their partner's families, isolating them and leaving them particularly vulnerable if widowed.[102]

Women comprise the majority of older persons around the world[103] yet they are often unable to accumulate sufficient retirement savings due to a variety of reasons including their role as primary caregivers, higher likelihood to be widowed, and

---

[96]*Macao Plan of Action, supra note 68*, 36–40.

[97]Id., 27.

[98]Id., 36–40.

[99]Id., 59.

[100]Id., 20.

[101]UNFPA, Population and Development Strategies, The Older Poor and Excluded in the Developing World, 2003, at 13.

[102]See S. Ghuman and M. Ofstedal, *Gender and Family Support for Older Adults in Bangladesh*, Population Studies Center, Report No. 04-563, 2004, at 14.

[103]UNDESA, Population Facts 2012/4, http://www.un.org/en/development/desa/population/publications/pdf/popfacts/popfacts_2012-4.pdf.

frequent lack of marketable vocational skills.[104] Older women are especially vulnerable workers for they, to a large extent, work in marginal economic sectors. Decisions about whether to retire may be beyond their personal volition[105] and they tend to be less well-off financially than older men due to lower education levels and interrupted or lower labor-force participation rates.[106]

Thus, the *Shanghai Regional Strategy* suggests enhancement of support for family caregivers so that they can combine work and family life, promote and support greater male responsibility in the family, and increase women's participation in the labor force through education and training.

F. Conclusion

Monitoring and reporting on the effects of eroding family support and the concomitant demand for assisted living in the region are of immediate and necessary concern.[107] Other related emerging issues include pension reform and management of retirement funds.[108] Similar to other regions, the longevity of women in Asia, in combination with patriarchal social constructs and prevailing gender-based discrimination, create additional challenges.[109] The evolution of cultures there are likely to exacerbate the challenges for the older population due to slowing economies and relatively fewer younger people.[110]

The region of Asia presents some amazing opportunities to support older persons. It is the epicenter of a boom in older populations and a blank slate of opportunity to create a human rights framework for older persons with the participation and support of older persons in the region. Even the agreed protection of the right to non-discrimination and equality will go a long way toward empowering older persons to claim their own rights in their own way.

---

[104]Ageing in South Asia: An overview and emerging issues for policy planning and research, Par 15 http://www.unescap.org/ESID/psis/meetings/Ageing_Change_Family/Ageing%20in%20South%20Asia%20An%20overview%20and%20emerging%20issues%20for%20policy%20planning%20and%20research%20by%20Mr%20Moneer%20Alam.pdf.

[105]Hyungsook Yoon, Jon Hendricks, Eds, *The Sweep of Asian Aging: Changing Mores, Changing Policies*, Handbook of Asian Aging, 2006, at 13.

[106]Economic Status, Work, and Retirement in China Economic status of older women, Chap. 5, at 153.

[107]Alam, Ageing in South Asia: An overview and emerging issues for policy planning and research http://www.unescap.org/ESID/psis/meetings/Ageing_Change_Family/Ageing%20in%20South%20Asia%20An%20overview%20and%20emerging%20issues%20for%20policy%20planning%20and%20research%20by%20Mr%20Moneer%20Alam.pdf.

[108]Id.

[109]The Rights of Older People in Asia, Niti Saxena, supra note 31, at 15.

[110]*See* Asian Development Bank, Impact of Population Aging on Asia's Future Growth, Donghyun Park and Kwanho Shin, October 2012 http://www.adb.org/publications/impact-population-aging-asias-future-growth; The Financial Times, *The End of Asia's Economic Dividend,* David Pilling, 12 March 2012 http://www.ft.com/cms/s/0/bd935806-6d00-11e1-a7c7-00144feab49a.html#axzz2UQYe5cgu.

Future Efforts

Advocacy for older persons in the future should build on the work done by experts in the field. More must be understood and shared about older persons' rights. A positive feedback loop of communication should exist between activists in the field and technical experts. The creation of the Independent Expert post brings this ideal one step closer to reality.

More dialogue will lead to improved outcomes at various levels addressing older persons' rights. At the grassroots level, an understanding of human rights can influence local policymakers, and enrich individuals' understandings of their own rights. At the regional and state level, a better understanding of the challenges older persons face in exercising their rights can lead to more inclusive policies and support of international norms. Internationally, both at the regional level and the global level, states can support each other in providing a basic framework of protections of this often-overlooked group.

Conclusion

This book has explored soft law and hard law forming the frameworks for older persons' human rights. It has also discussed how these laws *could* apply to older people, even if there has not been international agreement that these laws apply. What it is clear is that existing mechanisms fail to adequately address the rights of older persons. Laws do not adequately address the issues of older people, resulting in thin case law and even fewer references in the international system. Finally, social norms are informed by law, and vice versa. The positive externalities of stronger protection of older persons' human rights cannot be underestimated.

A Failed System

Older people face discrimination not just in their daily lives, local access to justice, and national legislation. They also face discrimination in terms of equal treatment with people of other ages at the international level. One analysis of the 124 Human Rights Committees' reports received from Member States found that between 2000 to 2008, "only three made specific reference to actions taken to address age discrimination, and just one highlighted the vulnerability of older people in long-term care homes."[1] The rights that could be protected by existing human rights mechanisms have been well-outlined in previous chapters. As the minimal case law cited shows, hundreds of millions of older people around the world are not claiming those universal rights at the national or international levels. This is the case despite documented, ongoing and systematic violations around the world.[2]

[1]Israel Doron, Itai Apter, *The Debate Around the Need for an International Convention on the Rights of Older Persons*, The Gerontologist Vol. 50, No. 5, 586–593.

[2]See, for example, the NGO Thematic Shadow Report on Older Women's Rights in Tanzania, 2008, submitted to CEDAW with 13 Tanzanian organizations, http://www2.ohchr.org/english/bodies/cedaw/docs/ngos/HAITanzania41.pdf (last visited Feb. 27, 2015).

C. Martin et al., *Human Rights of Older People*, Ius Gentium: Comparative Perspectives on Law and Justice 45, DOI 10.1007/978-94-017-7185-6_7

The inquiry of the UN's Open Ended Working Group toward Strengthening the Human Rights of Older Persons is predicated on the fact that the Secretary General's report in 2009 noting the failure of existing human rights instruments in addressing the rights of older persons. The region of Latin America and the Caribbean found the same; it is seeking to address this failure with a draft Inter-American Convention on the Human Rights of Older Persons at the Organization of American States.[3] The region of Africa has also sought to address these shortcomings with the African Common Position for the African Union[4] supporting the creation of an International Convention on the Human Rights of Older Persons at the UN. Even the countries of Europe, for all of their challenges to an international convention at the UN, have approved their own recommendation on older people's human rights.[5]

Why is this the case? One reason is that policies and soft law such as the Madrid International Plan of Action on Ageing (MIPAA) are not being taken up. Its stated aims were to call for changes in "attitudes, policies and practices" at all levels to fulfill the enormous potential of ageing in the twenty-first century. Its specific recommendations for action give priority to older persons and development, advancing health and well-being into old age, and ensuring enabling and supportive environments.[6]

The intent to better protect older persons' human rights is explicit in the 2002 Madrid International Plan of Action on Ageing[7] but has not achieved its intention. In 2011, the UN General Assembly recognized the limited awareness of MIPAA and its limited implementation.[8] It was intended to be a guide for national legislation. Countries have had ten years to implement it into their national laws, and most have not.[9] A lesson from MIPAA could be "[I]f the Madrid Plan of Action is not taken seriously, there is no real reason why another international document will be treated any differently."[10]

---

[3]Working Group on Protection of the Human Rights of Older Persons, at http://www.oas.org/consejo/cajp/personas%20mayores.asp.

[4]3rd Session of the AU Conference of Ministers of Social Development, *Africa's Common Position on Human Rights of Older Persons in Africa*, available at http://social.un.org/ageing-working-group/documents/africa-older-persons2012.pdf (last visited Feb. 27, 2015).

[5]Recommendation CM/Rec(2014)2 of the Committee of Ministers to Member States on the Promotion of the Human Rights of Older People, available at http://www.coe.int/t/dghl/standardsetting/hrpolicy/other_committees/cddh-age/default_EN.asp (last visited Feb. 27, 2015).

[6]UN website, Global Issues, Ageing: http://www.un.org/en/globalissues/ageing/.

[7]See Madrid International Plan of Action on Ageing, 12, available at http://www.un.org/en/events/pastevents/pdfs/Madrid_plan.pdf (last visited Oct. 27, 2014).

[8]UN, Follow-up to the Second World Assembly on Ageing, A/RES/65/182, available at http://www.un.org/ga/search/view_doc.asp?symbol=A/RES/65/182 (last visited Feb. 27, 2015).

[9]UNFPA, HelpAge International, Ageing in the 20th Century: A Celebration and a Challenge, 2012, "there is still scant evidence of resource allocation to support implementation of policies on ageing" at 107, available at http://www.helpage.org/download/5059f6a23af15 (last visited Feb. 27, 2015).

[10]Israel Doron, Itai Apter, *The Debate Around the Need for an International Convention on the Rights of Older Persons*, *supra* note 1.

These recommendations are not on their way to becoming customary. Non-compliance with soft law, and the blindness of hard law, galvanizes the need for a comprehensive international legal instrument.

Further Exploration

Another reason for this international movement toward an international convention are the gaps in the existing system; while a few rights may be covered under existing treaties, claiming human rights should not be a matter of identity politics. A person who experiences discrimination based on their age should not have to prove that they belong to another group, such as women, or persons with disabilities, in order to claim some sort of protection or to achieve some sort of redress.

Older persons face specific issues that require legal frameworks. One cannot read the existing treatment of older people's rights in international law and expect that the existing patchwork begins to scratch the surface. Gaps in norms are part of the problem. "A *normative gap* exists where persistent acts and circumstances depriving a person or people of their dignity are not provided for in existing human rights law."[11]

Other types of gaps contributing to these problems include *monitoring gaps* "where no independent monitoring of older people's rights takes place," and *information gaps* where limited information is systematically gathered about people once they reach a certain age.[12] As pointed out by HelpAge International, these two gaps contribute by inhibiting understanding of the extent of problems, and potential solutions for older people's human rights.[13]

These gaps all converge to leave blind spots around many areas of the human rights of older persons. Some issues lightly touched upon in this book, but requiring much further analysis include the uncharted legal territories of supported decision-making for people living with dementias, choices regarding living in communities versus institutions, the right to earn an income, violence against older men, access to justice, mandatory retirement ages, mandatory voting laws and age discrimination in all its forms.

[11]HelpAge *International, International human rights law and older people: Gaps, fragments and loopholes*, 2012, http://social.un.org/ageing-working-group/documents/GapsinprotectionofolderpeoplesrightsAugust2012.pdf (last visited Oct. 27, 2014).

[12]See OEWG Chairman's Report, April 2011, at 9, available at http://social.un.org/ageing-working-group/documents/chairmans%20-%205%20May.pdf (last visited Oct. 27, 2014).

[13]HelpAge International, *International Human Rights Law and Older People: Gaps, Fragments and Loopholes*, *supra* note 11, at 2.

Changing Norms

A social shift is necessary to protect older persons' rights that cannot be catalyzed without a more far-reaching process than what the limited engagement of social development has provided. Strong law informs societal change, and the ageing and changing of the makeup of societies will reinforce the importance of those laws.

Today, old age discrimination is a major cause of inequality and injustice. Some harmful societal expectations that older persons belong in the home, for instance, add to this stereotyping and discrimination. Other harmful stereotypes include the idea that older persons should be content with being dependent on others, and that older persons are useless in a modern economy. Age discrimination should be prohibited; this would go a long way to combatting negative stereotypes of older persons by giving them more fair and equal treatment and opportunities to participate in society.

With intentional societal change, it also represents an opportunity for older persons to continue to be productive and active in their communities, societies and nations. In order for this new role and contribution to be made to society, their rights need to be recognized and valued equally to other members of society. The idea that older persons do in fact have equal rights as others is an idea that has not found its way into national laws and policies.

In 2012, the UN High Commissioner for Human Rights reported to the Economic and Social Council noting that the demographic shift is enough to warrant dedicated attention to older people's human rights, noting that "Only a handful of international human rights mechanisms have devoted attention to older persons over the years or developed the guidance and specific tools to provide to Governments and other stakeholders that such a large population group would entail."[14]

An international convention on the human rights of older persons would ensure that older persons are treated as equal to persons of other ages. National legislation would change to ensure equal treatment of older persons. "Overall, the convention would define older people's rights as human rights and demonstrate that the abrogation of human rights is not acceptable. It would stipulate positive obligations on nations to realize equality and the enjoyment of rights by older people."[15] Protection of older persons' rights benefits society as a whole, and each and every one of us.

Overall, we hope that this book contributes to a better understanding of the current limitations and possibilities that international institutions offer to uphold the rights of older persons. We hope that this information will allow states and

---

[14]Report of the United Nations High Commissioner for Human Rights (E/2012/51), April 2012, pursuant to General Assembly resolution 48/141.

[15]Tang, K., & Lee, J., *Global social justice for older people: The case for an international convention on the rights of older people*, British Journal of Social Work, 36(7), 1143 (2006).

other policy makers to move forward with the international recognition of their human rights. We know this is only a first effort in compiling and reporting about the standards that are being produced from different international institutions. But we have no doubt that many others will follow with their expert analysis of these emerging standards, and that the ongoing discussion will finally crystalize in international human rights binding instruments explicitly recognizing the universal rights of older persons.

CPSIA information can be obtained at www.ICGtesting.com
Printed in the USA
BVOW06*0226300715

411098BV00002B/4/P

9 789401 771849